Constructing Worlds through Science Education

In the **World Library of Educationalists,** international experts themselves compile career-long collections of what they judge to be their finest pieces – extracts from books, key articles, salient research findings, major theoretical and practical contributions – so the world can read them in a single manageable volume. Readers will be able to follow themes and strands of the topic and see how their work contributes to the development of the field.

John K. Gilbert has spent the last 30 years researching, thinking and writing about some of the key and enduring issues in Science Education. He has contributed over 10 books and 175 articles to the field.

In *Constructing Worlds through Science Education*, John Gilbert brings together 16 key writings in one place. Starting with a specially written Introduction, which gives an overview of his career and contextualises his selection, the work presented supports the assertions that:

- students of science must be presented with a curriculum that is both philosophically valid and personally interesting to them
- informal educational provision has much to offer science education
- close attention to the development of thinking about the ideas of science is necessary if science education is to be effective
- there must be a close relationship between educational theory and practice in science education, underpinned by qualitative research in classrooms and laboratories
- international collaboration is an effective way of generating an active synergy of research involving diverse knowledge and skills.

For anyone seeking a clearer understanding of the developments in Science Education over the last 30 years, this book provides a useful route map to the thinking and work of one of the leading scholars in the field.

John K. Gilbert is currently Professor of Education at The University of Reading, UK, and Editor-in-Chief of the *International Journal of Science Education.*

Contributors to the series include: Richard Aldrich, Stephen J. Ball, John Elliott, Elliot W. Eisner, Howard Gardner, John K. Gilbert, ͺ ͺdson, David Labaree, John White, E.C. Wragg.

D1372186

World Library of Educationalists series

Constructing Worlds through Science Education

The selected works of John K. Gilbert

John K. Gilbert

Routledge
Taylor & Francis Group

LONDON AND NEW YORK

First published 2005
by Routledge
2 Park Square, Milton Park, Abingdon, Oxon OX14 4RN

Simultaneously published in the USA and Canada
by Routledge
270 Madison Ave, New York, NY 10016

Routledge is an imprint of the Taylor & Francis Group

© 2005 John K. Gilbert

Typeset in Sabon by
Newgen Imaging Systems (P) Ltd, Chennai, India
Printed and bound in Great Britain by
MPG Books Ltd, Bodmin

British Library Cataloguing in Publication Data
A catalogue record for this book is available
from the British Library

Library of Congress Cataloging in Publication Data
A catalog record for this book has been requested

ISBN 0–415–35217–7 (hbk)
ISBN 0–415–35218–5 (pbk)

For Julie Gilbert, who has known the actors, rehearsed the lines, provided an invaluable critique of the production, and whose insightful company over nearly 50 years has shaped the meaning of it all.

CONTENTS

ACKNOWLEDGEMENTS

The following articles have been reproduced with the kind permission of the respective journals

Davies, T. and Gilbert, J.K. 'Modelling: promoting creativity while forging links between science education and design and technology education'. *Canadian Journal of Science, Mathematics and Technology Education*, 2003, 3(1): 67–82.

Gilbert, J.K. 'The interface between science education and technology education'. *International Journal of Science Education*, 1992, 14(4): 563–578.

Gilbert, J.K. and Priest, M. 'Models and discourse: a primary school science class visit to a museum'. *Science Education*, 1997, 81(6): 749–762.

Gilbert, J.K. and Stocklmayer, S. 'The design of interactive exhibits to promote the making of meaning'. *Museum Management and Curatorship*, 2001, 19(1): 41–50.

Gilbert, J.K. and Watts, D.M. 'Concepts, misconceptions and alternative conceptions: changing perspectives in science education'. *Studies in Science Education*, 1983, 10: 61–98.

Gilbert, J.K., Osborne, R.J. and Fensham, P.J. 'Children's science and its consequences for teaching'. *Science Education*, 1982, 66(4): 623–633.

Gilbert, J.K., Boulter, C. and Rutherford, M. 'Models in explanations, part 1: horses for courses?' *International Journal of Science Education*, 1998, 20(1): 83–97.

Justi, R. and Gilbert, J.K. 'A cause of ahistorical science teaching: use of hybrid models'. *Science Education*, 1999, 83(2): 163–177.

Justi, R. and Gilbert, J.K. 'History and philosophy of science through models: the case of chemical kinetics'. *Science and Education*, 1999, 8: 287–307.

Justi, R. and Gilbert, J.K. 'Modelling, teachers' views on the nature of modelling and implications for the education of modellers'. *International Journal of Science Education*, 2002, 24(4): 369–387.

Reiner, M. and Gilbert, J.K. 'Epistemological resources for thought experimentation in science learning'. *International Journal of Science Education*, 2000, 22(5): 489–506.

The following chapters have been reproduced with the kind permission of the respective publishers

Gilbert, J.K. 'Research and development on satellites in education'. In J.K. Gilbert, A. Temple and C. Underwood (eds), *Satellite Technology and Education*. London: Routledge, 1991, pp. 205–218.

Gilbert, J.K., Watts, D.M. and Osborne, R.J. 'Eliciting student views using an interview-about-instances technique'. In C. West and A. Pines (eds), *Cognitive Structure and Conceptual Change*. Orlando, FL: Academic Press, 1985, pp. 11–27.

Gilbert, J.K., Boulter, C. and Elmer, R. 'Positioning models in science education and in design and technology education'. In J.K. Gilbert and C. Boulter (eds), *Developing Models in Science Education*. Dordrecht: Kluwer, 2000, pp. 3–17.

Pope, M.L. and Gilbert, J.K. 'Constructive science education'. In F. Epting and A. Landsfield (eds), *Anticipating Personal Construct Psychology*. Lincoln, NE: University of Nebraska Press, 1985, pp. 111–127.

Stocklmayer, S. and Gilbert, J.K. 'Informal chemical education'. In J.K. Gilbert, O. De Jong, R. Justi, D.F. Treagust and J.H. Van Driel (eds), *Chemical Education: Towards Research-Based Practice*. Dordrecht: Kluwer, 2002, pp. 143–164.

INTRODUCTION

Preface

An invitation to compose what amounts to a 'professional autobiography of ideas', based on published papers, is a most disconcerting experience. To address the task efficiently, one would need to be both many years away from all the relevant facts and to be emotionally detached from them, as is the case in the writing of all worthwhile histories. Neither of these conditions can be met here, for my professional life is still vigorously active. Perhaps more importantly, I have only kept sketchy records of ideas and output over the years, there being no expectation of such an invitation. I therefore do not wish to give an undue impression of coherence or progression in what is included. Scientific papers have been described as 'frauds', meaning that individual published papers have an internal logical structure that is never an accurate reflection of the actual conduct of the enquiries being reported. This must surely be true of a structure composed of many papers, produced over several decades, addressing a range of themes, albeit that they are allied. However, a structure had to be devised for this book. The established convention of publishing is that one or more storylines should run through it. I have decided to present the selected papers grouped by themes set in the context of national and international concerns in science education. The realities of decision-taking and action at the time were infinitely more complex however.

Signposts on a road with many turnings

The self-directed (as opposed to commissioned) educational research that anybody does depends, I suggest, on three factors. First, on the traditions of the field in which the work is set. Science education, being at the interface between science and education, draws on many disciplines: the sciences themselves (of which physics, chemistry, biology are prominent), the history and philosophy of science, cognitive psychology, social psychology, linguistics etc. The tradition has also been to relate this synthesis to the practice of science education, to what actually goes on in classrooms, lecture halls and laboratories, with a view to improving and developing that practice as well as theorising about it. Second, on those contacts with teachers, peers and students, which become personally valued. This is because of the influence that such people have both on the precise focus of what is done and the way that work is carried out. That valuation is often reflected in the co-authorship of papers, for diverse expertise in this extended field has been brought together. Most of my papers have been co-authored for this reason. Third, the 'climate of the times': that evolving spectrum of opportunities, concerns and issues, in the field of education in general, that shape enquiry on any particular theme. These three threads are always interwoven and can most readily be presented along a common time-line within particular physical situations.

At school

Like many children born and brought up in London, I regularly visited the Science Museum in Kensington, enjoying the objects and dioramas of yesterday's science and technology of which it was then largely composed. However, it was the Festival of Britain on the South Bank in 1951 and especially the Dome of Discovery, the subsequent destruction of which I still much regret, that had the most impact. Taken there as a reward for passing the 11+ examination and hence for gaining a much-prized grammar school place, I dragged my mother back for a further three visits (or so she always said). Set in the social context of bomb damage and food rationing, science and technology seemed to hold the key to an exciting and prosperous future for me and for many other young visitors. Images of giant models of the structure of substances, clearly represented as derived from Bragg's work on crystallography, have persisted in my mind down the years. The exhibition certainly provided the main intellectual drive for my adult life.

This initial interest was soon built on. Everybody who has ever had contact with a distinguished schoolteacher remembers for their whole life both the style of that person and the substance of what is taught. Gus Brooks, who taught me physics in the mid-1950s, was one such. In addition to being able to effortlessly and convincingly move between phenomena, abstractions and the technological exploitation of ideas, his intellectual elegance aroused an interest in 'teaching to promote thinking'. This was immediately picked up when I went to read chemistry at the University of Leicester.

At university

Whilst most of the chemistry syllabus consisted of largely unrelated facts – a situation that does not, alas, seem to have generally changed too much since – thermodynamics seemed to have the greatest explanatory potential of what was on offer. At the same time, it was a beast to understand. Eric Peeling, later at the University of Sussex, modelled clarity of thought as he cut paths through the jungle of terms and equations, leaving behind him, for me, an interest in what thinking actually entailed.

Like many other undergraduates, I learnt much from my peers. I was taken along by a social science student to hear a talk by Richard Hoggart, author of the best-selling book 'The Uses of Literacy', and became immediately interested in the nature of the social narratives of learning and personal action. This bore fruit a few years later when, as my postgraduate studies drew to an end, it was suggested that I bring my interests in science and social science together by becoming a schoolteacher. This was seen by my peers as a radical (and perhaps suicidal) occupational move for somebody with a newly acquired DPhil from the University of Sussex at a time (1965) when university chemistry posts were readily obtained.

At school again

By the late 1960s, I was Head of Chemistry at Banbury School, then one of the largest of the newly created comprehensive schools in the country. The Principal, Harry Judge, was nationally famous, so the school had a flow of visitors, such as Eric James, Richard Crossman, Margaret Thatcher, all of whom interacted energetically (and very differently!) with both staff and students. The educational context of the school had the effect of encouraging the staff to 'look for the big picture' and I started to think beyond the confines of the immediate classroom and laboratory. At the same time, the now-defunct Schools' Council published (in 1972) its influential Bulletin No.7 'Changes in School Science Teaching'. Pages 53–55 dealt with models and modelling, an approach to science education that, whilst just noted at the time, became central to my professional life much later.

The late 1960s was the period in which the Nuffield Science Projects were reaching their crescendo with the production of the 'A' level schemes. Although too late to be a 'trials school' proper, Banbury was one of the first schools to adopt Nuffield 'A' Level Chemistry. I was asked to contribute to a talk on the 'A' Level schemes to members of the Association for Science Education in Cardiff. The lecture room was packed, the excitement palpable, and the questions incisive. I came away convinced, perhaps for the first time, that science education could be made both a success for students and fun for the teacher at the same time, provided that it had a coherent intellectual basis. The Nuffield approach, did not, in the long run, turn out to provide that basis. However, it did leave a valuable legacy of ideas with me, including – a blindingly obvious point – that students learnt best when actively intellectually engaged with their work.

At university again

After a brief period at Keele University, 1974 saw me at the University of Surrey, where I would stay until 1988. Lewis Elton, a model of energy and application in professional work, was the Head of (what evolved into) the Institute for Educational Development. Lewis, an excellent physicist in his own right and therefore from a background of quantitative research methods, exemplified the need to be adaptable in the face of new demands. It was he who introduced me to qualitative research methods, opening the door to an under- standing of how people actually learn. He could also both pick and support staff and students: Diana Laurillard, Margaret Cox, Maureen Pope, David Boud, Mike Watts, Elizabeth Beatty and I, from that small group of staff and students, all went on to gain professorships.

It was in the period 1978–1984 that many of my professional commitments were firmed up. I enjoyed an extended period of collaboration with Roger Osborne, Reader in Physics at University of Waikato, New Zealand. Not only did we jointly develop the 'Interview-about-Instances' technique for eliciting individual's understandings of the concepts of science that has been so widely used throughout the world, but I also came to see the great value of international collaborations: the operation of an 'invisible college'. His untimely death in 1985 in a road traffic accident brought a life-long grief into the lives of many, including myself. Those who knew him still ask: What might have been? At about the same time, I worked extensively with Maureen Pope, who introduced me to George Kelly's theory of 'personal construct psychology' and who showed me the educational value of the history and philosophy of science, thus providing a multi-stranded coherence to my study of children's conceptions. The notion of building science education on the study of the class- room experiences of both teachers and students was reinforced by work with Mike Watts, who has always shown great insight into both. Martin Sweeting (later Professor Sir Martin Sweeting) aroused my interest in the relationship between science and technology through his ground-breaking work on small Earth-orbiting artificial satellites.

By the mid-1980s the vandals had arrived at the gates of Surrey – metaphorically speak- ing at least – and seemed intent on downgrading the hitherto emphasis on science educa- tion. It seemed, alas, a good idea to move. In 1988, I became Professor of Education at The University of Reading, where I have remained since. Most of my work there has been built around the further development of the professional relationships started at Surrey. Carolyn Boulter shared her enthusiasm for the study of the language in which science edu- cation is conducted. She was also strongly instrumental in getting my work focused on 'models and modelling in science education', a process that started in about 1995, and which continues to this day. Rosaria Justi, from Brazil, and I have worked on student-teachers' and experienced teachers' perceptions of what a model *is*. We have jointly evolved a perception of the scope and limitations of the complex field of models and modelling in science education. Miriam Reiner, from Israel, and I have looked at the role of thought experimentation in science education and at its conduct in the classroom. Beverley France, from New Zealand, and I have looked at adults' modelling of the processes and products of biotechnology and at its consequences for the 'public engagement in science and technology'. An allied area of work has been the learning that takes place in 'science and technology centres' as visitors use interactive exhibits, where my main collaborator has been Susan Stocklmayer from Australia.

In 1991, I became Editor-in-Chief of the *International Journal of Science Education*. What started as the small commitment of producing four Issues per annual Volume has grown greatly so that 15 Issues are now produced each year. It continues to provide me with an opportunity to support science education at world level and especially to provide professional development for new and/or poorly resourced researchers. More selfishly, it enables me to keep abreast of trends in the field at global level. This complex task would not have been possible without the support of a succession of Regional Editors: Chris Dawson, Beverley Bell, Gaalen Erickson, Kathleen Fisher, James Wandersee and currently Janice Gobert, David Treagust and Jan Van Driel.

In 2001, I was greatly honoured, and even more surprised, to be given the annual award for 'Distinguished Contributions to Science Education through Research' by the USA based National Association for Research in Science Teaching (NARST).

The future

The 'retirement age' for men of 65 years, introduced in Great Britain in 1911, was a great boon to most of the population who then did heavy manual work, it seems rather irrelevant to an academic in today's world. I propose to ignore it. The future looks bright, for I am now working ever-more extensively with Matthew Newberry and the teachers in the Cams Hill Science Consortium, who are conducting action research into the significance of 'models and modelling' for all aspects of the school science curriculum. I am also working on chemical education, especially on the role of visualisation in learning the ideas of chemistry. A full circle then, certainly starting from, and to a lesser extent finishing with, chemistry in the classroom. It would be wonderful, at the close of a career of 40 years, to be able to help the science teachers of England regain some sense of professional self-determination after many years in the wilderness of the 'Stalinist command economy' created by the educational policies of successive UK governments since 1988. There are glimmers of hope.

The themes addressed

What can be distilled from the above to characterise my professional commitments?

- An interest in the nature and development of thinking in respect of the ideas of science.
- The need to provide students with a curriculum that is both philosophically valid and personally interesting to them.
- The importance of taking advantage of opportunities for informal science education by both school-age pupils and adult visitors.
- The value of qualitative research conducted in or near classrooms, with close attention being paid to how ideas are actually expressed.

- The need for a boot-strap relationship between educational theory and practice. Without such a relationship, the former can become sterile and the latter unproductively pragmatic.
- The value of international collaboration as a way of generating an active synergy in research of diverse knowledge and skills.

Mixed to taste, illustrated with a selection from the 175 items (books, chapters, papers) I have published so far, they have produced a book divided into four parts.

Part 1: explanation, models, modelling in science education

Any curriculum in science education can only be successful if it has a clear relationship to the nature of science. Whilst science education can serve several purposes, one of them must be to provide insight into how science is conducted as a social enterprise. In other words, science education must be as 'authentic' as the peculiar circumstances of education will allow. This part consists of chapters on what I see to be key themes in the nature of 'authentic science' education: the provision of convincing explanations and the role of models and modelling in doing so. This work took place against a background of an ever-growing interest throughout the world from the mid-1980s onwards in the 'representation of the nature of science in science education'. This has perhaps been a response to the anti-science sentiment that has become progressively more apparent in most societies.

In Chapter 1, Carolyn Boulter, Margaret Rutherford and I discuss the notion of 'explanation' in science education. It is generally agreed that the purpose of science is to provide explanations of the natural world. Philosophers normally view 'explanation' as meaning causal explanations: the ascription of cause–effect relationships. However, in the case of science education, we argue that a causal explanation is preceded by statements of intention (why is this enquiry worth doing?), a description (how does the phenomenon behave?), an interpretation (why does the phenomenon behave as it does?), followed by one or more predictions (how might it behave under other circumstances?). We also talk about 'appropriateness' in an explanation: ensuring that the correct type of explanation is provided at a suitable level. Lastly, we discuss how these issues are viewed by the four major constituencies with an interest in the science curriculum: scientists, curriculum planners, teachers and students. Authenticity is all too often the casualty of the conflicting interests of these groups.

Whilst theories are often assumed to be the core of science, I have always found them hard to discuss and use. Of greater use are models: attempts to represent theories in particular situations such that they can be readily visualised. In Chapter 2, Rosária Justi and I argue the case for the role of models in science. We then identify, we think for the first time, the eight distinctive models that have played a major role in the study of 'chemical kinetics' over the years.

In Chapter 3, Rosária Justi and I take this argument one stage further and report on an enquiry into how these eight models of chemical kinetics are featured in textbooks and are used by teachers. It might have been expected that students were introduced to some or all of the models in the historical sequence of their development. What we found was that teachers select parts of different models to produce 'hybrid' models to serve a set of pedagogic purposes. Whilst this may be both convenient and efficient for the teachers, the practice completely nullifies any attempts to teach the nature of science. Hybrid models can never be replaced – the normal process of progress in science – because they never existed in science *per se*. So, whilst I would advocate the use of models in teaching, the crucial question is 'what models?'

One of the major roles for models in both science and science education is to form the basis of 'thought experiments': the design and conduct of experiments in the mind.

Whilst some eminent scientists and philosophers – Einstein, Mach, Kuhn – have written about their own experience of thought experimentation, there has been little enquiry into the use of this elegant technique in science education. In Chapter 4, Miriam Reiner and I discuss the field and report an empirical enquiry into what a group of students said and did when given problems that entail thought experimentation. Of greatest interest are the roles that visualisation and 'bodily knowledge' played in their work. Miriam and I became convinced that the central role of thought experimentation in authentic science education must be recognised.

The processes of modelling, the act at the heart of creativity in science, has been little studied. In Chapter 5, Rosária Justi and I propose a 'model of modelling', if only as a way of stimulating discussion of this complex phenomenon. We use this as a framework with which to review the various reports of 'developing modelling skills' that had been published up to 2002. We also report an enquiry into what teachers think modelling involves, for the broader introduction of modelling skills will entail them teaching such skills. There is much still to do before modelling, surely a cornerstone of authentic science education, is fully introduced into systems of science education.

Part 2: relating science education and technology education

I have never really understood why science education is so highly valued at a political level that it is mandatory for all school-age children in many countries. It is technology that is of interest to young people and that they have to deal with in the everyday world. Technology is the basis for so much employment in adult life, not science in isolation. Although there are periodic calls for science and technology to be more closely related in education, little practical activity, in the form of money for research and development, is made available. Even an honest, if limited, effort like the introduction of 'Design and Technology' into the UK National Curriculum is being allowed to wither and die. In the chapter presented in this part, I make an attempt to clarify the situation and to suggest some ways forward.

In Chapter 6, published in 1992, I present an analysis of the nature of technology derived from the work of Pacey. I then touch on the vexed – still unresolved – issue of the relationship between science and education. Having presented some ideas of what technology education might involve, I review the ways in which technology education might be associated with science education. Fifteen years on and these modes of association remain as unexplored as then. Surely technology – a key achievement of the cultural history of humanity – cannot go so comprehensively unrecognised in educational systems. Nor can it continue to be isolated from science education.

Given my ethusiasm for models and modelling, it is to be expected that I would advocate their use as a 'curricular bridge' between science education and technology education. In Chapter 7, Carolyn Boulter, Roger Elmer and I, spell out the language that is needed to discuss the range of ontological status that a 'model' may attain in science and the representational forms that may be used to express it. Against the background of brief analyses of the natures of science and of technology and of science education and technology education we suggest reasons why models could form an effective bridge between technology education and science education.

Part of that bridge rests on the central role that modelling plays both in science and in technology. In Chapter 8, Trevor Davies and I make the case for the commonality of processes in the creative act of modelling in science and in technology. We go on to suggest ways in which this creativity can be fostered and hence the bridge strengthened.

Most of the exemplars of the relationship between science and technology discussed in education are drawn from history, often the history of the nineteenth century. It is a truism to say that we must prepare young people to live in tomorrow's world, yet it is rarely done.

Why do we not look at today's science and technology, or, better still, at tomorrow's? Perhaps it is because the worlds of science, of technology and of education, are so separate from each other. In Chapter 9, written when I was at University of Surrey, I suggest the ways in which Earth-orbiting satellites – then just a relative novelty but today an everyday tool for society – could be used to link science education and technology education. Satellites are wonderful repositories of science and technology that have drastically changed the way we communicate with each other. It seems foolish to ignore their educational potential to the extent that has occurred: it is only one of their products, images of the surface of the Earth and of other planets, which is explicitly used in formal education, and then mainly in geography.

Part 3: informal education in science and technology

It is possible to look at the structures for the provision of education as being placed along a continuum. At one end is formal education, where the curriculum is inflexible and where participation is compulsory for school-age children. They have no choice over what is taught, when, how or why. They will actively participate in classes only if the material is of some direct interest to them or if the examinations that inevitably follow serve some instrumental purpose for them. At the opposite end is informal education, where the learner has complete choice over what is learnt, when, how and why. Informal educational provision is usually designed with the adult, or family, market in view. In between these two extremes lies non-formal education, where the resources developed for informal education are co-opted into formal education, if only for a short time. Over the last decade or so, the value of informal and non-formal science and technology education has come to be more recognised. Whilst most has been put into developing new provision, often on the basis of intuition, some research has also taken place. I have made a contribution to what is a very small body of research evidence.

In Chapter 10, Susan Stocklmayer and I discuss informal chemical education. We originally did this because, of all the three major sciences, chemistry is the least extensively represented in the informal provision on offer. We distinguish between the provision of a *situation*, which is an object, the development of a *context*, which the visitor creates by actively engaging with an object, and a *narrative*, where the visitor links a context to some ongoing aspect of their lives. To be most educationally effective, the situations provided must be as strongly linked to the narratives of 'the public' as possible. We then discuss the extent, scope and limitations of the various types of situation that are provided in informal science and technology education: books, newspapers and magazines, television, live shows and lectures, museums and interactive science and technology centres.

Chapter 11 follows up on this theme, not specifically focusing on chemistry but rather on the situations provided by interactive science and technology centres. Here we discuss the design of individual exhibits, a key element in deciding the quality of the experience to be had by a visitor. Drawing on theoretical work on the nature of analogy, we present a typology for exhibits. We suggest that the learning that results from the use of an exhibit is governed by whether the visitor understands the conventions underlying its construction and by the nature of the memories that it evokes.

Chapter 12, by Mary Priest and I, is a case study of non-formal education. Mary visited a primary (elementary) school and saw the preparations that were being made for a visit to the Science Museum in London. Those preparations involved locking the proposed visit into the science curriculum. The school party was then accompanied on the visit, with pupils being observed and interviewed as they went round an exhibition. Later, we visited the school to see what follow-up to the visit took place. Perhaps the most significant outcomes of this study was the realisation of the value of 'critical incidents': events in the visit when pupils related their museum experience to their school experience. To my mind, this case study illustrates just how non-formal science and technology education should be conducted.

Part 4: alternative conceptions and science education

It would be admirable if students received an authentic and relevant science education, facilitated by a suitable mix of formal, non-formal and informal provision. But what would be the nature of that education? How would students learn science and about science?

From the mid-1960s onwards, in the United Kingdom at least, the proportion of the school student population who study science has increased rapidly and today all students do so, willingly or not. With this explosion in numbers the diversity of prior attainment came to the fore and the need for a convincing psychology of learning became apparent: how were the abstractions of science to be effectively taught and learned to all?

The first approach considered built on the work of Jean Piaget. Based on an impressive programme of qualitative research by the Geneva School, Piagetian psychology represents an individual's capacity for thinking as developing through a series of invariant 'stages'. The position of an individual in this hierarchy at a given time governs the nature of what can be learnt at that time. It was turned into a technology for mass appraisal by science educators through its transformation into a quantitative methodology. The results seemed to show that relatively few students were capable of a full understanding of the ideas of science expected of them by society. On the other hand, some children showed an intellectual competence beyond that predicted for them by Piagetian psychology. It was, therefore, politically undesirable and pragmatically flawed. The search had to continue.

The first step taken was to stop looking for universal truths and to start to look at what students actually learn about the specific concepts that are of importance in science. The elegant interview technique that had been used by Piaget was the basis of the several research techniques that evolved in the late 1970s and early 1980s. Roger Osborne and I jointly developed the 'Interview-about-Instances' (IAI) technique that has since been used throughout the world, leading to scores of theses and hundreds, perhaps thousands, of research papers. Chapter 13 is surprisingly the only place where Osborne and I, together with Mike Watts who made a lot of use of the technique, wrote down what IAI actually entails. The use of the technique still has value as a way of sensitising trainee science teachers to the fact that students do know something, even if it is not what the teacher wants.

The late 1970s through to the mid-1980s saw an explosion in the number of studies of so-called 'alternative conceptions' (if viewed as genuine human creations) or 'misconceptions' (if viewed against the template of socially accepted science). It seems that most young people, before they are formally taught science, have and employ an understanding of words adopted by science, for example, 'force', 'energy' etc. In any population there is only a finite number of distinct understandings of a given word: typically between 5 and 12. In Chapter 14, Mike Watts and I discuss the nature of 'concept' and the ontological status of students' responses. We give an overview of the pattern of understandings then known for specific words and speculate on the nature of a model for the conceptual development that science education would still demand of all students.

There was no doubt that some definite action to promote conceptual development was needed, for it became evident that students' alternative conceptions interacted with what they were taught in often unexpected ways. In Chapter 15, Roger Osborne, Peter Fensham and I, present a series of models based on empirical evidence for the nature of that interaction.

Alas, even today, there is no accepted approach to promoting conceptual development in science for all students. For me, the most convincing theory of learning that I have come across is based on George Kelly's 'Personal Construct Psychology', outlined in Chapter 16 by Maureen Pope and myself. Kelly's core idea is that learning is the result of active mental engagement by an individual learner. Although the 'social construction of knowledge', after the ideas of Vygotsky, has become popular in science education in recent years, I am quite unable to see how a group of people actually learn something. Like Kelly, I believe that members of a group cause an individual to think, propose ideas that might be worth considering and evaluate each others' learning. However, mental activity is an individual art form and does not reside in some kind of inter-personal miasma. On the other hand, the 'social constructivist' approach does place great value on co-operative learning, which seems desirable.

And now, that title . . .

Scientists investigate the things in which they come to be interested – the consequences of which are sharply pointed out by feminists, ecologists and others whose interests do not seem to be served. In that sense then, decisions on the focus of scientific enquiry are social constructions. The product of such enquiries can only be called scientific if they are agreed with at a given time by a significant number of other scientists. However, this is not to say that the products of science are socially constructed, for all successful enquiry must relate convincingly to the world around us through the production of evidence against pre-established norms.

We must assume, in the absence of any convincing alternative view, that there is a common physical world – a real world – that we all share. However, that is not to assume that we all experience that real world in the same way. All the evidence available suggests that we all actively construct an interpretation of the real world, that construction depending on our focus of attention, prior experience and presumptions. So then, we individually 'construct worlds' in which to live. Science education is certainly about introducing students of all ages to the collective vision of that world currently held by scientists. But surely the long-term purpose for science education in respect of any individual must be to support that person in constructing their own world on the basis of empirical evidence supported by models.

I maintain that the task of science educators is to support people in 'constructing worlds through science education'. To do so requires that close attention be paid to the ideas of philosophy, psychology, sociology and linguistics. Given the fact that expertise in science education is thinly spread – and getting thinner – it seems necessary and desirable (as well as a lot of fun) to draw expertise from around the world. The testing of models against the reality of the classroom and laboratory can only lead to better models and, one would hope, to the better practice of science education. To do otherwise is to leave our fellow citizens at the mercy of superstitions. There are plenty of those about.

PART 1

EXPLANATION, MODELS, MODELLING IN SCIENCE EDUCATION

CHAPTER 1

MODELS IN EXPLANATIONS
Horses for courses?

Gilbert, J.K., Boulter, C. and Rutherford, M. 'Models in explanations: horses for courses?' *International Journal of Science Education*, 1998, 20(1): 83–97

This paper seeks to identify some of the issues associated with the role of models in scientific explanations. Starting from a broad definition, a typology of explanations is developed and the notion of 'appropriateness' in scientific explanations is explored. Some characteristics of explanations sought and provided for and by scientists, science curricula, teachers of science and students of science are identified. Finally, the nature of models and their contribution to explanations are explored.

What is an explanation?

Even if the most simple of definitions is adopted, that an explanation is the answer sought or provided to a specific question, it seems that no one explanation is appropriate in all circumstances and for all questioners. The notion of an appropriate explanation will vary as a function of relevant experience and the expectation perceived by those involved to underlie the question asked. When one of us (CB) was observing a class of 9-year olds, the teacher's general question, 'What is going to happen in the eclipse tonight?', produced many different answers. When another of us (JG) asked a group of 14-year olds in a science class 'What is a good explanation in science?', the answer was 'Good for what?'. When the third of us (MR) asked a number of science teacher educators 'How would you explain colour?', the first response was invariably 'To whom?'.

It is the latter investigation (Rutherford 1995) which is used as the source of the examples and exemplars in this paper. For that reason, the content area running through the text is 'light and colour'. Other areas could as appropriately have been used. The outcomes of this investigation, when combined with our previous work in the area of models (Boulter and Gilbert 1996), led us to conduct a broad enquiry into the role of models in scientific explanations. This study drew on a diverse literature, including that of the history and philosophy of science, as well as that of cognitive and social psychology. Thus, the references given for Newton and his work are only a few of the many available. The complexity of the inquiry was revealed by the fact that, when we showed interim versions of our ideas to other researchers in science education, their comments focused on many different aspects of them. The resulting papers are thus speculative, intended to raise issues rather than to arrive at definitive conclusions. This paper is an attempt to identify and explore some of the social-psychological issues involved in the relationship between questions and explanations.

An illustration of the complexity of the field is revealed by the range of types of relationships which are possible between a question asked and an explanation produced. One way of analysing these relationships is in terms of an implicitness–explicitness continuum. The most straightforward case is where the question is directly posed and the explanation provided is equally explicit. For example, where a chemistry teacher's question 'Why is chlorine gas green?' is answered by a student with 'Because the molecules are green' [*sic*]. In some cases, an explicit question can receive an implicit explanation, i.e. one where the actions of those questioned indicates that they are acting in response to their unvoiced answer. Thus, a physics teacher's question 'How does white light behave?' is followed by the students drawing a light ray being refracted by a prism in their notebooks, having decided that this is the explanation called for. Teachers' explicit questions have been characterized as being 'open', where a wide range of answers would be acceptable, or 'closed', where one specific answer is expected (Barnes *et al.* 1969). The above-mentioned drawing of a ray of light, rather than describing its colour characteristics, is an example of a closed type of response. Whether the expectation implied in a teacher's explicit question is either open or closed, it is very likely that the student will give a closed reply, having interpreted it in the narrow context of the topic being considered. Thus, the students mentioned above produced a 'physics' description of light in a physics class, rather than a 'chemistry' answer which should also have been known. The relationships can become more complex still when the question is implicit. For example, a teacher's provision of information about interference phenomena in optics can lead students to volunteer an explicit explanation for the implied question 'How does this come about?'. Although the observer could not perceive what is going on, there are cases where both the question and the explanation are implicit; for example, where a teacher provides tables of colour and wavelength, such that students seem to understand the patterns in their own data because they can predict data yet to be collected, having implicitly grasped the significance of wavelength. Alas, in many cases teachers provide explanations without students, or perhaps even also themselves, being aware of what the question is.

Martin (1972) has analysed the complexity of the field in another way, by identifying five meanings for the term 'an explanation' in science and science education, i.e.:

1   a clarification of what a phrase means in a scientific context. It is a description of how the phrase relates to a phenomenon;
2   a justification of some belief or action. It is the provision of reasons why a belief or action is reasonable;
3   a causal account of some state, event or process. It is a propositional statement stating why something is;
4   a citation of a theory from which a law may be deduced;
5   an attribution of function to an object.

The present paper is concerned with this genre of meanings, i.e. with the content of explanations. It is not concerned with that range of meanings for 'an explanation', which focuses on the activity of providing an explanation, or on the discourse involved in doing so.

One important pattern of relationships stems from the interactions between the nature of questions asked and the explanations which they elicit. A typology can be constructed from these relationships.

A typology of explanations

In a simple view, the conduct of science involves the posing of questions about phenomena in a natural world. This suggests that a typology of scientific explanations might be deduced from the expectations required by these questions. The empirical finding of Abrams and Wandersee (1995), that practising life scientists base their work on seeking answers to questions, rather than on trying to validate theory, supports this approach. A philosophically inclined person of scientific bent might ask the following range of questions.

Why is the inquiry to be carried out?

The first act of a scientific inquiry, often taken unconsciously, is to segregate out some aspect of nature from the continuum which constitutes the world in which we live. Some kind of boundary is drawn around it and the phenomenon is named. It is usual in so doing to give a reason for believing that the phenomenon chosen is capable of study in isolation (Lipton 1991). The making of a selection from the complexities of nature, the construction of that which is to be investigated, carries the assumed question of 'For what purposes is the study of this aspect of nature, this phenomenon, of importance?'. The explanation provided as an answer to this question is the *intention* underlying the conduct of the subsequent scientific inquiry.

Thus, in Newton's time, astronomy concentrated on the place and movement of the Earth amongst the planets as part of a general inquiry into 'the place of humanity in God's universe'. The lenses used in the telescopes were prone to chromatic aberration, so the images formed had coloured edges and were poorly focused. The intention of the work carried out by Hooke, Huygens and Newton at that time was to understand and eliminate this problem (Westfall 1994). Even though these three researchers had the same *intention*, their work differed in other respects, as will be noted later.

How does the phenomenon behave?

Most scientific inquiries, at least those into phenomena which have not hitherto been extensively investigated, always address the question 'How does this phenomenon behave?' at an early stage in their conduct. A *description* is an explanation which provides an account of its behaviour, initially as originally encountered and later under the effect of experimental manipulation. In both cases, behaviour often changes over a period of time. Newton, for example, collected together examples of many situations where the phenomenon of colour was found, e.g. glass, water droplets, paint and fabrics, and investigated their behaviour (Sabra 1981).

Of what is the phenomenon composed?

The naming of entities within the phenomenon, together with the identification of their relative spatial and temporal distributions, constitute an *interpretation* of its physical structure. This type of explanation often allows it to be ascribed to a group of similar phenomena and thus to be classified. Newton's early work on colour, like that of many other natural philosophers at the time, involved a naming of the colours of the visible spectrum, as revealed by the refractions of white light, and a demonstration that the same colours, in the same spatial relationship, were produced in different situations, e.g. in glass, in a water droplet (Sabra 1981).

Why does the phenomenon behave as it does?

The explanation produced in response to this questions is based on *causation*. A mechanism is proposed by means of which the phenomenon produces the observed behaviour through the operation of cause and effect on the entities of which it is composed, either deterministically or probabilistically. Newton, for example, initially tested Hooke's 'theory of pulses' as a way of explaining rainbows and chromatic aberration. He went on to conclude that white light is heterogeneous and may be split into the different colours because they are differently 'refrangible' (Finegold and Olson 1972: 31–35). Hooke had also provided a mechanism to explain colours (Westfall 1994). This mechanism used a scale of colours which related to how much 'darkness' had been added to white light. The 'scale of strength' took red (white light with the least addition of 'darkness') as the strongest colour, through dull blue, to black, which involved the total extinction of light by darkness. Hooke's mechanism, however, provided no causal explanation to underpin the 'scale of strength'.

Causality has, of course, been much studied. The deductive-nomoligical model takes a standard form for deterministic explanations (Hempel 1965):

- the statement of causal law;
- a statement about a phenomenon;
- a deducted causal explanation.

For example:

- Light is refracted by the boundary between two translucent media.
- Light is being refracted.

Therefore, there is a boundary between two translucent media.

The Statistical-Probabilistic model of causal explanation (Hempel 1965) is very similar in structure. The causal laws are statistical in nature, so that the causal explanation produced is probabilistic in form.

How might it behave under other conditions?

When quite a lot is known about the phenomenon, the way in which it may behave under different circumstances can be anticipated. The explanation provided is a *prediction* which, because it can be experimentally tested, is one of the most powerful, and indeed defining, tools of science.

Newton, having produced his causal explanation of the relationship between coloured light and white light, went on to predict and to subsequently test what would happen both qualitatively and quantitatively when light was passed through chains of prisms (Finegold and Olson 1972: 36–61). This led to his second proposition, second theory, that 'the light of the sun consists of rays differently frangible'. He stated that:

> sometimes I placed a third prism after the second, and sometimes also a fourth after the third, by all which the image might be often refracted sideways.
> (Newton 1952: proposition II, theory II, experiment 5)

The production of a descriptive explanation is often an early, if not the first, step in the production of a series of explanations of the other types in a *de novo* inquiry. Thereafter, each such inquiry takes place by seeking the other types of explanation, and indeed revisiting particular questions, in a pattern which is probably unique to it. In that the development of these other types of explanation is often facilitated by its production, it seems appropriate to include descriptive explanation within the listing for any given inquiry. This is contrary to the view of Bateson (1979: 81–83) who argues that behaviour is not a part of explanation because it makes no statement about causation in respect of the phenomenon under study. We, on the other hand, feel that it is a frequently ignored, but vital, step in the chain.

The elements of the above typology, derived by considering the questions being asked, map well onto Martin's (1972) types of explanation, i.e.,

- intentional explanation (Martin, Type 2 – justification of actions);
- descriptive explanation (Martin, Type 1 – clarification of meaning);
- interpretive explanation (Martin, Type 4 – citation of theory);
- causal explanation (Martin, Type 3 – causation);
- predictive explanation (Martin, Type 3 – deduction of future event).

Martin (1972: 69) himself casts doubt over whether his Type 5 explanation (attribution of function) is indeed an explanation. Its absence from a question-based typology is therefore not surprising.

Identifying the type of explanation that a given question requests is, on the face of it, relatively simple. It is much more difficult to decide on the substance of the explanation that seems called for, that is appropriate in the given circumstances.

Appropriateness in an explanation

In both science and in science education an appropriate explanation is one which adequately meets the needs of the questioner at the time that a question is asked. An appropriate explanation should facilitate and suggest directions for, as opposed to inhibiting, subsequent questioning. The value of an explanation, often described in terms of its 'level', 'power' or 'scope', is judged in respect of performance against four criteria of apparently equal status. The first is its 'plausibility': does it seem likely to solve problems that may arise in the future and does it fit with what is already known? The second is its 'parsimony': the fewer concepts it invokes in providing the explanation, the better. The third is its 'generalizability': the more contexts to which it can be applied, the better. The fourth is its 'fruitfulness': the greater the number of predictions which it supports, and, especially, the greater the proportion of these which are confirmed by experiment, the better. These criteria, originating in the work of Toulmin (1972), have been extensively explored by Strike and Posner (1985, 1992). Newton stated that Hooke's theory of colours (using only pulses) was not only insufficient, but in some respects unintelligible. In other words, it lacked power and scope (Westfall 1994). Descartes' account of the relative positions of primary and secondary rainbows, given in terms of the rotation of ether particles, continues to be plausible, although it is not fruitful in that it cannot explain why they are coloured (Fauvel *et al.* 1988).

An inappropriate explanation is one where the match with experimentally derived data is not 'close' in terms of the application of these judgemental criteria by the inquirer. Galileo's descriptive explanations of the movements of the planets were at too great a variance with predictive explanations based on Ptolemy's interpretive explanations. Extrapolating this still further, a 'non-explanation' is one which does not in any way answer the question asked, e.g. if a causative explanation is provided although a descriptive explanation was asked for. If a child asks an adult 'Where did I come from?' and receives an answer built around the notion of pregnancy, rather than 'From Birmingham', this may not be seen by the child as appropriate. Since the explanations provided in formal science education are, by implicit convention, drawn only from the pool available within orthodox Western science, they may well be seen as inappropriate, or even as non-explanations, in other contexts.

The actual application of these criteria of appropriateness will depend on the intentions which the inquirer has in respect of a particular explanation, or indeed in respect of the inquiry as a whole. This, in turn, will depend on the relationship of the inquirer to the scientific enterprise as a whole.

Who questions, who explains and why?

In the context of formal science education, four groupings may be seen to have distinctive perceptions about scientific questioning and explanations.

Scientists

Scientists undertaking *de novo* inquiries will themselves have the opportunity to ask the questions which will lead to all five types of explanation. However, in the realities of modern laboratories, most scientists work in teams on established lines of inquiry. This means that intentional and, probably, descriptive explanations will be supplied to them. The range of question types that may be asked by an individual scientist, and hence the range of explanatory types that may be produced, is restricted.

Formal scientific inquiry, into many phenomena and over a relatively short span of time, often remains locked into a cycle of producing ever more refined explanations of the latter types outlined above within the framework of an intention which has not been substantially changed. For a contemporary example, one has only to look at the current rapid-fire production of papers in the journal *Nature* on the phenomenon of AIDS. When an existing intention no longer poses challenging questions, the question 'How does this phenomenon relate to other phenomena?' produces a refinement, or more extensive change, in the intention being addressed. Certainly, discoveries are sometimes made, new models proposed, when somebody does not 'know' that what is being observed is 'unimportant', e.g. Bell and the inter-stellar material seen that led to the discovery of pulsars (Latour and Woolgar 1979). However, more commonly the focus often shifts either to a more microscopic level (e.g. in chemistry, from the composition of the AIDS virus to its spatial geometry) or to a more macroscopic level (e.g. in biology, from the classification of an individual organ to its function in an organism and how this may have evolved). Kuhn's (1970) vision of a wholesale and perhaps rapid change between paradigms has now been superseded by more gradualist representations, for example that due to Laudan (1984) in which one or more of the intentions of an inquiry, the theories and models being used to guide it, and the methodological rules within which data are collected and evaluated, are undergoing change at any one time. The work of Abrams and Wandersee (1995), based on that of Laudan (1984), suggests that society's broad aims for science, of which those adopted by scientists are a sub-set, and access to funding, provide the major dynamics for change, with the availability of novel methods and relevant theories playing a lesser role.

The questions asked and the explanations produced by scientists are intended to elucidate the nature of phenomena in the natural world. Inquiries which are successful, in the eyes of that select band who act as referees, are presented in the pages of scientific journals. For journal purposes, the most appropriate explanations are those that are, against an absolute measure, the most plausible, most parsimonious, most generalizable and most fruitful that can be devised.

Curriculum designers

The central purpose of science education is to introduce students to the processes and ideas with which scientists make sense of the world. If science education is to be 'authentic', that is to be related as closely as possible to the actual conduct and outcomes of science (Roth 1995), then it should be based on the questions asked by and the types of explanations produced by science itself. Too little is yet known about the nature of 'authentic science processes' for this aspect to be addressed with any confidence (Millar and Driver 1987). At the primary school end of science education, simple but scientific questions and explanations may just be implied through the requirement to study certain phenomena. At the other end of the age and experience spectrum, just before graduating with a science degree from a university, the explanations required by the curriculum would be the same as those currently being produced by scientists.

The particular discussions taking place during the development of a science curriculum, which will decide on the actual relationship between scientists' explanations (both historical and current) and those to be encapsulated in the curriculum, will be shaped by the range of professional experience and the ideological commitments of the discussants. These commitments will be concerned with the nature of the scientific enterprise, with the identity of particular valuable scientific knowledge, and with perceptions of what students of a given age and experience can and should do in respect of science. These commitments will be set within a framework of general educational aims, e.g. the development of broad 'skills of learning'. The quality of the judgements made will depend on the educational breadth and depth of insight into science and into education generally of those involved. To take the example of the curriculum that has been very influential, Fuller (1991) gives a vivid account of how a group of talented individuals, several very successful as scientists, schoolteachers, school equipment designers and teacher educators, were brought together to form the central team of the Nuffield A Level Physics Project.

Roberts (1982, 1995) has produced a framework, derived by the analysis of documentation, with which to discuss variation in the intended *emphases* of curricula, which represent the outcomes of these discussions. He identified seven emphases: the 'everyday applications' emphasis, where the intention is that students should acquire the understanding of science – its processes and outcomes – which are helpful in monitoring and controlling their own immediate physical environment; the 'personal explainer' emphasis, where the focus is on how individuals can and have sought the main explanations of science; the 'structure of science' emphasis, where the curriculum focuses on how the community of scientists functions intellectually, on a day-to-day basis, and how science progresses, on a longer-term basis; the 'science, technology and decisions' emphasis, which focuses on an understanding of the limits to science, in terms of the boundaries of its relationships with technology and also with other schemes of belief which purport to be scientific, i.e. the pseudosciences; the 'correct explanations' emphasis, where the focus is on introducing the student to the main explanations produced so far by science; the 'solid foundations' emphasis, where the focus is on preparing students to engage in more advanced science education and, ultimately, in science itself; and the 'scientific skill development' emphasis, where the focus is on the component skills which are said to contribute to the practice of science, e.g. experimental design, the conduct of experiment, observation, data collection and summary, data interpretation and the drawing of conclusions.

Given that curricula are usually designed by committees, it is likely that most will be a hybrid of the emphases. As each emphasis will be predicated on different intentional explanations, a broadly based hybrid curriculum would require students to be taught a wide range of explanations. The overall appropriateness of the collection of explanations contained within a given curriculum will be determined by the breadth of the hybrid and by the capabilities, prior achievements and anticipated futures of the students as perceived by the curriculum designers. However, it is important to note that most, if not all, of the questions for which explanations are to be presented will have been posed by the curriculum designers.

Science teachers

Science teachers work within a space shaped by the expectations of the curriculum, the demands of public accountability and the interpretations of both as set out in the textbooks adopted.

An analysis of the UK National Curriculum for Science (DFEE 1955) suggests that it places great stress on what Roberts (1995) would call the 'solid foundations', the 'correct explanations' and the 'scientific skill development' emphases. The range of intentional explanations to be taught is thus quite restricted. Bowe *et al.* (1992) found that, when implementing the National Curriculum for Science (England and Wales), teachers made the minimum changes to their existing practice. This suggests that the three, now required, emphases fitted their existing practice closely. This curriculum, like many others, states the descriptive interpretive and causal explanations which are to be taught without giving the questions to which they are a response. Teachers thus have to infer the scientific questions which are to guide their work.

Public accountability for teaching increasingly takes the form of requiring students to sit for examinations based on a prescribed curriculum. Attention is therefore focused ever more tightly on the three 'core' emphases. Recall and puzzle-solving test items, of the range of forms of assessment available, can most readily be produced with high reliability indexes. Thus, it seems likely that teachers will be aware that their pupils will be tested on an ability to remember a restricted range of explanations and to apply them in routine situations. They will teach accordingly, a decision which is usually justified. For example, the published marking scheme for the assessment of the National Curriculum for Science (England and Wales) at Key Stage 2 (age 11 years) (DFEE 1996), which poses a very constrained set of question types, places great reliance on the production of what we would characterize as 'unambiguously correct answers couched in a language very close to that of formal science'.

Textbooks represent an interface between the demands of the curriculum and the cognitive space created by individual teachers. In many cases, they do little to enhance the latter, for they are often written by teachers with, inevitably, an idiosyncratic range of teaching experience. Rutherford (1995) reviewed the explanations of colour contained in 10 widely used textbooks covering the range from primary school to undergraduate studies. Only descriptive explanations of colour were found in the biology textbooks whilst only two interpretive explanations were found in chemistry texts. Given that 'light' is a topic in most physics curricula, it is not surprising that descriptive, interpretive and causal explanations of colour were common, although no intentional or predictive explanations were found.

If the constraints on teachers in respect of the formulation of questions and the identification of appropriate explanations are severe, then they are as nothing compared with those experienced by science students.

Science students

The demands of the curriculum, the example set by textbooks and the realities of many classrooms converge to ensure that science students are largely taught descriptive and interpretive explanations, often without the antecedent questions being explicitly discussed. One main social task of young people is to understand how the natural world works and to gauge the extent to which they are able to control and change their environment. They will therefore value intentional causative and predictive explanations which, for the reasons set out above, tend to be sidelined if not ignored. The conventional classroom seems to offer science students little, if any, opportunity to design (or even to choose) their own intentional explanations. The only context in which they do have a real opportunity to do this is provided by extra-curricular project work (e.g. Watts and West 1992).

Models and representations

Each of the four social groupings discussed above is concerned either with different scientific explanations or with the same explanations at different levels of appropriateness. In both cases and in all instances, these explanations involve the modelling of the natural world. These models may be expressed in one or more different representational modes. The four modes most commonly used are: the material, where a physical object is used, e.g. a demonstration using a prism and a light source; the visual, where a diagram is used, e.g. a ray diagram of the incidence of light on a prism; the verbal, where some oral description is employed, e.g. the description of light in terms derived from that of wave in water; or the symbolic, where the conventions of mathematics are evoked, e.g. Snell's Law for refraction. The use of each of these modes is constrained by a generic set of 'scope and limitations' which defines its representational capability. Each model, expressed through these modes which can represent it, is able to provide a particular kind of explanation.

Models and how they explain

The interdependence of models and explanations in science is both evident and strong. As Giere (1990: 105) puts it:

> Explaining is a human activity whose practice long antedated the rise of modern science...all that is the distinctive about 'scientific explanations'...is that they deploy models developed in the sciences.... What science provides for 'scientific explanations' is a resource consisting of sets of well-authenticated models. How people deploy these models in the process of constructing or understanding explanations depends on the extra-scientific context.

Models can be of ideas, objects, events, systems or processes. Common experience, language and knowledge are all needed to build models in a specialist field (Arca *et al.* 1983). Their production involves the removal of features of the world-as-experienced which seem to be extraneous in the formulation of, or address to, a question. This process of simplification enables the purpose for an inquiry to be decided upon, attention to be focused on the behaviour of the phenomenon, on the entities which are postulated as forming its structure, on the spatial and temporal relations between them, on the causes of these relations and on the possible changes in behaviour which will result from changes to these entities. This attention is possible because the models, the simplified representations, are more capable of mental manipulation, often being image-like in form and always requiring the use of less memory capacity than the original.

Models play key roles in the explanations of science. First, they can provide the basis for all five types of explanation (intentional, descriptive, interpretive, causative, predictive) either singly or, as is common, in multiple combinations. Newton's model of white light as being composed heterogeneously of colours enabled a full range of explanations for common situations to be put forward, the most sophisticated being that for interference fringes ['Newton's Rings'] (Finegold and Olson 1972). Second, they can maximize the appropriateness of any explanation by matching a questioner's need with a particular type of explanation. For a younger student, the description of the production of coloured light from a ray of white light by a prism is often enough, whilst, for an older student, the notion of waves must be introduced to provide a causal explanation of interference patterns. Moreover, models can support the production of explanations which are plausible, parsimonious, generalizable and fruitful. The wave model and the practice model of light both meet these criteria, but have done so to different degrees in the course of their development. At any one time, each has been based on a few concepts and has been useful in the explanation of different phenomena. The wave/particle duality model, within Einstein's formulation, is only needed even today for scientifically advanced phenomena. Third, the inclusion of models in an explanation readily enables their appropriateness to be gauged, not least because they can be expressed in one or more of the range of modes of representation. They have this capability because, although they are all formed by a common process, they can attain a wide and interrelated range of epistemological statuses.

All models are formed by the processes of analogy, the seeking of similarities and differences. As Guidoni (1985: 137) wrote:

> For analogies to work we must 'recognise' something and then disregard but remember this first 'obvious' fit and 'force' another.

Analogies are thus drawn between a *source* (something which is perceived to be somewhat like the phenomenon under study), and the phenomenon itself, which is the *target* (see, e.g. Gilbert 1994). Hesse (1966) divided the entities and structural relationships which are possible between a given source and target into three sectors: the positive analogy, those where some similarity seems to exist; the negative analogy, where no useful similarity exists; the neutral analogy, where no clear decision over its similarity can be arrived at when the general comparison is made. Examples drawn from the wave model of light, based on a wave in water, are respectively: the notions of wavelength and frequency; the fact that a wave in water appears to be solid; the fact that the amplitude of a wave in water decreases as it moves. The key aspect to drawing a positive analogy is the decision on the degree of similarity between the chosen features of the source and target (Gentner 1983). This decision will undoubtedly have a strong bearing on the appropriateness of the model so constructed for the explanatory purposes being served.

The outcomes of these comparisons can have a wide variety of statuses. It is possible to talk about a *mental* model, which is a personal, private, representation of a target. By their very nature, mental models can only be fully appreciated (if then!) by the individuals having them. We all have our own model of what white light 'is like'. An *expressed* model is that version of a mental model which is expressed by an individual through action, speech or writing. The main characteristic of an expressed model is that it is in the public arena and therefore available for anybody to form a mental model of it. Textbooks often show a ray of white light entering a prism and a series of waves, each of different wavelength, leaving it. Each reader of the textbook will construct a mental model from that expressed model: the issue for textbooks' authors is to try and ensure that the mental models of different readers are as alike as possible. In passing, we note that the act of expressing a mental model often does seem, to the individual concerned, to change the associated mental model. Drawing the refraction of white light by a prism does (on at least some occasions!) cause a student to alter a prior mental model. A *consensus model* is an expressed model which has been subjected to testing by scientists and which has been socially agreed by at least some of them as having some merit for the time being. Consensus models are one of the main products of science. Newton's model of white light as being composed of a heterogeneous mixture of different coloured rays was one such. A *teaching model* is a specially constructed expressed model used by teachers to aid the understanding of a given consensus model; for example, the line-of-students-in-a-row teaching model of a wavefront used to explain refraction (Treagust *et al.* 1995: 25). Although there is much debate about the exact nature of mental models (Gentner and Stevens 1983; Johnson-Laird 1983), the other types of model may invoke the full range of modes of representation. A light wave may be presented, respectively, as: a line drawn on paper (the visual mode); an undulating entity (the verbal mode); something or specified wavelength and amplitude (the symbolic mode); or a sliver of crinkled cardboard (the material mode).

As a consequence of this flexibility of status and forum, models can make contributions to the explanations produced by all sectors of the scientific community.

For *scientists*, the models that they include in their published papers represent a simplified and accessible summary of their achievements. All scientists produce a plethora of mental and expressed models in the course of their work. While historians of science will, in the longer term, wish to unravel the relationships between expressed models and a given consensus model, it is only the published, and therefore consensus, model which counts as an achievement in the short term. Mathematical models are the most prized output of most scientists: they tend, at least in well-developed fields of inquiry, to play a major role in the most plausible, parsimonious, generalizable and fruitful explanations. These are often of the causal and predictive types. However, particularly where interpretive explanations are to be put forward, the visual or even material modes are used.

Curriculum designers will wish to select historically important consensus models in a given field of inquiry as being representative of the development of a particular science subject. Given that most fields of inquiry, over an extended period of time, produce models in many of the modes of representation, then curriculum designers often have a range to choose from. The problem is that the scope and limitations of each, i.e. the purposes which they might serve, are often not clearly established, which makes a rational choice amongst them difficult. Whilst the selection of consensus models will provide the main route to the definition of the content to be taught, the production of expressed (and hence, by implication, mental) models will be required if the processes of science are to be taught. Often little guidance is provided about which modes of representation are to be used and for what explanatory purposes expressed models are to be constructed.

Science teachers will value models for three reasons. First, as more accessible representations of inherently abstract theories. The more readily visualizable and memorable they are, the better. This means that models of the material or visual types will perhaps prove to be the most useful for class teaching purposes. The demonstration of colour phenomena is a powerful adjunct to the presentation of Newton's model, as Taylor (1994) showed. Second, an organizational framework through which to teach the large number of otherwise isolated facts which science has accumulated. Diagrams, as forms of visual model containing the symbolic representation of an array of facts, are particularly useful. For example, the spectrum chart of wavelengths, which is a consensus model in the visual mode, enables the relationship between significant wavelength groupings, e.g. visual, X-ray, UV, to be shown. Third, as a practical way of indicating the level of appropriateness at which they are required to learn. Each mode of representation carries with it an implicit set of scope and limitations to its use. Each historical model, within a given mode of representation, enables only certain purposes to be addressed. Combining the two perspectives enables what is to be learned to be framed. Teaching models will play a major role in the introduction of consensus models. Each one may, or may not, be in the same mode of representation as the relevant consensus model. This is a situation which calls for additional caution in their use as the basic conventions of interpretation for any two modes are different. Expressed models will play an equally major role in an authentic introduction to the processes of science. These are likely to be in any mode of representation, dependent on the purposes to which they are to be put.

Science students may well value models, and particularly consensus models, for two reasons. First, being simplifications and readily accessible, they reduce the load on both long- and short-term memory. The more parsimonious, the more highly generalizable and the more memorable they are, the better. Second, they provide a clear indication of the appropriateness required by the examiners.

In so far as students study science in order to find out more about nature, models should be valued because they provide a range of representations of what nature *might* be like. This notion of representation is often abandoned in favour of viewing models as nature *is*.

Little systematic work has been published so far on the particular merits of, and relations between, the various modes of representation of models – the visual, the verbal, the symbolic, the material. It may well be that individuals vary in the styles of thinking in which they engage and that the several modes of representation differentially facilitate these styles. Equally, there may be differences between the modes of representation in terms of the types of explanation which they can support. Lastly, the use of the different modes of representation of models can perhaps fine-tune the appropriateness of an explanation in a given field of inquiry. Linking an explanation, a mode of representation and a learner's mental framework together is self-evidently a complex task.

Horses for courses

This metaphor, drawn from the world of horse racing, states that particular horses will be better suited to racing than others on a course of given surface conditions ('the going'). Drawing some positive analogies from within it (Ortony 1979; Hesse 1996) enables the notions of 'appropriateness' in this paper to be revisited.

A typology of explanations does enable a discussion about what kind of scientific answer ('the horse') is required by a particular question ('the course'). The various social groupings engaged in science and science education seem to attach a particular importance to different sub-sets of the possible types of explanation (to 'different kinds of horse'). It has become evident that, of all these social groupings, there is currently least direct insight available into the workings of curriculum designers. Whilst all the types of explanation and all the social groupings do make use of all the types and modes of representation of models, it is evident that the interaction between these four factors is complex. Case studies would be a great help in modelling the relationships.

Appropriateness, from the point of view of the person answering a question in good faith, is gauged by the way in which the factors are reconciled in the particular explanation produced (how the 'going' is judged). Little is known about how a particular scientific explanation meets the criteria of plausibility, parsimony, generalizability and fruitfulness in a given context. Even less is known about how the questioner gauges the appropriateness of a provided explanation for their purposes.

References

Abrams, E. and Wandersee, J. (1995). How does biological knowledge grow? A study of life scientists' research practices. *Journal of Research in Science Teaching*, 32(6), 649–664.

Arca, M., Guidoni, P. and Mazzoli, P. (1983). Structures of understanding at the root of science education, Part 1: Experience, language and knowledge. *European Journal of Science Education*, 5(4), 367–375.

Barnes, D., Britton, J. and Rosen, H. (1969). *Language, the Learner and the School* (London: Penguin).

Bateson, G. (1979). *Mind in Nature: A Necessary Unity* (New York: Dutton).

Boulter, C. and Gilbert, J. (1996) Texts and contexts: framing modelling in the primary classroom. In G. Welford, J. Osborne and P. Scott (eds), *Research in Science Education in Europe* (London: Falmer Press), 177–188.

Bowe, R., Ball, S. and Gold, A. (1992). *Reforming Education and Changing Schools* (London: Routledge).

DFEE (1995). *Science in the National Curriculum* (London: Department for Education and Enterprise).

DFEE (1996). *Key Stage 2 Tests: 1996* (London: Department for Education and Enterprise).

Fauvel, J., Flood, R., Shortland, M. and Wilson, R. (1988) *Let Newton Be!* (Oxford: Oxford University Press).

Finegold, M. and Olson, J. (1972). *An Enquiry into the Development of Optics: Conceptions of Light and their Role in Enquiry* (Toronto, Canada: Dept of Curriculum, Ontario Institute for Studies in Education).

Fuller, K. (1991). Innovation, institutionalisation and renewal in the sixth form curriculum: a history of Nuffield A-Level Physics. Unpublished PhD thesis, University of Reading.

Gentner, D. (1983). Structure mapping: a theoretical framework. *Cognitive Science*, 7, 155–170.

Gentner, D. and Stevens, A. (1983). *Mental Models* (Hillsdale, NJ: Erlbaum).

Giere, R. (1990). *Explaining Science: A Congitive Approach* (Chicago, IL: University of Chicago press).

Gilbert, J. (ed.) (1994). *Models and Modelling in Science Education* (Hatfield: Association for Science Education).

Guidoni, P. (1985). On natural thinking. *European Journal of Science Education,* 7(2), 133–140.

Hempel, C. (1965). *Aspects of Scientific Explanation* (New York: Free Press).

Hesse, M. (1966). *Models and Analogies in Science* (London: Sheen & Ward).

Johnston-Laird, P.N. (1983). *Mental Models* (Cambridge: Cambridge University Press).

Kuhn, T. (1970). *The Nature of Scientific Revolutions* (Chicago, IL: University of Chicago Press).

Latour, B. and Woolgar, S. (1979). *Laboratory Life: The Social Construction of Scientific Facts* (California: Sage).

Laudan, L. (1984). *Science and Values* (Berkeley, CA: University of California Press).

Lipton, P. (1991). *Inference is the Best Explanation* (London: Routledge).

Martin, M. (1972). *Concepts of Science Education: A Philosophical Analysis* (London: Scott, Foresman).

Millar, R. and Driver, R. (1987). Beyond processes. *Studies in Science Education,* 14, 33–62.

Newton, I. (1952). *Opticks* (London: Dover Publications).

Ortony, A. (1979). *Metaphor and Thought* (Cambridge: Cambridge University Press).

Roberts, D. (1982). Developing the concept of 'curriculum emphases' in science education. *Science Education,* 66(2), 243–260.

Roberts, D. (1995). Junior high school science transformed: analysing a science curriculum policy change. *International Journal of Science Education,* 17(4), 493–504.

Roth, W.-M. (1995). *Authentic School Science* (Dordrecht: Kluwer).

Rutherford, M. (1995). Explanations of colour. *Proceedings of the Third International History, Philosophy and Science Teaching Conference,* Minneapolis, MN, October, 2, 979–988.

Sabra, A.L. (1981). *Theories of Light: From Descartes to Newton* (Cambridge: Cambridge University Press).

Strike, K. and Posner, G. (1985). A conceptual change view of learning and teaching. In L. West and A. Pines (eds), *Cognitive Structure and Conceptual Change* (Orlando, FL: Academic Press), 211–232.

Strike, K. and Posner, G. (1992). A revisionist theory of conceptual change. In R. Duschl and R. Hamilton (eds), *Philosophy of Science, Cognitive Psychology and Educational Theory and Practice* (New York: State University of New York), 147–176.

Taylor, C. (1994). *Colour in our Lives* (London: Royal Institution, Master Class Series [video]).

Toulmin, S. (1972). *Human Understanding* (Oxford: Oxford University Press).

Treagust, D., Venville, G., Harrison, A., Stocklmayer, S. and Thiele, R. (1995). *Teaching Analogies in a Systematic Way* (Perth, Australia: Science and Mathematics Education Centre, Curtin University).

Watts, D.M. and West, A. (1992). Progress through problems, not recipes for disaster. *School Science Review,* 73(265), 57–64.

Westfall, R. (1994). *The Life of Isaac Newton* (Cambridge: Cambridge University Press).

CHAPTER 2

HISTORY AND PHILOSOPHY OF SCIENCE THROUGH MODELS

The case of chemical kinetics

Justi, R. and Gilbert, J.K. 'History and philosophy of science through models: the case of chemical kinetics'. *Science and Education*, 1999, 8: 287–307

The case for a greater role for the history and philosophy of science in science education is reviewed. It is argued that such a role can only be realised if it is based on both a credible analytical approach to the history and philosophy of science and if the evolution of a sufficient number of major themes in science is known in suitable detail. Adopting Lakatos's Theory of Scientific Research Programmes as the analytical approach, it is proposed that the development, use, and replacement, of specific models forms the core of such programmes.

Chemical kinetics was selected as an exemplar major topic in chemistry. Eight models which have played a central role in the evolution of the study of chemical kinetics were identified by an analysis of the literature. The implications that these models have for the teaching and learning of chemistry today are discussed.

Introduction

History and philosophy of science in science education

The case for a greater role for the history and philosophy of science in science education has been explored in recent years e.g. by Matthews (1994). It rests on four basic arguments. Such a role would offer scope:

(i) To teach students about the nature of science.

The central argument here is that the evolution of scientific methodology has been one of the major cultural achievements of the last half millennium and, as such, should form part of the education of all young people. However, because of the explanatory power of modern science and because of the wealth which has been created through the use of technologies associated with science, it is often treated with undue reverence. In order to combat the resulting 'scientism', more emphasis has to be placed on the institution of science as a human endeavour.

(ii) To utilise any parallels between the development of subject matter per se and the development of an understanding of that subject matter by students.

There is no general agreement on the existence of a relationship between change in science and change in the understanding of specific content. There has been the assertion of 'strong' positive relationships between the two, through the attribution of progressively 'weaker' positive relationships, to a denial of any relationship at all. Piaget and Garcia (1989) propose the strongest relationship, that of identity between the two domains, because they believe that the same mechanisms, those of assimilation accommodation and equilibration, are at the heart of both. The cognitive-historical approach of Nersessian (1992) involves attempt to recreate the cognitive processes of great scientists by the analysis of their laboratory note-books. The relationship is seen by her as one of a strong analogy, with processes which were evidently used during the making of historical advances in science becoming a model for those involved in learning. Chin and Brewer (1993) have taken one issue which is important in science and which is thought to be important in learning, namely the status of 'anomalous data'. They showed that a common scheme of analysis could be applied to the use of anomalous data in both contexts, thus demonstrating their 'similarity' (p. 3). Many other researchers have just noted apparent parallels between the two domains, but without discussing what these might entail (e.g. Watts 1982; McCloskey 1983). Some, whilst noting the parallels in outcomes, believe that the social circumstances of the practice of science in history and of the learning of science today are too different to permit assumptions of parallels in processes between them to be drawn (McClelland 1984; Lythcott 1985).

(iii) To develop students' capabilities for critical thinking.

The study of the history and philosophy of science, because it entails an engagement with the core commitments which comprise the nature of science (AAAS 1989), offers a way to develop habits of critical thinking which, it is to be hoped, have transfer into everyday life.

(iv) To overcome practical problems in the production of schemes of work, classroom teaching and the facilitation of learning.

Two major practical problems in the provision of science education could be addressed. First, the history and philosophy of science might provide a more cogent basis to both teachers and students for the structuring of the curriculum than the content-themes approach which is currently widely used. The present approach tends to produce an apparently random collection of themes, often merely isolated facts, between which it is often hard to construct any defensible relations. The intellectual evolutionary links which lie at the heart of an historical and philosophical approach have a range of inbuilt structures, dependent on the way that they are approached. Second, because it relates very readily to other subjects in the curriculum e.g. history, first language, it could support a greater degree of cross-curricular linkage. This should help students acquire a coherent, rather than fractured, overall education.

If this greater role for the history and philosophy of science in science education is to be taken on and successfully addressed, three conditions have to be met. First, the history and philosophy to be addressed must have been subject to a reputable scheme of analysis which has been consistently applied: there must be an epistemological rationale for what is to be taught. Second, the analysis undertaken must be concerned with a common class of entities so that there is a consistent focus for study: there must be an ontological rationale for what is to be taught. Third, the epistemological and ontological analyses must have been applied to key curriculum areas: there must be a content rationale for what is to be taught.

Epistemological rationale: Lakatos's analysis of scientific programmes

The production of models of change in science is an on-going activity in the history and philosophy of science. Kuhn's (1970) model is 'course grained', identifying major 'paradigm changes' across broad sweeps of science over substantial periods of time. Lakatos's (1970) is 'medium grained', in that it also identifies major changes, but here within the context of specific areas of study and over shorter spans of time. Laudans (1977) is 'fine grained', in that it looks at specific areas of study in such detail as to detect continuous small changes in aims, theories and methodology. If the purpose is to inform educational provision and practice, the level of analysis of a specific subject matter area must roughly match that which can be incorporated within the existing structure of the chemical curriculum. Lakatos's (1970) approach is at an appropriate level of detail.

Lakatos (1970) was concerned to explain the dynamics of scientific programmes: the rise and fall of particular lines of enquiry, whether pursued by an individual, or, much more likely, by an institution or a movement (see also Gilbert and Swift 1985). A Lakatosian research programme has three elements. The 'hard core' consists of a series of fundamental assumptions, which provides a negative heuristic in that it is judged irrefutable by all those who operate within that programme. To abandon the hard core is to leave the programme. Programmes which share a common hard core but which have differences in their remaining component parts are different versions of a particular research programme. One of these other component parts is the 'protective belt' which consists of a changing set of auxiliary hypotheses which are philosophically compatible with the hard core. They both operationalise the hard core, by making specific predictions possible, and protect it because, if the predictions are not empirically confirmed, it is the protective belt rather than the hard core which is called into doubt. The 'positive heuristic' is a set of suggestions or hints on how to modify the protective belt.

Lakatos (1970) saw the comparative evaluation of research programmes as the central construct with which to explain scientific progress. This comparison could be between versions of the same programme, where different protective belts were erected around a common hard core, or between different programmes, those with distinct hard cores. The key to the evaluation between programmes lies in their relative placement on a continuum between 'progressive' and 'degenerate'. As Lakatos (1970: 134) himself put it:

> A research programme is said to be progressing as long as its theoretical growth anticipates its empirical growth, that is, as long as it keeps predicting novel facts with some success ('progressive problemshift'); it is stagnating if its theoretical growth lags behind its empirical growth, that is, as long as it gives only post hoc explanations either of chance discoveries or of facts anticipated by, or discovered in, a rival programme (degenerating problemshift).

It would be very attractive, for both scientific and educational reasons, if Lakatos's scheme could be applied to programmes today, with a forward projection in time, to provide guidance on choice between hard cores (programme choice) or between protective belts (programme variation choice). However, Feyerabend (1970) has pointed out that the lack of time limit for the definitive ascription of the adjective 'progressive' or 'degenerate' to a given research programme makes this impossible. Applying the scheme retrospectively, with a projection backwards in time, the task of historians of science, overcomes the difficulty and produces a description of the dynamics of programme change in the past.

Inevitably, Lakatos's (1970) approach has been criticised. Laudan (1977: 77–78) argues that Lakatos's approach: rests on the notion of increased empirical content, which is hard to quantify; portrays progress as resting solely on the expansion of empirical content, whilst the resolution of conceptual problems is left unresolved; pays undue attention to the succession of theories within a research programme, at the expense of the elimination of competitor theories; does not put forward a rational theory for theory choice; devalues the significance of the accumulation of anomalies as an engine of theory change; and that it is built around the idea of a 'hard core' which is seen as being too rigid too accommodate the realities of theory change. As this is not a paper in defence of Lakatos's (1970) ideas, we cannot enter this debate. Suffice it to say that, for the purposes on informing the practice of chemical education, Lakatos's (1970) ideas do place an emphasis on the expansion of empirical content, on a balance between a continuity and change in theory, and on the judgemental aspects of theory choice, which represent a potentially valuable additional dimension.

Ontological rationale: a focus on models

In his original work, Lakatos (1970) saw theoretical assumptions as the main, if not sole, constituents of the 'hard core', of a research programme. Giere (1990: 85) has suggested that:

> A theory – comprises two elements: (i) a population of models and (ii) various hypotheses linking these models with systems in the real world.

A theory may be taken as a set of abstractions which are mapped onto an imaginary world through the agency of models. These models are also mapped onto the world-as-experienced. Models are thus a 'visualisable' intermediary between the imaginary world of theory and the world-as-experienced: they are representations of an idea, object, event, process or system (Gilbert and Boulter 1995). Such representations are used, for instance, to make predictions, to guide enquiry, to summarise data, to justify outcomes and to facilitate communication. It is usual for a series of models to be associated with one theory. These different models can be within one representational tradition e.g. the 'molecular theory of matter' has associated with it the material models of the 'skeleton', the 'ball-and-stick' and the 'space filling' variety. They may also be drawn from different representational traditions e.g. the 'billiard ball' and the 'statistical' models of the 'kinetic theory of gases'.

The importance of models in the conduct of science, their status as one of the main outcomes of science and their property of being more readily visualisable than theories, make them central entities in the analysis of research programmes for educational purposes. However, models can have a variety of ontological statuses. A mental model is the product of a personal activity undertaken by an individual, whether alone or within a group (Gentner and Stevens 1983). Mental models, being incapable of direct access by others, can be expressed in the public domain through action, speech, writing or other symbolic form. It is likely that the process of expressing mental models changes them. Those expressed models which, following testing, gain social acceptance by the community of scientists, or more likely by a coherent sub-community of scientists, become consensus models (Toulmin 1953). This status persists for the time that they play an active role in the conduct of research. Several models may co-exist at any one time in the practice of scientific research in a given field of study e.g. the 'particle' and 'wave' models in the study of optical phenomena. In the longer run, competition between consensus models, in the Lakatosian manner, leads to their successive abandonment for research purposes.

A historical model is a consensus model which was developed in a specific historical context (Justi 1997a). It is not necessarily produced by an individual scientist. Even when one scientist can be identified as the author of an important set of ideas, it was the acceptance of such ideas in a scientific community that characterises it as a historical model. A sequence of historical models is not synonymous with a linear temporal succession of consensus models. As was founded in the study of the historical models of chemical kinetics, different contexts sometimes co-existed at the same time. This provides the basis for the development of several consensus models at any one time (Justi 1997b).

Teaching models are those expressed models specifically developed to assist in the understanding of historical or consensus models (Gilbert and Boulter 1995). Other researchers e.g. Treagust *et al.* (1992) refer to some of these as 'teaching analogies'. In passing it should be noted that the meaning of any expressed, consensus, historical or teaching, model can only exist, for an individual, in the form of a mental model.

Content rationale: the Lakatosian analysis of historical models

In our view, two types of ontological entity constitute the 'hard core' of a Lakatosian research programme in chemistry. The first of these is the 'theoretical background', the concept of matter on which the model is based and the analytical tools used in its construction. The second of these are the 'main attributes' of the model, the fundamental aspects of the phenomenon which the model has to explain. The 'protective belt' is comprised of 'secondary attributes', ideas about other aspects of the phenomenon which, taken together, provide a comprehensive characterisation of the model.

The relationship between 'progressive problemshifts' and 'degenerate problem-shifts' associated with the use of a given model is governed by the context in which it is used. A model is developed to address particular questions: explanatory success leads to it being seen as 'progressive'. In other research contexts, where other questions are being addressed, it may not be so successful and may be seen to be 'degenerate'. This shift in status may be identified by addressing the following questions:

(i) what features of the next earlier or competing model were modified and incorporated in the model being characterized?
(ii) how did the model overcome the explanatory shortcomings of its predecessor or competitor?
(iii) what unanticipated benefits did the model have?
(iv) what explanatory deficits did it have? (Justi 1997a).

The answers to these four questions give some indication of the 'positive heuristic' associated with each model.

Models of chemical kinetics

The topic of chemical kinetics was chosen to exemplify the outcomes of the analytical approach discussed. It is both a perennial theme in chemical research and is included in chemistry syllabuses at both school and university levels in most, if not all, countries. The elements of the models are stated by the use of modern words e.g. energetics. However, it must be recognised that the meanings attached to these words evolved over the years. The words only serve as indicators of equivalent functions in the successive models. It must be remembered in the following analysis that, although the same signifier is used, that which is signified changed between different historical contexts.

The main attributes of each of the models, identified by an analysis of original papers and texts on the history of science, are the meanings attached to the notions of chemical reaction, reaction rate and the determinants of these two. It is they that constitute the heart of the topic. The secondary attributes are, in each case, the ideas that it contains in respect to:

(i) 'Energetics': ideas related to the influence of energy on the rate of a chemical reaction.
(ii) 'Catalysis': ideas which explain the action of specific materials which change the rate of a chemical reaction.
(iii) 'Reaction Path': ideas concerning the way in which a chemical reaction is viewed at the 'molecular' level.

These secondary attributes can be discussed independently of each other, but not independently of the associated notions of chemical reaction and reaction rate.

Eight consensus models were identified and each allocated a title, derived from the theoretical background or main attributes involved, to serve as a descriptive label.

The Anthropomorphic Model

Early ideas about chemical kinetics were extremely vague and their precise authorship has been lost. The theoretical background of this model was the anthropomorphical conception of matter, that is the idea that things are like people. Thus, natural elements have human qualities such as love and hate. The main attributes were defined by the use of the vague term 'affinity', a concept whose meaning was based on that of 'human feelings'. Within this model, chemical reactions were viewed as transformations in materials that are controlled by the differences between them and the affinities of them. Therefore, it would be possible to transform any kind of matter into any other kind. The only idea about reaction rate in this model was that due to Aristotle: rate was synonymous with readiness for a transformation to occur (Mellor 1904; Mierzecki 1991). This would be a function of a balance between the similarity ('affinity') and the difference between the properties of the material involved in it. There were no secondary attributes within the scope of this model.

A comprehensive characterisation of this model entails a full explicitation of the meaning of 'affinity'. Greek philosophers assumed it was something like 'love' between the materials. Western alchemists had two ideas. The first of these was derived from Greek ideas: 'affinity' was a selective force related to love and hate between substances that caused them to undergo transformation when they were placed in contact. The second was that 'affinity' arose from a similarity in composition. Such a divergence in the meaning of one of the basic notions of the Anthropomorphic Model did nothing to improve its explanatory capability. Greek philosophers, as well as Western alchemists, were not concerned with predicting the rate of reactions.

The Affinity Corpuscular Model

From the seventeenth century on, when the meaning of 'affinity' had changed, ideas concerning chemical kinetics became more fruitful. At that time, R. Boyle and I. Newton, among others, proposed the idea that matter was constituted of invisible small particles. This corpuscular view of matter constituted the theoretical background of the Affinity Corpuscular Model.

In the seventeenth and eighteenth centuries, the gradual development of corpuscular views of matter meant that the concept of 'affinity' whilst being carried forward into the new model, changed in character. Boyle, for instance, thought 'affinity' was a result of corpuscles having appropriate shapes which permitted them to adhere together and which did not result of an attraction force. On the other hand, Newton thought attraction was a sort of force by which bodies tended towards one another, whatsoever be the cause. 'Affinity' and attraction were different approaches to physical explanation. However, although chemists knew about their different origins, they came gradually to use both words as having the same meaning: the tendency of substances to combine with each other as a result of the forces between them resulting from the distinctive characteristics of the particles involved (Duncan 1996). Thus, they used both words as if they were explaining the selective attraction they observed among substances. On account of this, the main attribute about chemical reaction in this model is the idea that it is a process in which elementary particles interact with one another as a result of their chemical affinity. This was the first time that the origin of a 'force' that brings about a chemical reaction was seen to be related to the characteristics of particles or originated from them. It was this major point that differentiated the Affinity Corpuscular Model from the Anthropomorphic Model.

By being somewhat more precise about the forces operating between reacting substances, the Affinity Corpuscular Model was the first to facilitate predictions about the likelihood and the rate of a reaction. Within this model, the rate of the transformation was related to the different degrees of affinity between the particles and depended on its readiness to occur. It was from a corpuscular view that the first ideas about the influence of temperature treated, in a qualitative way, in the Affinity Corpuscular Model. T. Bergman discussed the influence of temperature on the affinity between substances. For the first time, he showed that an increase in the temperature of the system could change the affinities of the substances present in it (Mierzecki 1991).

Berzelius' ideas, expressed in this model, were the first to be concerned with catalysis. Indeed, it was he who proposed the term 'catalyst' (Mellor 1904; Laidler 1986, 1995). For him, a catalyst was a substance that increased the rate of a chemical reaction and was regenerated in the final system. He also thought that a catalysis was a result of a kind of force that was not completely independent of the electrochemical affinities of matter. In effect, he noted the empirical impact of a catalyst and its non-consumption.

The Affinity Corpuscular Model was the first one that permitted predictions concerning the occurrence and the rate of a chemical reaction. Nevertheless, such predictions had a limited scope since they were based on affinity tables that were empirically developed as an attempt to quantify the force of affinity. No theoretical inference about the nature of affinity was derived from them. Another shortcoming of this model was the lack of any mathematical treatment of the relationship between the energy of a system and rate of reaction. Such a gap was addressed in the next model.

The First Quantitative Model

From the early years of the nineteenth century, when chemists began to observe, to analyse reactions involving, and subsequently to synthesise, organic substances, a very different approach emerged. This was because the relatively slow rates of many reactions involving organic substances allowed systematic measurement to be made throughout the whole process. The conditions for the occurrence of such reactions were susceptible to progressive modification. The study of the influence of such changes on the rate of a reaction could take place. This different type of reaction continued to be interpreted with a corpuscular view of matter. The new aspect in the theoretical background of this model was the use of the differentiation of exponential equations as the basis for the interpretation of chemical phenomena. The mathematisation of chemical kinetics began in Germany in 1850 with the quantitative study of the inversion of sucrose by Wilhelmy (Farber 1961; King 1981; Laidler 1985, 1987, 1995; Mierzecki 1991; Yablonskii *et al.* 1991). The importance of Wilhelmy's work cannot readily be reduced to the proposition that it produced the first quantified rate law for chemical reactions. By developing his law theoretically, based on physical assumptions, Wilhelmy changed the traditional reliance on empirical methods only (King 1980). This can be viewed as a new paradigm in chemistry, a new way of looking at chemical phenomena and of analysing them, which changed the development of the subject.

The scientists who contributed to the development of the First Quantitative Model (mainly L. Wilhelmy, P. Berthelot and L. St Giles) still based their ideas of chemical reaction on corpuscular views of matter. Notwithstanding, some changes were introduced: they believed that a chemical reaction was a process in which particles interacted with one another and that the cause of this interaction was not some kind of affinity.

The main attributes concerning reaction rate in this model are very distinct from those of the previous models. It was possible to state from the new theoretical background that the rate of a chemical reaction was proportional to the amount of substances that underwent transformations on each occasion. In the differential equation that represented such an assumption, there was a coefficient that was specific for each reaction. The new main attributes, the chemical reaction resulted from the interaction of specific particles and that reaction rate was proportional to the number of particles reacting in a given time, were jointly the key to the success of this model.

Progress in respect of the secondary attributes was mixed. The model had nothing to say about catalysis. However, the introduction, for the first time, of a simple theoretical mechanism (King 1980) enabled the relationship of rate (which implied the measurement of concentration) to temperature to be represented mathematically. This was included in the initial studies of Wilhelmy and in those carried out later by Berthelot and St Giles (Farber 1961; Laidler 1987, 1995). However the accuracy of the predicted values was found to be questionable. This was so because of limitations arising from the theoretical background: they were based on a very simple view of the nature of chemical reaction. By considering a chemical reaction only as a simple interaction of particles, the scientists who proposed this model did not take into account important factors that would modify the variables in the mathematical equations. It was this deficiency that led to the construction of that set of knowledge that we define as the Mechanism Model.

The Mechanism Model

The main attributes of this model arose from a collaboration between the chemist A. Harcourt and the mathematician W. Esson. Harcourt focused his attention on the process, *the course*, of a chemical reaction. He added the idea that such a process, in which particles interact with one another, occurred in distinct steps. He searched for a suitable reaction, one for which quantitative experimental measurement was possible. He also tried to establish some analogies between chemical reaction and the laws of mechanics, a procedure followed earlier by Wilhelmy (Partington 1964; King 1980, 1981, 1984). Assuming that the rate of a chemical reaction was determined by the mechanism by which the reaction occurred, Esson developed empirical equations which connected the amount of product formed and time by integrating the differential equations for rates of reactions. This was done for what we now call first-order, second-order and consecutive-first-order reactions (Harcourt and Esson 1865, 1866; Laidler 1984, 1987). Such main attributes conferred a significant explanatory benefit on this model.

As in the previous model, the association between chemistry and mathematics was shown to be essential to a more accurate model. The mathematical part of the Mechanism Model arose from two theoretical assumptions: the existence of different steps in the course of a chemical reaction, and the significance of the integration of differential equations. This type of equation provided considerably more accurate predictions concerning the rate of the chemical reactions which were studied and a better interpretation of the empirical data obtained.

The notion of 'step' also enabled the secondary attributes, for the first time, to acquire a level of qualitative and quantitative sophistication which permitted the making of credible predictions. In relation to energetics, for instance, Harcourt and Esson developed an empirical equation relating temperature and rate that fitted empirical data closely. The systematic treatment of a chemical reaction at the molecular level also permitted the discussion of the influence of temperature on molecular motion and, consequently, on the reaction rate. Harcourt and Esson also established, quite independently of thermodynamics, a kinetic absolute zero temperature ($-272.6°$C) at which chemical reaction would cease because molecules would be at rest (King 1980, 1981; Temkin and Bonchev 1992).

Ostwald worked on the secondary attributes concerned with catalysis. The treatment of this subject, within the previous First Quantitative Model, amounted only to an empirical recognition of the existence of the phenomenon. However, by being able to consider the existence of different steps in the course of a chemical reaction, it became possible to discuss the action of a catalyst. From his extensive investigations into catalysis, Ostwald proposed the existence of an alternative pathway in a catalysed reaction. For him, every reaction that could be catalysed was already proceeded by itself, perhaps immeasurable slowly. He also proposed that a decrease in a reaction rate could be caused by a negative catalyst or by a secondary effect of small amount of foreign substances. Ostwald's enquiries into catalysis enabled the mechanism of the effect to be discussed more theoretically (Ostwald 1909; Partington 1964; Laidler 1987). Nevertheless, this treatment only provided an overall view of the process by proposing an alternative pathway for a catalysed reaction. It did not provide any details of such an alternative pathway. Only when experimental techniques for detecting and identifying reaction intermediates were developed, at the beginning of the present century, did it became possible to formulate more consistent explanations of the way by which a catalyst acts (Laidler 1985, 1986).

The Thermodynamics Model

The development of the Thermodynamics Model represented a substantial step forward in the study of chemical kinetics. Within this model, the theoretical background of the previous model was significantly changed by the addition of the concepts and mathematical techniques of thermodynamics and by the use of the statistical distribution law.

In this model, the key development in the main attributes was in respect of the idea of 'interaction', the reason for the occurrence of a reaction given in previous models. This was specified more tightly in terms of collisions between molecules with 'sufficient' energy. The chemical basis of the main attributes remained as in the previous model, but the use of the new theoretical background produced equations to represent reaction rates which had greater explanatory power. The thermodynamic perspective came from Van't Hoff's argument that the rate of chemical reactions must have a temperature dependence which was analogous to that for equilibrium constants.

The secondary attribute of energetics was substantially improved within the Thermodynamics Model. Van't Hoff believed that temperature was not the reason for the occurrence of a reaction, but that it was responsible for a change in its rate. He proposed the most consistent equations up to that time for the influence of temperature on rate. This was done by analysing the energy necessary for the occurrence of forward and the reverse reactions and by postulating that the standard change in internal energy (a variable involved in equilibrium equations) was not necessarily temperature independent (Van't Hoff 1986). Nevertheless, it was only when Arrhenius gave a new interpretation to one of those equations that it became widely accepted. Arrhenius concluded that it was not possible to assume that the increasing frequency of collisions of the reacting molecules was the origin of the increase in reaction velocity with temperature. He proposed that the minimum energy necessary for the occurrence of a reaction was associated with an '*energy barrier*' for that reaction. This relationship between one of the variables in a mathematical equation (the activation energy) and a physical entity (an energy barrier) gave rise to a completely new and powerful level of explanation (Arrhenius 1889; Mellor 1904; Laidlar 1987). The simultaneous consideration of two points, the association of a physical quantity (activation energy) with one of the variables of the equations and the utilisation of a statistical distribution law as the basis for the equations proposed to explain reaction rate and the relationship between temperature and rate, distinguish the Thermodynamics Model from previous models.

In the Thermodynamics Model, the equations were obtained by differentiation, a process that has been used earlier in the Mechanism Model. However, as the concepts of chemical reaction on which the initial ideas were based were different, the equations were also completely different. Notwithstanding, it is not possible to assert that the equations of the Thermodynamics Model fitted the data better than the equations of the Mechanism Model. Both of them showed good results, but were not perfect. The great acceptance by scientists of the equations of the

Thermodynamics Model was due to the provision of the relationship between the variable 'energy' and the physical quantity 'energy barrier', which had not been present in the previous model.

The definition of the activation energy also provided a new way to explain the action of a catalyst. In the Thermodynamics Model, the preferred pathway (defined in the Mechanism Model only in a general way) was specified as a pathway with a lower activation energy. Even with the acceptance of this definition, mechanisms through which catalysts act were not satisfactorily explained.

In respect of the reaction path, the distinction between order and molecularity of a reaction was proposed in this model (Laidler 1995). This was a very important step in facilitating the prediction of the mechanism of reactions and, therefore, of the equations that represent their rates. Moreover, Van't Hoff derived the integrated rate equations for first-, second- and *n*-order reactions in the forms that are still used.

The Thermodynamics Model was the most complete model proposed up to that time. It was the first time that a chemical reaction was thought of as a whole process. This was not done by chemists before Van't Hoff. However, this model had two main shortcomings. First, it did not provide an explanation of how molecules acquired the necessary activation energy (King and Laidler 1983). Second, it was not possible to obtain accurate values of activation energies, since the Arrhenius Equation was not obeyed, even for simple reactions (Logan 1982). These two points motivated studies that resulted in several different models at the beginning of the twentieth century.

The Kinetic Model

Arrhenius' ideas about the rate of a chemical reaction, together with the acceptance of the equation first proposed by Van't Hoff and interpreted by Arrhenius to explain the influence of temperature on reaction rates, represented major advances. This led to the problem of how molecules acquired excess energy becoming one of the main issues that motivated research. By changing the main focus of the explanations from the energy involved in a collision to the overall view of how collisions occur, scientists were able to provide an alternative way of looking at a chemical reaction. Assuming the kinetic theory of gases, M. Trautz and W. Lewis, working separately, addressed this question by proposing that the behaviour of molecules was analogous to that of hard spheres. They also related the 'frequency factor' to the frequency of collisions between reacting molecules and calculated the magnitudes of frequency factors (Lewis 1918; Glasstone *et al.* 1941; Christiansen 1964; Laidler and King 1983). Thus, the use of the kinetic theory of gases as a theoretical background provided a shift towards a qualitative level of analysis. A completely new main attribute was defined: a chemical reaction, involving the breaking and making of bonds, would be caused by collisions between molecules that not only had sufficient energy (as stated in the Thermodynamics Model) but which also came together with an appropriate 'spatial orientation'.

The acquisition of an appropriate orientation as a necessary factor for the occurrence of effective collisions did not comprehensively explain the source of the activation energy. However, it did lead to the introduction of a 'steric factor' into the Arrhenius Equation. This was done in order to explain why only a proportion of the collisions occurring were effective. The steric factor was the most important contribution to the secondary attribute of energetics within this model.

In respect of the relation between chemical reaction and energetics, the rate of a reaction was assumed in this model to be proportional to the Arrhenius pre-exponential factor and to the frequency of collisions. Although helpful, the actual evaluation of the magnitude of the steric factor proved to be problematic (Rice and Ramsberger 1927; Glasstone *et al.* 1941; Laidler 1987, 1995).

Despite these difficulties with the steric factor, the Kinetic Model contributed to a better understanding of the process of a chemical reaction and of the reasons why different reactions took place at different rates. This model could have provided a qualitative basis for studies into catalysis and into the mechanisms of chemical reactions. Probably some scientists did think about these issues based on assumptions of the Kinetic Model. However, no specific study has been found which was based only on the Kinetic Model. This was because of the contemporary development of the Kinetic and the Statistical Mechanics model. The Kinetic Model continued to explain the action of a catalyst in terms of a decrease in the activation energy of a reaction, as had been proposed by Arrhenius. Scientists whose studies were concerned with the relationship between catalysis and mechanism benefited from the co-existence of the two models.

The Statistical Mechanics Model

At the same time that the Kinetics Model was being developed, a parallel group of workers were adopting a quantitative approach through the application of statistical mechanics to systems of reacting molecules. The theoretical background of this model stemmed from L. Pfaundler's qualitative discussion of chemical equilibrium and reaction rates in terms of the statistical distribution of molecular speeds.

Van't Hoff (1896) had also based his ideas, expressed within the Thermodynamics Model, on those of Pfaundler. However, he took a deterministic view of collision theory. Scientists at the beginning of the twentieth century, on the other hand, used the ideas of the statistical distribution of molecular speeds and of the necessity of a critical energy in order that a chemical reaction could occur. The completely new perspective proposed an explanation for reaction based on the notion of a 'potential energy surface'. Thus, R. Marcelin suggested that a chemical reaction be conceived of as a motion of a point in phase space whose co-ordinates were the distance between the molecules and their momentum. This much more complex and abstract approach to explanation provided the elements for the discussion of the reaction rates with greater precision, thus overcoming the deficiencies of the Thermodynamics Model.

In 1932, H. Pelzer and E. Wigner put forward a mathematical equation for the rate of a chemical reaction obtained from the consideration of the passage of systems through the 'col point' of the potential energy surface. Their ideas made a great contribution to the more general model that would be presented three years later (Laidler and King 1983; Laidler 1987, 1995). Another important assumption of the main attributes about reaction rate was that it was proportional to the concentration of 'activated complexes'. An expression for the concentration of species present at a critical surface in the phase space was reached by Marcelin when he applied the statistical–mechanical methods of Gibbs. Later, this idea was extended by J. Race by proposing a more precise formulation for the activated state in a chemical reaction.

In the field of secondary attributes, the major contributions about energetics within this model were the introduction of equations that led to a temperature dependence of rate which was consistent with the Arrhenius Equation. It was also demonstrated that, for a single step reaction, the activation energy was equal to the difference between the energy of the activated state and the average energy of the reactant molecules. Knowledge about catalysis was also improved by studies of reactant adsorption on surfaces. In respect to reaction path, the study of specific chemical reactions with the aid of this model led to the development of advanced notions about reaction. For instance: M. Bodenstein suggested, and J. Christiansen developed, the idea of 'chain reaction'; W. Nerst proposed the participation of atoms as intermediates in some specific mechanisms; and H. Taylor considered the possibility that organic free radicals might be involved in some reactions (Christiansen 1919; Semenoff 1927; Rice and Herzfeld 1934; Laidler 1995).

Since this model essentially only took a quantitative approach, it did not yet represent an integrated, comprehensive, treatment of reaction rates. This shortcoming was addressed in the Transition State Model.

The Transition State Model

The Transition State Model was an attempt to overcome the several shortcomings of the Thermodynamics, the Kinetics and the Statistical Models, by integrating the theoretical backgrounds of all three. However, the comprehensiveness of the Transition State Model was derived not only from its multiple basis, but also from an assumption that reaction rate was a way of understanding *how* chemical reactions occurred.

By considering a chemical reaction as a process in which reactant molecules are distributed across a potential energy surface, the Transition State Model proposed that only some of the molecules form 'transitional complexes' at the col of the surface. The path of a reaction on a potential surface is defined in terms of inter-atomic distances, that is, by considering 'the continuous series of configurations through which the system must pass in order that the chemical change may occur' (Evans 1938: 49). A system is regarded as moving on the surface with a kinetic energy and position which was particular to the surface. Thus, it was possible to derive the concentration at which the species passed through the critical configuration of the transition state by statistical methods. Moreover, the product of these quantities with the frequency with which the species were transformed into products was regarded as equal to reaction rates. In this way, this model took into account the statistical properties of reactive systems as well as the microscopic details of molecular collisions (Evans and Polanyi 1935; Eyring 1935, 1938a,b; Evans 1938; Glasstone *et al.* 1941).

The great contribution of this model, in respect to earlier models, was not the provision of a more accurate calculus, but rather than the establishment of a deeper relationship between thermodynamics and kinetics variables. In fact, the calculation of absolute reaction rates was not an easy process because this approach required a very accurate knowledge of potential energy surfaces. Nevertheless, by applying such an approach it was possible to obtain some insight into how complicated chemical and physical transformations took place. The simultaneous consideration of quantitative and qualitative approaches made a comprehensive view of chemical reactions possible. The Transition State Model provided more accurate proposals for mechanisms and credible predictions of the behaviour of systems in their entirety. As a consequence of the utilisation of new experimental techniques e.g. in respect of fast reactions, there has been significant progress in the detailed understanding of the mechanisms of catalysed reactions.

In the meantime, the great attraction of the model is that it forges a strong and concise link between different aspects that had been treated as quite isolated in each one of the previous models. This explains why this research programme superseded the previous ones and has become the basis of the currently accepted scientific knowledge concerning chemical kinetics.

Discussion

The Lakatosian analysis of the models which have been important in the evolution of key areas of scientific enquiry offers one route to the greater inclusion of the history and philosophy of science in the science curriculum. It can help in respect of the following.

Science as a human endeavour

Most, if not all, models used in key areas of science were the outcome of the work of several individuals, often contemporaries of each other. This, in itself, can be used to dispel the idea that scientists work alone and without reference to each other. The nature, extent and mechanisms of the collaboration and competition which led to the evolution of particular models can be demonstrated by an analysis of the personal papers of the individuals concerned, where these exist. Nersessian's (1992) analysis of Michael Faradays papers on electromagnetism is an exemplary case study.

Parallels between the development of science and that of individuals

This is a very complex issue about which cannot be argued in full here. Nersessian (1989) has argued for the pedagogic value of case studies which reconstruct the mental struggles and self-educational strategies through which great scientists went as their major ideas emerged. Her view is that, if students can be persuaded of some degree of parallel between their own ideas and those held by scientists prior to the acclaimed discovery/invention, then they may use somewhat similar lines of argument and action themselves. The use of such an approach to personal conceptual development could throw some, probably very partial, light on the 'ontogeny recapitulates phylogeny' issue, outlined earlier. Nersessian (1992: 54) is of the definite view that:

> 'recapitulating' the historical approach is not possible, feasible, or desirable... (rather) the history of science (might) be viewed as a repository of knowledge of how to go about constructing, changing, and communicating scientific representations.

In this sense, the characterisation of different models and the discussion about the transition from one model to another one can be of a great importance.

The development of critical thinking skills

The conduct of science is often taken to represent an exemplar, but certainly not the sole utiliser, of critical thinking skills. 'Science for All Americans' (AAAS 1989) lists some possible skills. These are seen to emerge from an acceptance of the significance of: realism, fallibilism, durability, rationalism, antimethodism, demarcationism, predictability, objectivity, moderate externalism, ethics and so on. These skills could be developed by the careful evaluation of case study material, as outlined earlier, built around these issues.

Practical problems in science education

If the models which have been used in a field of scientific enquiry are to be used to structure the science curriculum, then a strategy for their use will be needed. Hoeve-Brouwer (1996) has outlined three approaches. First, by introducing the successive models, in the chronological order in which they evolved, over the span of the years during which the specific topic is studied. Anecdotal evidence suggests that this is a widespread practice, often carried through badly, with models being introduced and later abandoned without discussion of why this is needed. However, the approach can be implemented with due attention paid to historical circumstances, the progress of science and current curricular needs e.g. as by Ben-Zvi *et al.* (1990). This approach does have the advantage that it gives time for students to develop an understanding of subordinate concepts. Second, to introduce the full range of models during one segment of the curriculum and their application to obtain solutions to particular problems. This approach does require that students acquire mastery over quite advanced concepts and probably means that topics of contemporary importance (e.g. chemical kinetic) have to be left until late in a course. However, a more authentically historical view can be developed through this approach, as has been demonstrated by Hoeve-Brouwer (1996) and by Harrison and Weaver (1989). Thirdly, by ignoring historical models and only teaching the currently consensus model. This approach does not meet most of the aims being discussed here, represents the most severe cognitive challenge to students, but does have its advocates e.g. Niedderer *et al.* (1990).

The full use of a models-based historical approach to the science curriculum, which maximises cross-curricular linkages, can only be realised if the original analysis of the topic in question is undertaken appropriately. This implies an 'externalist' analysis, that is one which embeds the development of scientific ideas within the sociological, economic and political circumstances in which the work took place. In this study we have characterised the attributes of the models which were landmarks in the development of chemical kinetics and have discussed the transition between such models by identifying some major failures and successes of each of them. This was done from a Lakatosian point of view and with the emphasis on changes in declarative scientific knowledge rather than on the social processes that accompanied such changes. This approach will enable rational decisions to be taken about which models of chemical kinetics to include in the curriculum and when and how they should be included.

The outcomes of this analysis, if included in the curriculum, would make the notion of change in scientific knowledge more understandable. It could provide links between the sciences, here between physics and chemistry, and to other subjects in the curriculum, here mathematics, history, social geography, perhaps even languages. The analysis that we have presented is a necessary pre-requisite to a series of 'cognitive reconstructions' (Nersessian 1992) which would enable the processes of change to be elucidated, assuming that the necessary data and personnel were available.

References

AAAS (American Association for the Advancement of Science) (1989). *Project 2061: Science for All Americans*, AAAS, Washington, DC.

Arrhenius, S. [1889] (1967). 'On the reaction velocity of the inversion of cane sugar by acids', in M.H. Black and K.J. Laidler (eds), *Selected Readings in Chemical Kinetics*, Pergamon Press, Oxford, 31–35.

Ben-Zvi, R., Silberstein, J. and Mamlok, R. (1990). 'Macro–micro relationships: a key to the world of chemistry', in P. Lijnse, P. Licht, W. De Vos and A. Waarlo (eds), *Relating Macroscopic Phenomena to Microscopic Particles*, University of Utrecht, CB-Press, Utrecht.

Chin, C. and Brewer, W. (1993). 'The role of anomalous data in knowledge acquisition: a theoretical framework and implications for science education', *Review of Educational Research* 63(1), 1–49.

Christiansen, J.A. (1964). 'Some features of chemical kinetics', in O. Bastiansen (ed.), *The Law of Mass Action – A Centenary Volume 1864–1964*, Det Norske Videnskaps-Academi I Oslo, Oslo, 47–78.

Christiansen, J.A. [1919] (1967). 'On the reaction between hydrogen and bromine', in M.H. Back and K.J. Laidler (eds), *Selected Readings in Chemical Kinetics*, Pergamon Press, Oxford, 119–126.

Duncan, A. (1996). *Law and Order in Eighteenth-Century Chemistry*, Oxford University Press, Oxford.

Evans, M.G. (1938). 'Thermodynamical treatment of transition state', *Transactions of the Faraday Society* 34, 49–57.

Evans, M.G. and Polanyi, M. (1935). 'Some applications of the transition state method to the calculation of reaction velocities, especially in solution', *Transactions of the Faraday Society* 31, 875–894.

Eyring, H. (1935). 'The activated complex in chemical reactions', *Journal of Chemical Physics* 3, 107–115.

Eyring, H. (1938a). 'The calculation of activation energies', *Transactions of the Faraday Society* 34, 3–11.

Eyring, H. (1938b). 'The theory of absolute reaction rates', *Transactions of the Faraday Society* 34, 41–48.

Farber, E. (1961). 'Early studies concerning time in chemical reactions', *Chymia* 7, 875–894.

Feyerabend, P. (1970). 'Consolations for the specialist', in I. Lakatos and A. Musgrave (eds), *Criticism and the Growth of Knowledge*, Cambridge University Press, Cambridge, 197–230.

Gentner, D. and Stevens, A. (1983). *Mental Models*, Lawrence Erlbaum, Hillsdale, NJ.

Giere, R. (1990). *Explaining Science: A Cognitive Approach*, University of Chicago Press, Chicago, IL.

Gilbert, J.K. and Boulter, C.J. (1995). 'Stretching models too far', paper presented at the Annual Meeting of the America Educational Research Association, San Francisco, CA, 22–26 April.

Gilbert, J.K. and Swift, D.J. (1985). 'Towards a Lakatosian analysis of the Piagetian and Alternative Conceptions research programs', *Science Education* 69, 681–696.

Glasstone, S., Laidler, K.J. and Eyring, H. (1941). *The Theory of Rate Processes – The Kinetics of Chemical Reactions, Viscosity, Diffusion and Electrochemical Phenomena*, Erlbaum, Mahveh, NJ, 3–34.

Harcourt, A.V. and Esson, W. (1865). 'On the laws of connection between the conditions of a chemical change and its amount', *Proceedings of the Royal Society* 14, 470–475.

Harcourt, A.V. and Esson, W. (1866). 'On the laws of connection between the conditions of a chemical change and its amount', *Proceedings of the Royal Society* 15, 262–265.

Harrison, A. and Weaver, E. (1989). *Chemistry: A Search for Understanding*, Harcourt Brace Jovanovich, San Diego, CA.

Hoeve-Brouwer, G.M. (1996). *Teaching Structures in Chemistry*, Unpublished PhD thesis, Department of Chemical Education, University of Utrecht, Utrecht.

Justi, R.S. (1997a). *Models in the Teaching of Chemical Kinetics*, Unpublished PhD thesis, Department of Science and Technology Education, The University of Reading, Reading.

Justi, R.S. (1997b). Historical models of chemical kinetics, in J.K. Gilbert (ed.), *Exploring Models and Modelling in Science and Technology Education: Contributions from the MISTRE Group*, The University of Reading, Reading, 245–256.

King, M.C. (1980). *Time and Chemical Change: The Development of Temporal Concepts to Chemistry with Special Reference to the Work of Augustus Vernon Harcourt*, Unpublished PhD thesis, Open University.

King, M.C. (1981). 'Experiments with time: progress and problems in the development of chemical kinetics', *Ambix* 28 (Part 2), 70–82.

King, M.C. (1984). 'The course of chemical change: the life and times of Augustus G. Vernon Horcourt (1834–1919)', *Ambix* 31 (Part I), 16–31.

King, M.C. and Laidler, K.J. (1983). 'Chemical kinetics and the radiation hypothesis', *Archives for the History of the Exact Sciences* 30, 45–86.

Kuhn, T. (1970). *The Structure of Scientific Revolutions*, University of Chicago Press, Chicago, IL.

Laidler, K.J. (1984). 'The development of the Arrhenius Equation', *Journal of Chemical Education* 61, 494–498.

Laidler, K.J. (1985). 'Chemical kinetics and the origins of physical chemistry', *Archives for the History of the Exact Sciences* 32, 43–75.

Laidler, K.J. (1986). 'The development of theories of catalysis', *Archives for the History of the Exact Sciences* 35, 345–374.

Laidler, K.J. (1987). *Chemical Kinetics* (3rd edn), Harper and Row, New York.

Laidler, K.J. (1995). *The World of Physical Chemistry* (2nd edn), Oxford University Press, Oxford.

Laidler, K.J. and King, M.C. (1983). 'The development of transition-state theory', *Journal of Physical Chemistry* 87, 2657–2664.

Lakatos, I. (1970). 'Falsification and the methodology of scientific research programmes', in I. Lakatos and A. Musgrave (eds), *Criticism and the Growth of Knowledge*, Cambridge University Press, Cambridge, 91–196.

Laudan, L. (1977). *Progress and its Problems*, Routledge and Kegan Paul, London.

Lewis, W.C.McC. (1918). 'Studies in catalysis, Part IX: the calculation in absolute measure of velocity constants and equilibrium constants in gaseous systems', *Journal of the Chemical Society* 113, 471.

Logan, S.R. (1982). 'The origin and status of the Arrhenius Equation', *Journal of Chemical Education* 59, 279–281.

Lythcott, J. (1985). ' "Aristotelian" was given as the answer, but what was the question?', *American Journal of Physics* 53, 428–432.

Matthews, M.R. (1994). *Science Teaching: The Role of History and Philosophy of Science*, Routledge, London.

McClelland, J. (1984). 'Alternative frameworks: interpretation of evidence', *European Journal of Science Education* 6, 1–6.

McCloskey, M. (1983). 'Naive theories of motion', in D. Gentner and A. Stevens (eds), *Mental Models*, Erlbaum, Hillsdale, NJ, 299–324.

Mellor, J.W. (1904). *Chemical Statics and Dynamics*, Longmans Green, London.

Mierzecki, R. (1991). *The Historical Development of Chemical Concepts*, Kluwer, Dordrecht.

Nersessian, N. (1989). 'Conceptual change in science and in science education', *Synthese* 80, 163–183.

Nersessian, N. (1992). 'Constructing and instructing: the role of abstraction techniques in creating and learning physics', in R. Duschl and R. Hamilton (eds), *Philosophy of Science, Cognitive Psychology, and Educational Theory and Practice*, State University of New York Press, New York, 48–68.

Niedderer, H., Betge, T. and Cassens, H. (1990). 'A simplified quantum model: a teaching approach and evaluation of understanding', in P. Lijnse, P. Licht, W. De Vos and A. Waarlo (eds), *Relating Macroscopic Phenomena to Microscopic Particles*, University of Utrecht, CB-Press, Utrecht, 67–80.

Ostwald, W. (1909). *The Fundamental Principles of Chemistry* (translated by H.W. Morse), Longman Green, New York.

Partington, J.R. (1964). *A History of Chemistry* (Vol. 4), MacMillan, London.

Piaget, J. and Garcia, R. (1989). *Psychogenesis and the History of Science*, Columbia University Press, New York.

Rice, F.O. and Herzfeld, K.F. (1934). 'The thermal decomposition of organic compounds from the standpoint of free radicals. VI: the mechanism of some chain reaction', *Journal of the American Chemical Society* 56, 284.

Rice, O.K. and Ramsberger, H.C. (1927). 'Theories of unimolecular reactions at low pressures', *Journal of the American Chemical Society* 49, 1617.

Semenoff, N. [1927] (1967). 'The oxidation of phosphorus vapour at low pressures', in M.H. Back and K.J. Laidler (eds), *Selected Readings in Chemical Kinetics*, Pergamon Press, Oxford, 127–153.

44   *Explanation, models, modelling*

Temkin, O.N. and Bonchev, D.G. (1992). 'Application of graph theory to chemical kinetics', *Journal of Chemical Education* 69(7), 544–550.
Toulmin, S. (1953). *The Philosophy of Science*, Hutchinson, London.
Treagust, D., Duit, R., Joslin, P. and Lindauer, I. (1992). 'Science teachers' analogies: observations from classroom practice', *International Journal of Science Education* 14, 413–422.
Van't Hoff, J.H. (1896). *Studies in Chemical Dynamic*, Frederick Miller and Williams & Norgate, Amsterdam and London.
Watts, D.M. (1982). 'Gravity: dont take it for granted', *Physics Education* 28(9), 116–121.
Yablonskii, G., Bykov, V., Gorban, A. and Elokhin, V. (1991). 'Kinetic models of catalytic reactions', in R. Crompton (ed.), *Comprehensive Chemical Kinetics* (Vol. 32), Elsevier, Amsterdam.

CHAPTER 3

A CAUSE OF AHISTORICAL SCIENCE TEACHING
Use of hybrid models

Justi, R. and Gilbert, J.K. 'A cause of ahistorical science teaching: use of hybrid models'. *Science Education*, 1999, 83(2): 163–177

This chapter discusses the desirability of an approach to the teaching of chemical kinetics based on models. The historical development of the subject was analyzed and eight historical consensus models were proposed. In a case study conducted in a class of 15–16-year-old students in Brazil, the models of chemical kinetics put forward by the teacher and the textbook were analyzed and discussed in the light of the historical models. The models expressed by the teacher and the textbook were found not to be any of the previously defined historical models, but rather what is termed a *hybrid* model. The existence of hybrid models in science teaching is proposed as a new component in science teachers' training courses.

Introduction

Since the 1980s, the emphasis in science education has been gradually shifting toward a greater concern with an overall understanding of science and its role in society. From this perspective, the purposes of science education have been proposed as being: learning science, that is, understanding scientific conceptual knowledge; learning about science, that is, understanding what the conduct of science involves; and doing science, that is, taking part in activities that contribute to the development of skills with which to obtain reliable scientific knowledge (Hodson 1992). The citizens of the twenty-first century are being educated now. It is of pivotal importance that they became scientifically literate in the sense just described. They will need not only to understand scientific phenomena, but also to have the ability "to generate fruitful and relevant questions and frame them in an effective way for investigation" (Burbules and Linn 1991: 228), and to be able to acquire and extend scientific knowledge on their own.

The introduction of history and philosophy of science (HPS) into science teaching has been advocated in recent years as a way of placing such an emphasis on science education, of making it concerned not only with the products, but also with the processes of development, of science (Allchin 1995; Brush 1978; Carson 1992; Hodson 1985, 1992; Matthews 1994; Monk and Osborne 1997). In many present-day science courses: (i) science is presented as a collection of "agreed upon facts"; (ii) the methodology for the production of scientific knowledge is presented as homogeneous and based on empiricism, leading to a static and context-independent creation/discovery view of outcomes; and (iii) students memorize the specific facts presented to them without questioning either their development or relationship to other scientific or nonscientific knowledge.

Where they have viewed the introduction of HPS into science education as important, or even the most important, way to change the situation just described, many researchers have proposed ways in which this could be done. For some of them, a unit about the history of science should be added after the teaching of a specific aspect of a course: Kauffman (1989) reported on the use of such an approach. We agree with Brush (1989) in believing that this could be a naive approach because it involves presenting students only with the conclusions reached by scientists and, sometimes the successful experiments that were done. It does not require students to think about the process of construction of scientific knowledge. It does not contribute to the establishment of those relationships between the history of science and science itself that could help in fostering meaningful learning. Such a view is shared by other researchers, who proposed the introduction of HPS into science education in a more integrated way. This means through activities such as, for instance, the reproduction of historical experiments (Ellis 1989; Kipnis 1995), the discussion of historical situations that provided initial steps in the development of a specific theme (Arons 1988), and the dramatization of historical debates (Solmon 1989). Such an integration may be also be achieved by taking the history of science as the basis for the organization of a course, as it was done in the Project Physics course (Brush 1978; Matthews 1988).

Independently of the approach adopted, science teachers are an essential element in the introduction of HPS in science education. It is of pivotal importance that they both understand and agree with the arguments in favor of the introduction of HPS into science education and that they have appointed knowledge to be able to effect a successful introduction.

By probing the teaching of a particular topic in chemistry through an approach based on models, the research presented in this chapter suggests a new reason for introducing HPS into science teaching.

Chemical kinetics, that branch of chemistry concerned with the study of the rate of chemical reactions, was chosen as the scientific theme because:

1   The main objectives of chemical kinetics are to describe and explain the observed relationships between reaction rate and the variables that exert influence on it. Also it must provide critical support for the mechanism proposed for any chemical reaction (Yablonskii *et al.* 1991). It can thus provide a basis for the comprehension of important chemical processes, which is essential for the education of thoughtful citizens.
2   We could not find any research report in which the teaching of chemical kinetics was analyzed from the perspective of either teachers' or students' understanding.

Models in science and science education

Models have been recognized as essential elements in the process of the development of theories (Forcese and Richer 1973; Gilbert and Boulter 1995; Lind 1980; McMullin 1968; Nersessian 1992; Norman 1983). As human constructs, models initialy exist in the mind of a person, independently of whether this person is thinking alone or within a group. These individual constructs are called *mental models* (Duit and Glynn 1996; Glynn and Duit 1995; Norman 1983; Vosniadou 1994; Vosniadou and Brewer 1992). Due to the impossibility of direct access to mental models, and to limitations in forms of expression, it is necessary to distinguish mental models from *expressed models* (Gilbert and Boulter 1995). The latter are what are placed by an individual into the public domain through any form of expression (e.g. speech, writing). When expressed models are discussed and accepted in a specific community through publication in a refereed journal, they become *consensus models* (Gilbert and Boulter 1995, 1998; Toulmin 1953). A consensus model developed in a specific context may be called a *historical model* (Justi 1997). In such a definition, a context is a system of philosophical, scientific, technological, and social beliefs. This implies that a historical model is not necessarily a model developed by an individual scientist, nor that its creation and use were tied to a specific time period, but rather that it achieved consensus status over some period in the past.

Teaching models also play an important part in science education. They may be defined as those models specially developed to help students understand consensus models and to support the evolution of mental models in specific scientific areas (Gilbert and Boulter 1995). Due to its function, a teaching model has a special level of complexity. That is, it should preserve the conceptual structure of a consensus model, demonstrate the constant and dynamic interplay of thoughts and actions in science, and deal with students' previous knowledge by providing ways to build on their personal understanding of science.

In the research reported here, teaching models were assumed to be not only concrete objects produced by a teacher to help students to learn, but also analogies and illustrations that lead students to visualize what is intended. This is the case, for instance, when a teacher says to students, "imagine yourselves as molecules and...," or when a teacher uses pictorial representations of molecular models to explain the meaning of a diagram about processes involving molecules. The important role of analogies in the understanding of scientific concepts has been intensively studied in recent years (see e.g. Collins and Gentner 1990; Dagher 1994, 1995; Glynn *et al.* 1989, 1995; Harrison and Treagust 1993; Treagust *et al.* 1992; Wong 1993).

Aiming to educate critical citizens, it is important to make students think about not only scientific knowledge itself but also about issues of HPS and the relationship between science and technology. Within the models approach, these concerns can be explored through the development of an understanding of the consensus model(s) of specific themes (Duit and Glynn 1996; Gilbert and Boulter 1994, 1998; Glynn and Duit 1995; Millar 1996). Understanding is viewed here as the establishment of meaningful relationships between the mental models of both the teacher and students and the consensus models (both current and historical). By understanding a specific theme, people are able to discuss it and to have doubts about it as well as to establish relationships and apply this knowledge in different contexts.

Analysis of research field for historical models

Philosophical framework

The definition of historical models used in this research was based on Lakatos's analysis of scientific research programs. According to Lakatos, a scientific research program has three elements: the "hard core"; the "protective belt"; and the "positive heuristic." The first comprises the major assumptions that guide all who work within a given research program. The second comprises the auxiliary hypotheses that protect the hard core from refutations. Finally the third is a set of suggestions about how to modify the protective belt (Lakatos 1978).

In this research, two elements play the role of the hard core: the "theoretical background" and the "main attributes" of the historical models. The theoretical background corresponds to the general philosophical and scientific ideas on which the model is or was based, as well as the analytical tools used in its construction. In other words, it characterizes the context in which the model is or was developed. On the other hand, the main attributes are the fundamental scientific ideas specific to the subject of the model. The protective belt is associated here with the "secondary attributes," ideas that complement the main attributes to permit a comprehensive characterization of each model. Each secondary attribute can be discussed independently of the other, but all of them are directly related to the main attributes.

In Lakatos's view, a research program is overthrown when its hard core has to be changed. This is a consequence of a kind of "competition" between the "progressive problemshifts" of one research program and the "degenerating problemshifts" of another (Lakatos 1970). In delimiting historical models, the following points have been systematically investigated to characterize such problemshifts:

1   The deficiencies in the explanatory capability of the previous model.
2   The features of that former new model that have been modified and incorporated into the new model.
3   The way by which the new model overcame the explanatory deficiencies of its antecedents.
4   The unanticipated explanatory benefits of the new model.
5   The explanatory deficiencies of the new model (Justi 1997).

Historical models of chemical kinetics

The history of the development of chemical kinetics was analyzed in light of the system just described. In this case, the main attributes were considered to be the meanings of "chemical reaction" and "reaction rate," for these are the fundamental ideas in this subject. The secondary attributes were the ideas concerned with:

1   *Energetics*: Ideas related to the influence of energy on the rate of a chemical reaction.
2   *Catalysis*: Ideas that explain the action of specific substances that change the rate of a chemical reaction.
3   *Reaction path*: Ideas about the way in which a chemical reaction is viewed at the molecular level.

Such an analysis resulted in the definition of eight historical models of chemical kinetics. These models have been identified by name, as they relate to their theoretical background or main attributes: anthropomorphic, affinity corpuscular, first quantitative, mechanism, thermodynamics, kinetic, statistical mechanics, and transition state. For a complete presentation and discussion of such models, see Justi (1997) and Chapter 2 of this volume. They can be characterized briefly as follows.

Early ideas of chemical kinetics were extremely vague and based on an anthropomorphic view of matter and on the equally vague idea of affinity. Therefore, within the *Anthropomorphic Model*, rate was synonymous with "readiness" for transformation to occur (Mellor 1904; Mierzecki 1991).

In the seventeenth and eighteenth centuries, the gradual development of atomic theories caused affinity to come to mean a tendency of substances to combine with each other as a result of the forces between them resulting from the distinctive characteristics of the particles involved (Duncan 1996). From such a view, a chemical reaction was considered a process in which elementary particles interacted with one another as a result of their chemical affinities. In the *Affinity Corpuscular Model*, the rate of a reaction was related to the different degrees of affinity between the particles and depended on their readiness to occur.

The *First Quantitative Model* originated from the introduction of mathematics into the study of chemical kinetics. This made possible the establishment of the first proportionality relationships between reaction rate and the number of particles reacting in a given time (Partington 1964).

The introduction of the notion of "steps" in a chemical reaction provided the basis for the development of the *Mechanism Model*. In this model, the first ideas about the relationships between mechanism and reaction rate were established. These ideas formed the basis for the discussion of the action of catalysts and for the development of more elaborate quantitative expressions of reaction rates (Laidler 1987, 1995).

The *Thermodynamics Model* was based on the idea of a chemical reaction as a process that involved reacting molecules colliding with sufficient energy. Therefore, the rate of a reaction was said to be proportional to the amount of substances that underwent transformation in a given time; that is, to the number of molecules that acquire more than a "critical" energy (Arrhenius 1889; Laidler 1995).

The analogy of molecules acting as if they were hard spheres provided a new way to analyze chemical reactions. In the *Kinetic Model*, such an idea led to the introduction of the Arrhenius preexponential factor and of the frequency of collisions between reacting molecules into the study of the rate of chemical reactions (Glasstone *et al.* 1941).

Concurrent with the development of the kinetic model, the application of statistical mechanics provided the basis for a different model. In the *Statistical Mechanics Model*, a chemical reaction was viewed as the motion of a point in phase space, the coordinates of which were the distances between the molecules and their momentum. The expression for reaction rate was thus obtained from the consideration of the passage of systems through the col point of the potential energy surface (Laidler 1987, 1995).

The *Transition State Model* was an attempt to overcome the several shortcomings of the thermodynamics, the kinetics, and the statistical mechanics models, by forging a strong and concise link between thermodynamics and kinetic variables. By considering a chemical reaction as a process by which a system is passed over the top of the energy barrier between the initial and final states (Evans 1938), this model proposed that rate could be calculated by focusing attention on the molecular complexes at the col of the surface. Moreover, the application of statistical methods made possible the development of equations that related the concentration of the species involved to the rate of a reaction (Evans and Polanyi 1935; Eyring 1935).

The issues examined in the delineation of each particular model in the historical sequence of models that represent the development of the study of chemical kinetics illustrate the dynamism of the construction of scientific knowledge. The change from one model to another model is an evolutionary process. Therefore, sometimes: (i) a model completes a former one; (ii) a model focuses some of the explanatory deficiencies of the previous model by using different theoretical assumptions; or (iii) a model provides a completely new overall view of the phenomena (Justi 1997).

Analysis of teaching of chemical kinetics

About the conduct of the research

The purpose of the research reported here was to explore the teaching of chemical kinetics from the point of view of the models used. A case study was conducted in a Brazilian chemistry class (15–16-year-old students). All the 20 meetings of the class concerned with chemical kinetics were observed and video-recorded. The teacher was interviewed six times for 40–90 minutes on each occasion. From the transcriptions of both the class and the interviews with the teacher it was possible to characterize the expressed model of chemical kinetics used by the teacher. The textbook used in the teaching was analyzed and its expressed model was also characterized. The criteria used in the analysis of the teacher's and the textbook's expressed models of chemical kinetics were the same as those used in the definition of the historical models; that is, by identifying the respective theoretical backgrounds and main and secondary attributes involved. Each one of these expressed models was then compared with the eight historical models previously identified to investigate the nature of any relationships between them. The teacher that took part in this case study has a good reputation in the city where he lives and the textbook he used in the classes observed is a typical Brazilian chemistry textbook.

Textbook expressed model and historical consensus models of chemical kinetics

Attempts to associate the model of chemical kinetics expressed by the textbook with any one of the historical consensus models previously defined were unsuccessful. This was because the authors had developed a completely different model in which they merged characteristics of several distinct historical models.

The theoretical backgrounds of their expressed model included empirical observations and simple ideas derived from the corpuscular theory through to the principles of statistical mechanics and thermodynamics. It could be argued that this amalgam could contribute to the development of an *overall* view of chemical kinetics by students. However, this does not seem to be the intention of the authors, as is argued in what follows.

There are no specific discussions about the historical development of scientific ideas as an evolutionary and continuous process. The sole reference the authors make to a specific scientist is to what they called "van't Hoff's law" (i.e. a 10° C increased in the temperature of a system doubles the rate of the chemical reaction). The authors do not try to discuss the context in which such a "law" was proposed, to establish any kind of relationship between what the scientist was proposing and the scientific knowledge of his time. They simply present "the law" as a truth to be believed. Moreover, they make frequent reference to "chemists" in an impersonal and generalized way. They use expressions such as "chemists concluded" (p. 258), "chemists represent" (p. 270), "chemists make the following definitions" (p. 279) in a way that means "chemists are authorities, all they say is true." This can be viewed as a subtle way of simply giving information rather than trying to "explain an interpretation (a product of human figuring)" (Sutton 1993: 1220). This also constitutes a misrepresentation of the process of development of scientific knowledge (Sutton 1996).

In no part of the textbook is there any reference to the fact that the chemical content presented is a model proposed in a specific context to explain specific phenomena. The authors seem to consider the chemical content from an empiricist point of view. Scientific knowledge is dealt with in an absolutist way; that is, it is presented as true and confirmed. There is no discussion concerning any aspect of the knowledge from distinct points of view. On the contrary, they use different theoretical backgrounds simultaneously in an attempt to explain and emphasize different aspects of this knowledge.

Among the main attributes of hybrid model presented in the textbook, it is possible to see that some are characteristics of different historical consensus models, as can be seen in Table 3.1.

The significant difference is that, in each of the historical models, the secondary attributes are clearly and closely related to the main attributes, so that there is a coherence between secondary attributes and the protective belt of a hard core. This is not seen in the hybrid model given in the textbook. This is because the authors discuss each specific attribute as they think is suitable for their immediate purposes. The authors, for instance, discuss the concept of activation energy from the point of view of the collision theory. For them, "activation energy is a minimum value of energy that reactant molecules must have so that a collision between them is effective" (p. 270). However, when they say there is a species called "the activated complex," they add elements of the transition state theory to the explanation, and assert, while discussing the occurrence of a reaction between NO_2 and CO, that: "The chemical species that exists at the instant of the collision, when the NO bond is partially made, is called activated complex or transition state" (p. 258). However, "activated complex" and "transition state" are different concepts, formulated from different theoretical backgrounds. The "transition state" was defined within the transition state model as the configuration of atoms whose potential energy increases with deformation in all directions except along the reaction path (Evans 1938). It has no similarity to the "activated complex," as defined by Arrhenius, as a real molecule in which the potential energy increases for all types of deformation.

Table 3.1 Identification of the origin of the main attributes of the hybrid model used in the chemical kinetics textbook

Main attribute		Historical model
Chemical reaction	Process that can occur in steps	Mechanism
	Process that results from effective collisions between particles of reactants	Thermodynamics
	Process in which the collisions occur not only with sufficient energy, but also with an appropriate orientation to the necessary bonds to be broken and made	Kinetic
Reaction rate	It is proportional to the concentration of the reaction substances	First quantitative
	It is determined by the mechanism through which the reaction occurs	Mechanism
	It is related to the activation energy of the reaction	Thermodynamics
	It depends on the frequency of collisions	Kinetic

Another aspect that illustrates the lack of connection between the main and secondary attributes in the model expressed in the textbook is related to the utilization of the energy profile diagrams. Such diagrams were originally developed from potential-energy surfaces in the context of the transition state theory. In these diagrams, the energy of the system is plotted against the interatomic distances to give "potential surfaces." Thus, different positions on this surface mean species that have particular energies. Moreover, from the lowest point over these "energy hills" it is possible to obtain a measure of the activation energy (Eyring 1935; Eyring and Polanyi 1931) these potential-energy surfaces thus provide a valuable pictorial representation of the course of a chemical reaction. Perhaps because of this usefulness, the authors of the textbook present such diagrams as a representation of the reaction path without explaining them or making any reference to the origin. On the contrary, they present them *as if they were a diagrammatic representation of what is predicted by collision theory.* This is done by introducing pictorial representations of molecular modules as being associated with an energy profile diagram. Such a picture strengthens not only the idea that the energy profile diagram is a representation of the predictions of the collision theory, but also that a transition state is a concept within the scope of the collision theory. This means that the authors merge elements of different models as if they were within the same hard core.

Teacher expressed model and historical consensus models of chemical kinetics

The teacher expressed model was characterized by the analysis of the transcriptions of all the 20 classes concerning chemical kinetics by using the same criteria as in the definition of the historical models. As in the case of the textbook, the theoretical background that underpinned his model also include some empirical observations from daily phenomena, simple ideas derived from the corpuscular theory, the collision theory was introduced and discussed broadly. However, the Maxwell–Boltzmann distribution law and some principles of thermodynamics were introduced in a very particular way. The meaning of the Maxwell–Boltzmann diagram, for instance, was not explained, nor were the reasons given for its use. The teacher simply presented a diagram showing two curves for different temperatures and said that, at a higher temperature, "the number of molecules with energy higher than the minimum energy, that is, those that are above the medium value, increases." It seems that the diagram was presented just because it is "usual" to present it in such a context, and not because it was very necessary to the understanding of the influence of temperature on reaction rate.

It was also impossible to associate the model of chemical kinetics expressed by the teacher with any one of the historical models. Similar to textbook case, the teacher used a mix of completely distinct theoretical backgrounds as basis for main and secondary attributes that were presented without any discussion about the differences in the contexts in which they were developed and in which they are valid. At no stage did the teacher make any historical references.

The comparison of the theoretical backgrounds and the main attributes of the models expressed by the teacher with those of each of the eight historical models resulted in a situation similar to that presented in Table 3.1. As the teacher did not base his classes on the textbook (by his own declaration and as a matter of classroom observation), this cannot be interpreted as a consequence of "the teacher repeating the textbook." We surmise that the teacher's own scientific knowledge may have been derived largely from school-level textbooks. This is likely to occur, because, since leaving the university (about 25 years ago), this teacher had not attended course or taken part in organized teacher development activities that could have influenced his scientific knowledge. The only difference between the model expressed by the teacher and that expressed in the textbook is that, to the main attributes regarding chemical reaction presented in Tabe 3.1, the teacher added that: "A chemical reaction is a process that requires affinity between reactants."

Until the eighteenth century, chemists used the word "affinity" to express what they observed in their experiments. They did not provide explanations for such observations. At that time, before Dalton, and under the influence of the corpuscular theories of Boyle and Newton, scientists could hypothesize causes for affinity only in terms of shapes and sizes of particles. In the classes, the teacher did not make such suppositions, but he used the word "affinity" in a manner similar to earlier chemists – that is, as if it provides an explanation. The meaning he attributed to affinity was the same as was attributed by scientists in the eighteenth century: a tendency of chemical substances to combine with each other. During an interview, he said that substances have affinity with each other when they "present a tendency of reactivity, a tendency to desire the other substance in order to complete itself." This indicates that the main attributes of the teacher's expressed model are also comprised of ideas of the anthropomorphic model integrated with those from other theoretical backgrounds.

As in the case of the textbook, elements of the secondary attributes were not coherently related to corresponding main attributes in the teacher's expressed model. At times, a secondary attribute defined in one specific historical model was presented as a natural consequence of an idea expressed in a main attribute of another historical model. Such findings were observed when:

1   The teacher used the idea of affinity between substances while discussing the collision theory. This happened when he discussed affinity between reactants as one of the conditions for the occurrence of a chemical reaction. This point was emphasized strongly by the teacher. Indeed, he introduced the idea of collision between molecules from an analogy used to explain the meaning of affinity:

> If I feel affinity with someone but I stay here and that person stays over there, this will result in nothing. Molecules must be put in contact. Chemistry called this "collision." Molecules must collide.

From this it follows that affinity between molecules is the original cause of the occurrence of their collisions. Assuming both that the teacher uses "affinity" meaning "a tendency to attract" and that the collision theory deals with "molecules as hard spheres colliding with sufficient energy and appropriate orientation," without discussing any force of attraction, such an analogy merges main attributes of different historical models. Thus, making one becomes the origin of the other.

2   The teacher introduced the concept of transition state as synonymous with activated complex within the scope of the collision theory. While discussing the idea of an effective collision, he said:

> This instant is the so-called activated complex. It is the highest energy instant of the reaction. It is a transition state: it is the intermediate of the reaction.... If the reactants reach the transition state, the activated complex, it is because the collision was good.

In an interview, when he was asked whether there is a difference between activated complex and transition state, he answered:

> When we say activated complex, it is what is mixing, reacting. It is just because it is reacting that there is not an outcome yet. Activated means with a sufficient quantity of energy to make the reaction occur. When I say transition state I am thinking of what will result. The molecules are undergoing change, but it is not completed.... I do not know whether it is just a question of nomenclature to call this an activated complex.... I do not know if we should call activated complex the moment when the molecules have the maximum energy, when the transition state is formed, or if we should call activated complex the moment when the break occurs.

Such quotations make us think that the activated complex and the transition state are indistinguishable in the teacher's model of chemical kinetics.

3   The teacher used energy profile diagrams as if they were representations proposed by the collision theory. Such diagrams were introduced during the initial discussions about "effective collision." This, as well as the previous point, was emphasized by the use of pictorial representations of molecular models associated with specific stages of an energy profile diagram. Such representations underpinned most of the explanations about effective collisions and activated complex.

4   The teacher used the Maxwell–Boltmann diagram to justify predictions arising from the collision theory. The meaning of such a diagram was not explained, nor were the reasons for its introduction. While discussing the influence of temperature on the rate of a reaction, the teacher said that:

> The higher the temperature, the higher the number of collisions. When the number of collisions is increased, the kinetic energy of the constituents of the system is also increased. When the number of collisions is increased, the number of effective collisions is also increased and, then the rate of the reaction. Now, pay attention here. [The teacher drew a diagram showing two curves for different temperatures.] At a higher temperature, the number of molecules with energy higher than the minimum energy, that is, those that are above the medium value, increases.

The diagram was introduced just to "show" the differences in the energy of molecules at different temperatures. In an attempt to explain this, following the excerpts just presented, the teacher introduced an analogy in which he associated both different average marks required from students at different times with different values of activation energy at different temperatures and the number of students who have marks above the average with the number of molecules that react. It seems that, by assuming the familiarity of students with the analog domain, the teacher hoped that students would naturally accept the relationship between the two domains. However, while the teacher presented the elements of the energy profile diagram, he discussed the process from the perspective of the collision theory. Thus the analogy implies that the energy that molecules acquire as a consequence of collisions is similar to the acquisition of marks by students.

All these points involve the collision theory. It seems that other sorts of ideas were added to the general analogy of molecules acting as if they were hard spheres colliding with sufficient energy and adequate orientation (proposed by Trautz and Lewis in 1916 and 1918, respectively), independently of the coherence of their backgrounds or predictions.

Models expressed in the teaching as hybrid models

The aforementioned points shows that the models of chemical kinetics expressed by both the textbook and the teacher represent no single one of the historical models previously defined. Moreover, these models cannot be viewed as a sequence of elements of different historical models, but rather as an aggregation of such elements. If elements of different models were presented sequentially, it would be possible to characterize the progressive and degenerating problemshifts of each of these models, as was possible in the discussion of the delimitation of the historical models. This means that they are not based on the subsumption of different consensus models. It is proposed that they be called *hybrid models*: models constituted of elements of different historical models treated as if they constituted a coherent whole (Justi 1997).

We are not advocating that textbooks and teachers should present an expressed model that coincides with a given historical model, nor that they should present a linear progression of historical models of a given subject. What we consider important is that teachers and textbooks make the backgrounds of their expressed models clear. They should state the context in which they are valid. When we say that chemical kinetics was taught as a hybrid model in this case study we are emphasizing that there are serious inconsistencies in the relationships between the hard core and the protective belt of the models expressed by the teacher and the textbook. Therefore, scientific knowledge is not presented as provision using models that are both developed in and are valid in specific contexts. This implies that the hybrid relationships are presented as natural ones – that is, as if there are no gaps between them, no questions requiring different ways of thinking about the phenomenon, and no alternate approach to interpreting a phenomenon. This contributes to a view of scientific knowledge as something established, true, and that must be accepted. By presenting chemical knowledge in this way, the hybrid model does not help students to understand the process of the development of such knowledge.

Discussion

The identification of both the textbook's and the teacher's expressed models of chemical kinetics as hybrid models provides an insight with which teaching can be discussed. The existence of a hybrid model in teaching means that no history of science is possible because it implies that scientific knowledge grows linearly and is context independent. It leads students to have misconceptions in their mental models of the theme being discussed and/or to have difficulties in understanding points for which hybrid relationships are introduced. It is difficult to identify general suppositions about the reasons for the existence of hybrid models in teaching. However, by looking more closely at the points that characterized the hybrid models of chemical kinetics identified here, it is possible to hypothesize some reasons. Such reasoning may, in time, lead to a more general discussion.

All the hybrid relationships found in this research involve the collision theory. Ideas from different models were added to such an approach for interpreting the occurrence of a chemical reaction, as if natural and coherent relationships among them existed. From the point of view of teaching chemistry to adolescents, the collision theory is an interesting approach to chemical reactions. The association with simple and well-known mechanical systems makes it easy to be understood. Historically, other ideas were developed to explain specific aspects of reaction rate (as, e.g. those involved in energy profile diagrams). Such ideas, by involving more abstract entities and by not being based on a mechanical model, cannot be "visualized" as can those derived from the collision theory. On the other hand, they provide more valuable explanations of some aspects. The hybrid model seems to represent an intention to provide explanations that are simultaneously comprehensive and simple. However, to simplify a set of ideas to present them to students must not mean to mix ideas from different backgrounds as if they constituted a coherent whole, as if the concepts and relationships that underpin different historical consensus models were the same. It seems that, in establishing a hybrid model, parts of different historical consensus models are being viewed only as different "language," or different "forms of expression," with which to talk about the same things. The theme of all the historical models defined here is the same (chemical kinetics), but the ideas within each models are related to different targets and backgrounds, which make their meanings different from one another.

It seems obvious that, depending on both the scientific theme and the grade level of the students, the latter should be presented with a model that is simpler than the current consensus model. However, this does not mean that a new model should be produced by adding parts of different historical consensus models according to conveniences that converge on a simple outcome. This attitude is not consistent with proper teaching of science and it does not contribute to students' understanding of either scientific reasoning or how science evolves.

Implications for science education

What factors favor the use of hybrid models? The issue is about the education of textbooks authors and teachers. The use of hybrid models in teaching points to the need for an urgent reconsideration inside chemistry departments about what knowledge has been taught, as well as how such teaching has been conducted. It leads to a discussion about whether university courses are the reasons why graduates produce and teach hybrid models. Three points comprise such a discussion:

1   The level at which basic chemical concepts are taught in university courses. When students enter universities they may hold misconceptions about some concepts or they may not comprehensively understand some basic chemical concepts. If university lecturers deal only with the current and most complex ideas, they would favor the construction (and consolidation) of many conceptual gaps and misunderstandings in further teachers' mental models.

2   The level at which university courses attribute importance to the history of science. This may vary from the absence of discussion about this subject or a mere superficial report of dates and names, at one extreme, to the introduction of the history of science as an intrinsic part of all subjects (i.e. permeating the discussion of all chemical themes, at the other extreme). However independently of the position adopted in such a continuum, it seems implausible to introduce the history of science into university courses if students are continuously expected to memorize facts or mathematical relationships, to uncritically accept what lecturers or textbooks present to them, and to rarely ask why (Kipnis 1996). The history of science must be part of chemistry courses in a way that contributes effectively to future teachers' understanding of either the process of construction of scientific knowledge or the identification of ideas produced in different backgrounds or contexts.

3   The level at which philosophical discussions about the nature of scientific knowledge taken place in university courses. This means discussion about both the nature of models and theories and the relationships between them

from different points of view. Also, it is essential that future teachers view scientific knowledge as provisional, that they understand how scientific knowledge changes, and that they develop a more critical view of scientific knowledge if they are to present an authentic view of science to their students.

By critically and coherently considering these three points, universities would provide future teachers with far superior chemistry courses; they would provide education and not training in chemistry (Matthews 1992, 1994); they would disallow conditions for the production of hybrid models; and they would make chemistry more understandable.

The existence of hybrid models in teaching is not an isolated case found only in Brazilian teaching. From an informal survey among British chemistry textbooks hybrid models for other concepts were also identified. Therefore, a systematic investigation to identify hybrid models in both other scientific themes and in different teaching contexts is important. This may:

1   Make authors of textbooks and teachers conscious of hybrid models, which is certainly the first step in the removal of such models from science teaching.
2   Make authors of textbooks and teachers more interested in the history and philosophy of science so that they can realize the importance of its introduction into teaching, which aims at a comprehensive understanding of science.
3   Provide teachers and educational researchers with an interesting explanation for some of the students' misconceptions and difficulties in learning of specific themes.

References

Allchin, D. (1995). How NOT to teach history in science. In F. Finley, D. Allchin, D. Rees, and S. Fifield (eds), *Proceedings of the Third International History, Philosophy, and Science Teaching Conference* (vol. 1, pp. 13–22). Minneapolis, MN: University of Minessota.

Arons, A.B. (1988). Historical and philosophical perspectives attainable in introductory physics courses. *Educational Philosophy and Theory*, 20, 13–23.

Arrhenius, S. (1889/1967). On the reaction velocity of the inversion of cane sugar by acids. In M.H. Black and K.J. Laidler (eds), *Selected Readings in Chemical Kinetics* (pp. 31–35). Oxford: Pergamon.

Brush, S.G. (1978). Why chemistry needs history – and how it can get some. *Journal of College Science Teaching*, 7, 288–291.

Brush, S.G. (1989). History of science and science education. *Interchange*, 20, 60–70.

Burbules, N.C. and Linn, M.C. (1991). Science education and philosophy of science: congruence or contradiction? *International Journal of Science Education*, 13, 227–241.

Carson, R.N. (1992). If science is presented as historical narrative: a sample chapter. In S. Hills (ed.), *The History and Philosophy of Science in Science Education* (vol. 1, pp. 149–155). Kingston, ON: Queens University.

Collins, A. and Gentner, D. (1990). Multiple models of evaporation process. In D.S. Weld and J. Kleer (eds), *Readings in Qualitative Reasoning about Physical Systems* (pp. 508–512). San Mateo, CA: Morgan Kaufmann.

Dagher, Z.R. (1994). Does the use of analogies contribute to conceptual change? *Science Education*, 78, 601–614.

Dagher, Z.R. (1995). Review of studies on the effectiveness of instructional analogies in science education. *Science Education*, 79, 295–312.

Duit, R. and Glynn, S. (1996). Mental modelling. In G. Welford, J. Osborne, and P. Scott (eds), *Research in Science Education in Europe* (pp. 166–176). London: Falmer Press.

Duncan, A. (1996). *Laws and Order in Eighteenth-Century Chemistry*. Oxford: Oxford University press.

Ellis, P. (1989). Practical chemistry in a historical context. In M. Shortland and A. Warwick (eds), *Teaching the History of Science* (pp. 156–167). Oxford: Basil Blackwell.

Evans, M.G. (1938). Thermodynamical treatment of transition state. *Transactions of the Faraday Society*, 34, 49–57.

Evans, M.G. and Polanyi, M. (1935). Some applications of the transition state method to the calculation of reaction velocities, especially in solution. *Transactions of the Faraday Society*, 31, 875–894.

Eyring, H. (1935). The activated complex in chemical reactions. *Journal of Chemical Physics*, 3, 107–115.

Eyring, H. and Polanyi, M. (1931/1967). On simple gas reactions. In M.H. Black and K.J. Laidler (eds), *Selected Readings in Chemical Kinetics* (pp. 41–67). Oxford: Pergamon.

Forcese D.P. and Richer, S. (1973). *Social Research Methods*. Englewood Cliffs, NJ: Prentice-Hall.

Gilbert, J.K. and Boulter, C.J. (1994, August). Modelling across the curriculum: the demands presented at the 7th Symposium on Science and Technology Education in a Demanding Society, De Koningshof Veldhoven, the Netherlands.

Gilbert, J.K. and Boulter, C.J. (1995, April). Streching models too far. Paper Presented at all the annual meeting of the American Educational Research Association, San Francisco, CA.

Gilbert, J.K. and Boulter, C.J. (1998). Learning science through models and modelling. In B. Frazer and K. Tobin (eds), *The International Handbook of Science Education* (pp. 53–66). Dordrecht: Kluwer.

Glasstone, S., Laidler, K.J., and Eyring, H. (1941). *The Theory of Rate Processes – the Kinetics of Chemical Reactions, Viscosity, Diffusion and Electrochemical Phenomena*. New York: McGraw-Hill.

Glynn, S.M. and Duit, R. (1995). Learning science meaningfully: constructing conceptual models. In S.M. Glynn and R. Duit (eds), *Learning Science in the Schools: Research Reforming Practice* (pp. 3–34). Mahwah, NJ: Erlbaum.

Glynn, S.M., Britton, B.K., Semrud-Clikeman, M., and Muth, K.D. (1989). Analogical reasoning and problem solving in science textbooks. In J.A. Glover, R.R. Ronning, and C.R. Reynolds (eds), *Handbook of Creativity* (pp. 383–398). New York: Plenum.

Glynn, S.M., Duit, R., and Thiele, R.B. (1995). Teaching science with analogies: a strategy for constructing knowledge. In S.M. Glynn and R. Duit (eds), *Learning Science in the Schools: Research Reforming Practice* (pp. 247–274). Mahwah, NJ: Erlbaum.

Harrison, A.G. and Treagust, D.F. (1993). Teaching with analogies: a case study in grade-10 optics. *Journal of Research in Science Teaching*, 30, 1291–1307.

Hodson, D. (1985). Philosophy of science, science and science education. *Studies in Science Education*, 12, 25–57.

Hodson, D. (1992). In search of a meaningful relationship: an exploration of some issue relating to integration in science and science education. *International Journal of Science Education*, 14, 541–562.

Justi, R.S. (1997). Models in the teaching of chemical kinetics. Unpublished doctoral thesis, Reading, UK: University of Reading.

Kauffman, G.B. (1989). History in the chemistry curriculum. *Interchange*, 20, 81–94.

Kipnis, N. (1995). Blending physics with history. In F. Finley, D. Allchin, D. Rnees, and S. Fifield (eds), *Proceedings of the Third International History, Philosophy, and Science Teaching Conference* (vol. 1, pp. 612–623). Minneapolis, MN: University of Minnesota.

Kipnis, N. (1996). The 'historical-investigative' approach to teaching science. *Science and Education*, 5, 277–292.

Laidler, K.J. (1987). *Chemical Kinetics* (3rd edn). New York: Harper & Row.

Laidler, K.J. (1995). *The World of Physical Chemistry* (2nd edn). Oxford University press.

Lakatos, I. (1970). Falsification and the methodology of scientific research programmes. In I. Lakatos and A. Musgrave (eds), *Criticism and the Growth of Knowledge* (pp. 91–196). London: Cambridge University Press.

Lakatos, I. (1978). *The Methodology of Scientific Research Programmes*. Cambridge: Cambridge University Press.

Lind, G. (1980). Models in physics: some pedagogical reflections based on the history of science. *European Journal of Science Education*, 2, 15–23.

Matthews, M.R. (1988). A role for history and philosophy in science teaching. *Educational Philosophy and Theory*, 20, 67–81.

Matthews, M.R. (1992). History, philosophy, and science teaching: the present rapproachment. *Science and Education*, 1, 11–47.

Matthews, M.R. (1994). *Science Teaching – the Role of History and Philosophy of Science.* New York: Routledge.

McMullin, E. (1968). What do physical models tell us? In B. van Rootselaar (ed.), *Logic, Methodology and Philosophy of Science* III (pp. 385–396). Amsterdam: North-Holland.

Mellor, J.W. (1904). *Chemical Statics and Dynamics.* London: Longman Green.

Mierzecki, R. (1991). *The Historical Development of Chemical Concepts.* Warsaw-Polish Scientific Publishers.

Millar, R. (1996). Towards a science curriculum for public understanding. *School Science Review*, 77, 7–18.

Monk, M. and Osborne, J. (1997). Placing the history and philosophy of science on the curriculum: a model for the development of pedagogy. *Science Education*, 81, 405–424.

Nersessian, N.J. (1992). How do scientists think? Capturing the dynamics of conceptual change in science. In R.N. Giere (ed.), *Cognitive Models of Science* (pp. 3–44). Minneapolis, MN: University of Minnesota Press.

Norman, D.A. (1983). Some observations on mental models. In D. Gentner and A.L. Stevens (eds), *Mental Models* (pp. 7–14). Hillsdale, NJ: Erlbaum.

Partington, J.R. (1964). *A History of Chemistry* (vol. 4). London: Macmillan.

Solomon, J. (1989). Teaching the history of science: is nothing sacred? In M. Shortland and A. Warwick (eds), *Teaching the History of Science* (pp. 42–53). Oxford: Basil Blackwell.

Sutton, C. (1993). Figuring out a scientific understanding. *Journal of Research in Science Teaching*, 30, 1215–1227.

Sutton, C. (1996). Beliefs about science and beliefs about language. *International Journal of Science Education*, 18, 1–18.

Toulmin, S. (1953). *The Philosophy of Science.* London: Hutchinson.

Treagust, D.F., Duit, R., Joslin, P., and Lindauer, I. (1992). Science teachers' use of analogies: observations from classroom practice. *International Journal of Science Education*, 14, 413–422.

Vosniadou, S. (1994). Capturing and modeling the process of conceptual change. *Learning and Instruction*, 4, 45–69.

Vosniadou, S. and Brewer, W.F. (1992). Mental models of the earth: a study of conceptual change in childhood. *Cognitive Psychology*, 24, 535–585.

Wong, E.D. (1993). Understanding the generative capacity of analogies as a tool for explanation. *Journal of Research in Science Teaching*, 30, 1259–1272.

Yablonskii, G., Bykov, V., Gorban, A., and Elokhin, V. (1991). Kinetic models of catalytic reactions. In R.G. Compton (ed.), *Comprehensive Chemical Kinetics* (vol. 32). Amsterdam: Elsevier.

CHAPTER 4

EPISTEMOLOGICAL RESOURCES FOR THOUGHT EXPERIMENTATION IN SCIENCE LEARNING

Reiner, M. and Gilbert, J.K. 'Epistemological resources for thought experimentation in science learning'. *International Journal of Science Education*, 2000, 22(5): 489–506

Thought Experiments (TEs) are reasoning processes that are based on 'results' of an experiment carried out in thought. What is the validity of an experiment – that has not been actually executed – for knowledge about the physical world? What are the features that make it distinctive and how do we integrate it into learning environments to support such thought processes? This study suggests that a thought experiment draws on three epistemological resources: conceptual–logical inferences, visual imagery and bodily motor experience. We start by stating how students' TEs are related to recent research on learning science and then proceed to describe the nature of TEs. The central part of the paper deals with cognitive theories and empirical examples of visual imagery and bodily imagery. It also deals with how these enable implicit knowledge about the world to be retrieved. Students may have, but are not aware of, such knowledge for it is hidden when learning is only based on formal representations. We show that imagination is structured, goal-oriented, based on prior experiential imagery and internally coherent. Students can, for example, mentally rotate objects at constant velocity. Students can 'zoom in and out' to inspect imaginary situations, transfer objects, predict paths of imaginary moving objects and imagine the impact of forces on mechanical systems. We show that the TEs are powerful because of these capabilities. We further claim that these are not exploited by school learning environments and offer a first step towards understanding imagery in science learning.

Introduction

In 1818, the French Academy of Science offered a prize for an explanation of the diffraction of light. The underlying motivation was to resolve the apparent contradiction between Fresnel's wave theory and the particle theory as argued for by Laplace, Poisson and Biot. Poisson argued against the wave theory by suggesting an imagined experiment: suppose one places a circular obstacle perpendicularly to the axis of a beam of light from a point source: What would the centre of the shadow look like? Poisson suggested that the most reasonable answer is that the centre, like the rest of the shadow, should be dark. However, he argued that, if one assumed the wave theory, then the waves should arrive in phase at the periphery and thus recombine at the centre of the shadow to create a bright spot. Poisson thought this an unlikely outcome and felt that it would seal the fate of the then doubtful wave theory. This imaginary experiment was then actually carried out and the bright spot was indeed found at the centre of the shadow. The light wave theory became the accepted explanatory framework for diffraction (Sorensen 1992).

Two points are of interest. The first relates to the imaginary part of this argument – Poisson used a strategy of conviction that led the audience through an imaginary experiment, carried out in thought alone. Experiential imagery, along with logical and conceptual knowledge, was used as an epistemological resource. The second was the conviction of the thought experimenter in the validity of an imagined result. What is the source of such a conviction about the result of the experiment that led Poisson to make such a professionally high-risk empirical claim without having actually performed the experiment? It is this capacity for producing conviction in scientists and the readiness of the scientific community to accept its outcomes that has made the TE such a powerful tool in learning and in argumentation, in this and many other cases (see Brown 1991; Sorensen 1992; Wilkes 1989).

Any strategy that leads to a deep conviction about the processes of coming to know something must have a place at the centre of classroom science learning. Thought Experiments (TEs) are used in science teaching because they are part of accepted scientific practice (Nersessian 1992; Sorensen 1992), because they play a major role during periods of conceptual change in science (Kuhn 1977), are seen as 'authentic' by the scientific community (Roth 1995), thus being a crucial component in the enculturation of students into the methodology of science. A major reason for studying how TEs are used by students is because they tap into the fundamental processes of learning – the processes by which students use non-propositional prior knowledge, such as imagery, along with logical inference and conceptual knowledge, in generating new knowledge (Reiner 1998). In this paper we focus on the relationship between mental imagery and TEs.

Imagery is considered a central component in historical TEs in physics. For instance, Gooding (1992) argues that TEs in physics work because of a process of visualization in the learner that leads to embodied inferences which do not necessarily obey the rules of logical derivation. Any effort to understand physical phenomena by using thought experimentation is thus dependent on our experience of mental images (Teng 1992). Furthermore, imaginary cases, being an inner reflection of the outside world, are sometimes as good as actual cases for learning about science, although the former become more compelling only as their descriptions grow more elaborate (Dancy 1985). Based on the central role of images in scientists' experiments in thought, we suggest that one component that makes TEs a powerful tool for learning is the use of intuitive imagery in constructing experimental situations in thought.

Imagery may be of two kinds: visual-pictorial or bodily. Visual imagery often deals with imagined pictorial situations in an animation-like manner. The recombination of waves to create a bright point in Poisson's TE, or Einstein's imaginary ride on a ray of light (see Reiner 1998) are examples of visual imagery. Bodily knowledge rises from physical experience, such as riding bikes or playing basketball and provides us with tacit knowledge about the dynamics of objects and motor performance. The potential power of TEs in learning lies in retrieving experiential knowledge, in the form of visual or bodily images, of which students are not necessarily aware.

The purpose of this study is to analyse how imagery is used by students in constructing a TE. We look at the imaginary situations that students construct, explore how imaginary objects are manipulated in this mentally constructed environment, identify the tacit rules that dominate the events in the imaginary world, and the process of integrating imagery and conceptual–logical thinking in a TE while solving physics problems.

The paper evolves in two parallel tracks: theoretical and empirical. The first draws on cognitive theories about imagery and image schemata in the context of science learning. The second is based on the result of an empirical learning experiment intended to identify the imagery used by students in building TEs in a context of collaborative problem solving. We thus support our theoretical analysis with examples taken from students' protocols of problem-solving activities.

Thought experiments and research in science learning

Two types of research questions have played a major role in enquiries into the learning of science over the last two decades. The first has addressed what the students know and gave rise to an extensive amount of research on students' concepts in science (Driver 1983; Scott *et al.* 1993). The second has addressed the mechanisms that lead to particular ways of generating knowledge, for example, questions concerning the role of analogies (Clement 1988; Dagher 1995; Gentner and Stevens 1983) or of cognitive conflict in learning (Hewson and Hewson 1984).

In this paper we integrate both concerns: the study of students' knowledge and the mechanisms that lead to such knowledge. As we have argued above, these mechanisms arise from the practice of the scientific community (Hennessey 1993).

What is a thought experiment?

A TE is design of thought that is intended to test and/or convince others of the validity of a claim. It is a special type of mental window through which the mind can grasp universal understandings (Brown 1986, 1991). For instance, we may wish to test a theory by identifying its implications, or to test the coherence of a theory by confronting claims based on it with those based on another theory, or to test the plausibility of a rule. In these circumstances, it may, according to Wilkes:

> Be appropriate to ask a 'what if' question. To answer such a question the thought experimenter needs to imagine a possible world in which the 'what if' situation actually occurs. Then, the results and implications of the 'what if' situation are examined. These results may support or weaken the tested claim.
> (Wilkes 1989: 2)

The 'what if' situation can be one of two types. It may be a 'merely imagined' situation, such as Poisson's TE: an experiment which could have been carried out in the laboratory, but which, for various reasons, is executed in thought only (Brown 1986). The other type of TE, the 'truly imagined' situation, is a construct of imagination only and thus could not under any circumstances be carried out in the laboratory. Einstein's TE about riding on a ray of light is truly imagined.

Any TE, merely or truly imagined, consists of five components (Reiner 1998). In Poisson's TE, for instance, the components are: an imaginary world of a light source and an obstacle; a problem (or a hypothesis) about whether light is of a wave or of a corpuscular nature; an experiment and a result concerning light waves that are in phase and hence recombine to create a bright spot at the centre of the shadow; and a conclusion – there is a bright spot, hence light has a wave nature.

Even though some components of a TE are imagery, not always accessible in propositional forms, it is possible to analyse the structure of TEs after they have been represented in verbal or written form. This is especially valuable in studies based on learning with classical TEs of science, what we have called consensus TEs. However, such an analysis gives little insight into the pro-cesses involved in spontaneously generating a TE. Nersessian (1992) believes both that linguistic analysis undervalues the experimental nature of TEs and that setting out TEs as propositional representations cannot provide an account of the pro-cesses of their generation. At the heart of the educational value of TEs lies the need to understand how they are generated.

The imagined dynamic world, animation-like, involved in generating, 'running', and evaluating the outcomes of a TE involves both Thought and Experiment. A TE must, then, simultaneously be a THOUGHT experiment and a thought EXPERI-MENT, for it is a heuristic for knowing something that cannot always be derived solely by logical argumentation (Brown 1986). The THOUGHT aspect involves creating an imaginary world. Imagining, however, is never undertaken de novo. The components of imagined worlds, no matter how wierd and unfamiliar these may be, are often based on familiar components, which are reconstructed, rearranged, and constrained, by the experimenter's intention, the goal which is sought (Ward and Scott 1987). The construction made is thus related to the experi-menter's prior knowledge and experience and to the reservoir of existing mental models. Some of these may be literal 'pictures', some may be the dynamic images of events, some may be based on tacit intuitions, or on other implicit types of knowing (Reber 1992). The EXPERIMENT aspect means that the thought process is subject to the constraints of experimentation, such as the need to manipulate variables, to achieve internal coherency between the experiment and the imaginary world, and to achieve external coherency with the theory being tested.

A TE is thus constructed both by the use of tacit components and of logical arguments. Nersessian's (1992) argument that TEs have a non-propositional component does seem to be substantiated. We support her claim that these tacit, implicit, ways of knowing, are overlooked or ignored when the analysis undertaken is based on the verbal representation only. These aspects are analysed in the following from two points of view: theoretical and based on actual TEs performed by students.

The learning experiment

To support our arguments, we analyse a series of TEs performed by groups of teachers and of students in a collaborative learning experiment. The experiment was carried out after normal class sessions in the laboratory, the subjects having volunteered for a problem-solving session of about three hours.

Purpose

The overall purpose of this study was to explore how students use TEs for learning. More specifically we aimed to identify whether students generate TEs when solving problems. Then we identified the kinds of imagery and bodily knowledge used by students in order to imagine a physics world, to 'see with their mind's eye' the results of an experiment carried out in thought and to draw conclusions about the sequence of events in the physical world.

Three problems that may give rise to TEs were introduced to four groups: three groups each of three physics students who were in their final year of a degree in physics and in physics education and one group of five serving biology teachers who were attending an in-service course. The groups were asked to solve the problems collaboratively and to provide written solutions. The discussions were video taped and protocols were subdivided into episodes. TEs were identified according to the five components: imaginary world, problem, experiment, results and conclusion. Each sector of the discussion that included the structure of a TE was analysed to identify its imagery components.

Description of the tasks

1   A system is designed to produce energy. It includes capillary tubes that dip into water in a valley. The capillary tubes are sufficiently narrow as to enable water to rise up to the top of the hill. When falling down from the top of the hill, the water rotates a wheel which is connected to a generator with the intention of producing electrical energy. Will the system work? Explain your answer.
2   Some people like canoeing in the narrow parts of a river. In order to avoid possible problems they need to predict changes in the velocity of the water. How would they do that? The phenomenon is like that in a pipe that gets narrow and then wider again.
3   Vinegar and oil mixed as a salad dressing separate after a while and the oil floats on the surface of the vinegar. Some people claim that when a polar fluid such as vinegar in water is mixed with a non-polar fluid such as oil the molecules attract each other with very weak forces. Others claim the opposite: that the attractive force between the oil and the water molecules is very strong. Which of the two satisfactorily explains the fact that the two separate?

The results are reported here within the framework of the discussion of the imagery and bodily components of a TE given earlier.

The construction of a Thought Experiment

Two components of a TE are analysed in the following: the role of visual imagery in a TE; and the role of bodily knowledge in predicting behaviour of an object in an imaginary world.

The role of imagination and visual imagery – structured imagination

Imagination and intuition are necessary basic tools for a thought experimenter. In a sense, they fulfil the roles in a TE which are equivalent to those of an activity design for a learner who is physically experimenting. This equivalence, like any analogy, has scope and limitations, giving rise to several questions: if the world created is entirely a product of imagination, how can what we learn from it relate to the physical world? What are the criteria of coherency in that imagined world? What are the processes by which one creates such a world? What are the rules of thinking in such a world?

As Murray (1986) points out, it is easier to describe what the nature of imagining is not than what it is. It is not creating an image of a known but invisible object or situation, for seeing a known physical situation in 'the mind's eye' is not just the recall of an image or event stored in the memory. Imaginative thinking is also not the experiencing of strange irrational realities (fantasies) as may happen during hallucination.

We suggest that constructing a TE does not involve random, undisciplined, imagination but rather a spontaneous operation of structured imagination which is constrained by three criteria: the design must support the attainment of a particular goal; the design must be based on prior experience and concepts; and the design must be internally coherent.

The imaginary world has to be designed in a manner that supports the attainment of the goal being addressed. Finke *et al.* (1996) state that imagination invariably involves the generation and experience of ideas and products directed at some goal. The goal may be concrete such as a design of an artefact; artistic such as the production of a painting; or scientific such as imagining a 'what if' situation to test a theory. The imagined situation if focused on an activity that results in some kind of product. In the first task, for instance, students 'looked' at the path of a stream of water:

> D: The water is pulled upwards by the pipes...then pushed out from the pipes, [it] falls down and rotates the wheel.
>
> G: What do you mean 'the pipes pull the water'?
>
> D: The molecules...I guess of the glass, pulls the molecules of water... (While talking, he points at the lower part of the pipe, draws small circles on the pipe, on the water, and vectors to show the direction of forces.)
>
> D: These (points at the circles on the pipe) pull these (points at the circles in the water).

D and G created two imaginary worlds. The first is based on the description of the situation. The second derives from the first and is aimed at the testing the plausibility of the situation described. The external drawing seems to be a reflection and trigger for extending the imaginary world.

The imaginary world will consist of features which are dominated by prior experience. Imagined situations are linked to the would-be thought experimenter's existing pool of known objects and situations. For instance, although riding on a ray of light is entirely an imagined act, the concepts of 'riding', 'velocity' and 'frame of perceptual reference' are all based on everyday experiences of travelling. Constructing such situations, although an act of imagination, depends fundamentally on the accumulated memory of everyday experience. Ward and Scott (1987), for instance, asked fifty college students to imagine creatures that live on a planet similar to Earth and describe their diet, sensory system and appendages. The majority of the students described symmetric creatures with sensory organs which were similar to those of humans (e.g. 92% of proposals had eyes) and with at least one major type of appendage (e.g. 88% of proposals had legs). Most of the imagined creatures (86%) had a mouth. In short, the imagined creatures had a remarkable similarity to Earth creatures. Rather than being the idiosyncratic product of wild imagination, unpredictable and random, it seems that the use of the imagination to construct exemplars of a category is highly structured by the prior known features of other members of the same category.

In a similar way, the undertaking of many tasks in science learning is affected by prior experience of such tasks and their associated knowledge (Resnick 1987; Ward and Scott 1987). For instance, in an attempt to explain the changes in the pressure between the large part and the narrow parts of the river, E described an imaginary situation in which one part of the canoe is in the wider part of the river and the other part is in the narrow part of the river. The canoe is described as moving towards the narrow part. He suggested that:

> G: If the canoe moves fast (draws a canoe) it exerts force on the water here (points at the narrow part)...I mean in the narrow part of the river. You can think of each infinitely small element of water...a very small cube, as a 'canoe' that exerts force on the water in the front...so I guess the pressure in the narrow part is higher than in the wider...because there is all that amount of water from the wide part to exert pressure.
>
> D: You mean...like the accumulated gravitational pressure of fluid in the bottom of a can?
>
> G: Not really...(he hesitates)...the force...it comes from the velocity of the water...not the mass...more like a ball that has velocity.

G uses prior knowledge of various types: the strategy he acquired in calculus to imagine 'infinitely small' cubes of water exerting force on other cubes of water, and the impetus idea 'the force...comes from the velocity of the water' D picks up the imaginary would but associates it with his own knowledge: the strategy of exploring the changes in the hydrostatic pressure in a can. E did not accept that:

> E: It just does not make sense. It means that the further you go in the river the more pressure accumulates. For big rivers you can get to an infinitely big pressure this way...have storms in the river....

and, after a while

> also, look at the velocity – the canoe moves faster in the narrow part. It accelerates, which means the force exerted on the canoe here (points at the wider part) is bigger than here (points at the front). So the pressure (points again at the wider part of the river) is bigger in the narrow...when the velocity is bigger.

E knows (as do the others) that rivers do not get more stormy further away from the source and therefore reject G's explanation. He also recruits his prior knowledge about the relations between force and acceleration as the basis for conclusions about changes in pressure of the water. The other interesting point is the language E uses: 'look at' the velocity, as if the river, canoe and velocity are an imaginary dynamic world that is shared and viewed by all the members of the group.

An imagined world has to be internally coherent if the thinker is to achieve a goal. For instance, a heavier object is always so, a position is attained, so there are no internal contradictions.

These three constraints on imagination are not necessarily explicitly obeyed by the thought experimenter. They may be tacit, playing an implicit role in the construction of such a world (for a broad analysis of Structured Imagination see Finke and Shepard 1986). Such is the case with the TE performed by the subjects in this study. While working on the narrowing river, H suggests that the velocity has to be bigger in the narrow part of the canal:

> H: Like you are a particle going with the current, here (in the wide part).
>
> R: Like in a river...that turns narrow.

H: Suppose we look at a particle of fluid here (the wide part of the river). You are the particle...like you get out from a crowded room into a narrow corridor. You are pushed from behind...because behind it is wider, so you have more particles to push you. The front is narrow so you have less particles. The net force is towards the narrow pipe. In the front, particles move forward so there is no force exerted on you. Because these particles are pushed...you push them. When you get here (where the narrowing of the river begins) you move faster. Here (points to the narrowest part of the river) all the pushes from behind accumulate and you move fast. It is very crowded (in the narrowest part) until suddenly you get to quiet waters, here (the broad river again). As if you suddenly have space. Why...because there is a lot of space.

H generated an imaginary dynamic world that can be observed. It is designed in a manner that supports the attainment of the goal being addressed, by setting particles in motion inside a pipe so that this motion could be analysed and 'seen'. He provides an explanation based on prior concepts, on known imagery and embodied past experience. Concepts such as 'velocity', 'direction', 'net force' are known to H and used accordingly. Both imagery and the 'feel' of 'a push' are used to describe the events. The world is coherent in the sense that there are no internal contradictions in the world that H described. The behaviour of the particles was consistent, always slower in the wide part and faster in the narrow part. The relations between the velocity in the narrow part of the pipe and the wide part are justified by H's 'theory' that elements of water apply pressure.

The role of imagination and visual imagery – manipulation of imaginary objects

'Running' a TE requires the thought experimenter to judge properties of objects such as size, position or shape. It involves 'seeing in the mind's eye' as direct sensory input is, by definition, absent. It involves the 'perception' of situations which are retrieved from memory rather than from the gathering of new information. These 'perceptions' are the components from which imaginary worlds are constructed. This perspective on TEs entails a number of questions. What are the processes involved in constructing these imagined objects? How do these imagined objects relate to objects that are physically perceived? How do we move the components of an image around? How can we see details in an imagined object? How are dynamic events visualized? Do such events obey the rules of the physical world, which would allow inferences about them to be readily made? Are they, on the other hand, the subject of unknown laws of dynamics, ignorance of which does not allow us to draw a conclusion about size, position or shape? These questions can be explored with the aid of some simple examples. Suppose we visualize a ball thrown with angle α, to the ground and with intial velocity v. Can we visualize where it will fall? How does the arc relate to a physical arc? Stated differently, how physically accurate is the visualized world? Suppose we 'see' two objects moving. The first moves with a higher velocity than the second. After a time t units, does the system of images 'know' to position them at a distance that relates to their corresponding velocities? Suppose that, in a particular TE, a wheel is rotating. The person is required to 'see' the system in a new position after t units of time. If asked to make this prediction in a hands-on experiment, one would multiply the angular velocity by the time of rotation in order to know what the new position is.

How does the mental system of images 'know' how to position the object? What if the person has never learnt the relation between velocity and distance in the world-as-commonly-experienced, does the system of images still position the rotating wheel correctly, as it would have been in the physics laboratory? The answers to these questions have profound implications for the use of TEs for learning. Is a system of images capable of 'knowing' something that the person does not explicitly, verbally, know? Is access to this type of knowledge the key to the 'learning power' of a TE?

The answer that researchers of cognitive imagery provide is that 'yes, imagined objects are treated by the mind in a way similarly to that for physical objects'. There are numerous sources of evidence that imagined objects share common processing mechanisms with physically perceived objects (Finke and Shepard 1986). The rotation of a wheel in the mind's eye is identical to rotation of a sensory perceived wheel (Hagarty 1993). In this sense there is some 'knowledge' that is accessed, becomes available, in imagined situations, but which is hidden when verbal representations are required. When dealing with the dynamics of the river for instance, the students' conclusions about currents were based on the details of the visualized world only. M visualized the river as if being covered with leaves. He could 'see' the behaviour of the system:

M: Suppose the river is covered with leaves...only the wide part, when these (the leaves) get to here (points at the narrow part), they do not pile up. You can see that they just move faster...so the water moves faster.

K: Sometimes the leaves at the edges move slowly or hardly move at all. The currents in the middle are stronger.

The picture K and M shared was based on an imaginary world each of them created, in which they 'saw' that the leaves move faster in the middle of the river and hardly move at the edges. Based on that they concluded that currents in the middle are stronger than at the edges, without any logical or empirical account for this 'fact' other than their shared visualized world. In a similar way, inspecting the dynamics of the imaginary system described in Task 1, G saw relations between changes in the velocity of the flowing water and number of rotations of the rotor:

G: The water falls faster...the water completes a whole cycle faster... (draws in the air the path of the water, a kind of a parallelogram). If the water is faster (moves his hand faster) the rotor rotates faster (imitates with his hand the movement of the rotor).

After a while E compared the collision of the water with the rotor to a fast and a slow ball colliding with two corresponding balls (probably identical) at rest. He draws a mental animation in which the collisions happen and the results are observed:

E: Because each drop of water falls faster.... Like when a faster ball hits a ball in rest. The ball at rest moves faster...than the one that was hit by a slow ball.... The balls that hit stop.

The process of engaging images in a TE involve both their inspection and their transformation.

(a) *The inspection of images*   The process of looking at objects in images shares many properties with that of actual perception (Kosslyn 1995). If asked to inspect a picture or an ensemble of laboratory equipment, the time needed for a full inspection is longer when the number of details to inspect is bigger. Similarly, when subjects are shown the parts of a pattern and asked to assemble them into the whole, the time needed for the assemblance increases with the number of details (Palmer 1977; Thompson and Klatzky 1978). The time needed to complete the mental synthesis increases with the number of parts involved, but the time taken to recognize the whole pattern is independent both of the number of parts and of the manner in which they are assembled (Glushko and Cooper 1978). This evidence suggests that people immediately form a complete image rather than store its parts.

When one needs additional information about equipment in the laboratory, the thing to do is to scan it again. A similar process happens with stored images. Once formed, mental images can be scanned to retrieve the information they contain (see the 'inspection' of the river to 'see' the motion of the leaves, or the 'inspection' of the 'perpetual motion' system earlier in the paper). The time needed to scan between two components of an image increases proportionally with the distance between the components, just as does the scanning of two separated points in a laboratory set up (Kosslyn *et al.* 1978). Similar findings were found for three-dimensional patterns. Cooper (1976) found that when subjects were given two orthographic views of an object and had to judge whether a third view was consistent with the two given, they mentally constructed a three-dimensional image of the object and then mentally scanned it from the new perspective. For additional details of a part of a picture (e.g. subjects were asked to identify the colour of the stripes on the head in an image of an imaginary honeybee). Kosslyn (1995) reports that subjects 'zoom in', which means to increase the size of the mental object, or 'zoom out', to decrease the size of the mental object. They would do the same thing in respect of a physically perceived object. For instance in discussing the currents in the river, M and K first observed the overall motion of the leaves, but then 'zoomed in' in order to 'watch' the velocity of the leaves in the middle of the river. Similarly, G, E and D zoom into the water in order to 'see' and draw the 'elements' of the water that exert pressure.

(b) *Transformation of images*   Research into the transformation of images shows that the time required to verify that two perspective drawings depict different views of the same three-dimensional objects increases with the rotational distance between them. This suggests that the subject constructs a three dimensional image which is mentally rotated to bring it into alignment with the given image (Shepard and Mettzler 1971, 1988).

For instance, in discussing the molecular forces between the water molecules in Task 3, D claimed that:

> D: The forces between the water molecules make the molecules arrange themselves in space in a particular pattern... water molecules are like this (drew a banana shape roughly in the air). ...here (points at the left edge) you have H's (hydrogen atoms) so it's positive. Here too (points at the right edge). In the middle is an O (oxygen atoms) negative.

and, after a while:

> D: If you have two molecules...
> E: This is not a minimal packing... for minimal packing you need to put the molecules on top of each of each other?

D: Obviously you can't put them on top of each other (creates two half-bananas with her palms in the air, her palms on top of each other...why?...because these (points to the edge of her fingers, 'the hydrogen poles') reject each other. This too (points at the root of her hand). It (the hand/molecule) will move. The negative, I mean the O, will be nearest the H. Another O, from another molecule is near to the other H. It moves so as to put the H nearer to the O. And then another one (molecule) moves. So it is in equilibrium. They (the molecules) move until all the forces in all directions are equal, then they stop.

We can also mentally rotate objects, the time taken increasing with the amount of rotation undertaken, whilst the velocity of rotation seems to remain constant (Shepard and Cooper 1982). These results seem surprising because one would not intuitively expect mental manipulations to obey the same laws as physical objects, yet they do (Finke and Shepard 1986). This surprising overlap between the physical and imaginary world provides TEs with a validity as a way of coming to know. Experiments in thought have more in common with physical experiments than expected. Both obey some similar basic rules. In this sense, the validity of experiments in thought stems from similar considerations to those associated with experiments in the laboratory.

There is a general correspondence between spatial relations depicted in a mental image and those present in actual physical situations (Finke 1989). This principle also asserts that imagery and perception share many types of the same neural mechanism, which is supported by neuro-psychological data. The part in the brain activated is similar in both cases (Farah *et al.* 1988).

The non-propositional nature of images

A key issue for TEs is whether images are stored as propositional or as non-propositional knowledge. This is of major importance in the classroom: how do we communicate with students about images? There are two views that may be taken: images are like pictures in the mind, of a depictive non-propositional kind (Finke 1990; Kosslyn 1995; Nersessian 1992); or are like a set of ordered propositions stating facts and relationships that are retrieved in order to create images (Pylshyn 1973, 1981). Based on the principle of spatial equivalence between the imagined worlds and physical worlds, we assume that a picture is depicted as a whole, rather than in parts. The tendency today is to treat imagery as depictive non-propositional information (Kosslyn 1995). If this is so, it shows that imaginary worlds correspond in many ways to physical worlds. The use of imagery in a TE enables the learner to recruit and make explicit knowledge that exists somehow in a very tacit, implicit, manner. This tacit knowledge is made explicit and visible, empowering the learning process with TEs.

The role of bodily knowledge

Bodily knowledge is the kind of knowledge reflected in motor and kinaesthetic acts. When imagining an experiment, such as being in an accelerating car, riding a bike on a curve, being pushed by a crowd towards the exit, or the forces exerted on each other by 'particles' that move into a narrow passage in a river, the learner recruits her/his knowledge embodied in bodily movements. So the learner knows, for instance, what way his/her body moves if the car would accelerate forward.

A person playing tennis, whilst observing the flight of the ball in the air, will be manipulating the racket so that it hits the ball at exactly that angle and velocity to direct the ball towards a particular, predetermined, place on the court. The player is probably responding directly to visual and other types of input sensory information. But how does that person know exactly how to manipulate the racket so that it imparts the correct direction and magnitude of velocity to the ball? If a robot was performing such an act, it would probably have some means to describe the position and velocity of the ball in space built into the controlling intelligent software. The robot would process incoming information in order to predict the time position in space and the velocity of the ball, when it hits the racket. The 'correct' velocity of the racket, when it connects with the ball, would also be calculated together with the initial acceleration of the ball and force applied by the racket. This type of processing would require the use of the laws of kinematics and dynamics, such as the change in velocity due to gravitational force, the properties of projectile motion, and the law of conservation of momentum. This would lead to an application of complex mathematical models of constrained motion to the connected set of body, hand and racket. Most competent tennis players are effective without knowing all this physics. Moreover, knowledge of the physics of the motion does not necessarily improve the level of expertise of even professional tennis players (Allard 1993).

This phenomenon is widely recognized in the world of athletics. When achieving at a particularly high level, an athlete seems to disconnect bodily performance completely from overt cognitive control and the body 'takes over' (Starkes and Allard 1993). It seems as if the body 'knows' something the player 'does not'. Rather than rational, propositional, knowledge being used, some sort of imagistic, embodied, form of knowledge, which is not 'registered' in the conventional manner is being employed.

Clement's (1988) findings support this view in that he showed that embodied intuitions about forces have a role in understanding physics situations. He suggests that knowledge embodied in perceptual motor intuitions are used for physics problem solving by experts. Furthermore, these intuitions are implicit in the sense that some aspects of an action are controlled without being consciously differentiated perceptually or articulated verbally. He showed that such intuitive knowledge is spontaneous and often imagistic in form. His subjects used a 'kinaesthetic sense' which involved the conception of the situation as muscle movements or as the sensation of muscular effort.

A similar effect is involved in the students' attempts to explain the change in pressure when moving from a wide into a narrow passage in the river. They apply the feeling of being pushed to the particles. In discussing the change in pressure in a narrow passage, students use sentences which reflect a kinaesthetic and force image pattern. They sometimes use their own 'feeling' in order to evaluate the forces exerted, making remarks such as:

You are the particle.
You are pushed from behind.
    more particles push you.
    there is no force exerted.
Because these particles are pushed...as if you push them.
When you get here you move faster.
    all the pushes from behind accumulate and you move fast.
It is very crowded...until suddenly you get to quiet waters, here.
    as if you suddenly have space.

This kind of intuitive knowledge is found in both 'novices' and 'experts' with respect to formal knowledge of physics (Reif 1985).

DiSessa (1993) refers to 'phenomenological primitives' (p-prims) as kinds of basic, in a sense axiomatic, intuitions that govern the understanding of physics phenomena on which students reflect even before undertaking any formal learning of physics. P-prims are relatively minimal abstractions of simple common phenomena which need no explicit justification for their existence. He proposes that both 'naive' and 'expert' physics are partially built out of phenomenological primitives, being modified and reorganized into coherent structures of knowledge. Such 'physics intuition' is thus an epistemological resource which is used at all levels of learning about physical phenomena. It is an elemental knowledge structure which is activated by an imagined or actual situation and which is context situated, self evaluating and embodied.

The sense of touch and forces exerted on different parts of the body are so deeply integrated within our everyday interaction with the environment that we are hardly aware of them. Acts such as walking and writing involve the sensation of force. The sensation of force is not just part of such acts, it is crucial to our ability to perform them. Imagine walking on an absolutely smooth field of ice, where the force exerted by friction is very low. Imagine writing on a very smooth piece of paper, so smooth that the pen encounters no resistance. Many of the TEs that are conducted in everyday life make use of an embodied representation of force.

In an interview with the biology teachers we asked them to imagine riding a bicycle at a constant velocity along a variably curved road and to describe what they would do during the ride. Some curves on the road had a very large radius, while some had a very small radius. All five teachers had some experience of cycle riding but had no formal knowledge about the dynamics of riding on a curved road. All of them described how they would lean their body and cycle towards the centre of the curve. They also described how the angle of the body to the ground decreased with a decrease in the radius of the curve.

> T: You have to bend your body to your left.
> E: Draw the road here (on a piece of paper).
> T: (Draws the curved road so that it curves towards the left.)
> T: Otherwise you fall. As if it threw you out of the road...
> E: What threw you out?
> T: When you ride, to maintain balance you have to bend. I know that from experience.... Everybody knows that. I don't know how everybody knows that. You just know.... If you stay straight...you fall.

This particular teacher imagined the sensation of force in order to justify his imagined action. He knew what the 'correct' situation was by retrieving the bodily force sensations related to riding from his memory and using that to describe the movements that would be involved. A propositional-only approach to knowledge would require knowledge of the physics as a basis to solve the problem: this is obviously not the case.

There seems to be no sound theory to account for the relations among the forces felt and our cognitive interpretations attached to them. Individuals know how to attach meanings to forces felt such as the vibrations of an engine, the tremors of an earthquake, the gentle swing of a boat on quiet water. This is done unconsciously. There is no syntax or dictionary of meaning to attach to patterns of forces (Johnson 1987). We suggest that images of forces are triggered and used without any external stimulus, but through the intention and needs of the TE alone. These images carry tacit knowledge used in a TE. Learners know what the results of an experiment in thought may be by using this innate non-verbal knowledge, that 'comes to life' through the images of forces associated with the particular situation in the TE.

Johnson (1987) in an attempt to explain this type of knowledge, claims that:

> Human bodily movements, manipulation of objects, and perceptual interactions involve recurring patterns.
>
> (Johnson 1987: xix)

These patterns are termed 'image schemata':

> Because they function primarily as abstract structures of images.... They are gestalt structures consisting of parts standing in relations and organised into unified wholes, by means of which our experience manifests discernible order.
>
> (Johnson 1987: xix)

According to this view, understanding the order of our environment is based on such bodily image schemata. These may emerge as a result of embodied acts and can then be extended into imaginative structures that have a meaning. Repeated to-and-fro pushes, for example, are interpreted as a swing. By applying the schema, we can imagine similar situations and can make predictions about events in the future. The schema is a structure of embodied imagination in the sense that we can imagine the movements associated with it. The patterns within image schemata are figurative, pictorial and dynamic. They are also non-propositional, in the sense that it is hard to put the feel of swing into words that convey the sensation accurately. This approach suggests that students construct meaning on the basis of mental structures of embodied imagination of a figurative dynamic, non-propositional character. These mental structures are used to construct mental 'realities' in a TE.

Conclusion and discussion

We have showed that thought experimentation in which students construct imaginary situations are a frequently-used strategy for problem solving. These imaginary worlds are enriched reflections of the physical world, including invisible components in the visible world such as particles and events that have never happened in the physical world, but which make sense to the learner. By using images of a visual nature, and images of bodily experience, the thought experimenter accesses tacit knowledge, which the person is not necessarily aware of, and of which only a small portion can be articulated in a verbal manner. Such tacit knowledge, when coupled with logical processes in a TE, is unconsciously recruited to generate new knowledge. These types of tacit, non-explicit type of knowledge act as the basis for generating new states of knowing, which are put in a propositional form at the conclusion of a TE.

This work suggests that TEs have a justificatory power based on their reflection of the behaviour of physical objects in thought integrated with conceptual inference processes. We rotate, displace and overall manipulate objects in thought in a similar manner to their physical manipulation. Furthermore, we have a type of implicit knowledge that concerns the behaviour of objects under force. We construct interpretations based on force sensations (and vice versa) and have the ability to access in memory an image of the impact of these forces. We can imagine the feeling of these forces on the body and construct an interpretation. These inner images of the interrelations among the forces, objects and their behaviour are the basis for constructing cause and effect events in thought, mainly because the imagined objects are manipulated similarly to objects in the physical world, and thus are valid for learning about the physical world. The inner world seems to be a reflection of the outer world, constrained also by the nature of senses. The TE relies on these capabilities of mental object manipulation. Mach coined the term *Gedunkenexperimente* to describe a process of thought that triggers inner knowledge...(a sort of 'animal faith') to generate scientific principles (Sorenson 1992). According to Sorenson (1992), in his convincing analysis of TEs, Mach:

> Grounded the similarity between our inner private world and the outer public one in the 'biological necessity of conforming thought to environment'.
>
> (Sorenson 1992: 51)

This process is unconscious. It is a survival necessity, such as the need to predict physical events in order to respond to danger, which drives us to mimic natural patterns.

> Iron fillings dart towards a magnet in imagination as well as in fact, and when thrown into a fire they grow hot in conception as well.
>
> (Sorenson 1992: 51)

Whilst Faraday's description and analysis of electrical and magnetic phenomena were couched in the mental manipulation on the dynamics of lines of force (Wise 1988).

What is the nature of this inner knowledge? How do we access it and recruit it for TEs in school learning? Is it developmental? Is it 'imprinted' and static? In an attempt to support the validity of TEs as reasoning tools in physics, Mach (1983) termed the inner private world as instinctive knowledge, a kind of knowledge that is similar to the one imprinted in instincts:

> Everything which we observe in nature imprints itself uncomprehended and unanalysed in our percepts and ideas, which then in their turn mimic the process of nature in their most general and most striking features. In these accumulated experiences we possess a treasure store which is ever close at hand and of which only the smallest portion is embodied in clear articulate thought.
>
> (Mach 1983: 36)

This stored knowledge is a basis for the structure and events in an imaginary world in which 'experiments' take place (Polanyi 1958, 1962).

Although there is an essentially individual nature to TEs, experimentation in thought can be constructed in a collaborative classroom set-up (Reiner 1998). When students are allowed to work together to solve meaningful problems, TEs evolved out of non-organized, non-predictable, in a sense chaotic, students' discussion. Each student's narrative had no meaning as a TE, but the accumulation of all the narratives – without any of the participants so intending – gradually emerged as a collaborative TE. A TE is thus a tool for both the personal and social construction of meaning.

The study of students' ideas in science education research has so far broadly focused on the analysis of verbal information. Tacit knowledge is hardly reflected at all in a symbolic form such as speech. It may be described in forms of visual images and in the construction of the feel of force exerted on the body. But these are not easily put into words. This implies that a comprehensive picture of students' ideas needs to rely on an integration of the analysis of bodily knowledge, mental imagery and conceptual–logical beliefs. It raises a major methodological research problem: what are appropriate methods to uncover that which may be hidden? A possible source for these might be the methods employed by psychologists who explore implicit, unintended, learning and knowing (for a review see chapters 2 and 3 in Reber 1992).

Another interesting implication is for the design of learning environments. The intellectual power of a TE is in the integration of the three components suggested. Two of them, visual imagery knowledge and bodily knowledge, recruit tacit knowledge for generating science concepts. Most computer environments include mainly visual information. The new generation of learning environments improve the visual information by positioning the learner inside a virtual world. Yet most of the information is still visual only. There is a need to include more than one modality of sensory information in the learning environment. Not just visual imagery, but also force sensations and support for conceptual logical schemes. Future learning environments may include olfactory information as well as video, audio and other forms.

References

Allard, F. (1993). Cognition, expertise, and motor performance. In J.L. Starkes and F. Allard (eds), *Cognitive Issues in Motor Expertise* (North Holland: Elsevier), 17–33.

Brown, J.R. (1986). The structure of thought experiments. *International Studies in The Philosophy of Science: The Dubrovnik Papers*, 1–15.

Brown, J.R. (1991). *The Laboratory of The Mind* (London: Routledge).

Clement, J. (1988). *Imagery in Problem Solving in Physics*. Paper presented at the Workshop on Cognition in Science Learning, Tel-Aviv, Israel.

Cooper, L.A. (1976). Demonstrations of mental rotations to an external analog of an external rotation. *Perception and Psychophysics*, 19, 296–302.

Dagher, Z. (1995). Analysis of analogies used by science teachers. *Journal of Research in Science Teaching*, 32, 259–270.

Dancy, J. (1985). The role of imaginary cases in ethics. *Pacific Philosophical Quarterly*, 66, 141–153.

DiSessa, A.A. (1993). Towards an epistemology of physics. *Cognition and Instruction*, 10, 105–225.

Driver, R. (1983). *The Pupil as a Scientist?* (Milton Keynes: Open University Press).

Farah, M.J., Peronnet, F., Gono, M.A. and Giard, M.H. (1988). Electrophysiological evidence for a shared representational medium for visual images and visual precepts. *Journal of Experimental Psychology (General)*, 117, 241–246.

Finke, R.A. (1989). *Principles of Mental Imagery* (Cambridge, MA: MIT Press).

Finke, R.A. (1990). *Creative Imagery: Discoveries and Inventions in Visualisation* (Hillsdale, NJ: Erlbaum).

Finke, R.A. and Shepard, R.N. (1986). Visual functions of mental imagery. In K.R. Boff and J.P. Thomas (eds), *Handbook of Perception and Human Performance* (New York: Wiley), 37, 1–55.

Finke, R.A., Ward, T.B. and Smith, S.M. (1996). *Creative Imagery Theory: Research and Applications* (Cambridge, MA: MIT Press).

Gentner, D. and Stevens, A. (1983). *Mental Models* (Hillsdale, NJ: Erlbaum).

Glushko, R.J. and Cooper, L.A. (1978). Spatial comprehension and comparison processes in verification tasks. *Cognitive Psychology*, 10, 391–421.

Gooding, D.C. (1992). What is experimental about thought experiments? *PSA*, 2, 280–290.

Hagarty, M. (1993). Constructing mental models of machines from texts, and diagrams. *Journal of Memory and Language*, 32, 717–742.

Hennessey, S. (1993). Situated cognition and cognitive apprenticeship. *Studies in Science Education*, 22, 1–41.

Hewson, P.W. and Hewson, M.A.G. (1984). The role of conceptual conflict in conceptual change and the design of science instruction. *Instructional Science*, 13, 1–13.

Johnson, M. (1987). *The Body in the Mind* (Chicago, IL: University of Chicago Press).

Kosslyn, S.M. (1995). *Image and the Brain* (Cambridge, MA: MIT Press).

Kosslyn, S.M., Ball, T. and Reiser, B.J. (1978). Visual images preserve metric spatial information: evidence from studies of image scanning. *Journal of Experimental Psychology: Human Perception and Performance*, 4, 47–60.

Kuhn, T.S. (1977). The function of thought experiments. In T. Kuhn (ed.), *The Essential Tension* (Chicago, IL: University of Chicago Press).

Mach, E. (1983). *The Science of Mechanics* (9th edn), trans. Thomas J. McCormack (London: Open Court), 36.

Murray, E.L. (1986). *Imaginative Thinking and Human Experience* (Pittsburgh, PA: Duquesne University Press).

Nersessian, N. (1992). Constructing and instructing: the role of 'abstraction techniques' in creating and learning physics. In R. Duschl and R. Hamilton (eds), *Philosophy of Science, Cognitive Psychology and Educational Theory and Practice* (New York: State University of New York Press), 48–68.

Palmer, S.E. (1977). Hierarchical structure in perceptual representation. *Cognitive Psychology* 9, 441–474.

Polanyi, M. (1958). *Personal Knowledge: Towards a Post-critical Philosophy* (Chicago, IL: University of Chicago Press).

Polanyi, M. (1962). *The Tacit Dimension* (New York: Doubleday).

Pylyshyn, Z.W. (1973). What the mind's eye tells the mind's brain: a. critique of mental imagery. *Psychological Bulletin*, 80, 1–24.

Pylyshyn, Z.W. (1981). The imagery debate: analogue media versus tacit knowledge. *Psychological Review*, 87, 16–45.

Reber, A.S. (1992). Implicit learning and tacit knowledge: an essay on the cognitive unconscious, *Oxford Psychology Series*, 19 (Oxford: Oxford University Press).

Reif, F. (1985). Acquiring an effective understanding of scientific concepts. In L.H.T. West and A.L. Pines (eds), *Cognitive Structure and Conceptual Change* (New York: Academic Press), 133–151.

Reiner, M. (1998). Thought experiments and collaborative learning in physics. *International Journal of Science Education*, 20, 1043–1058.

Resnick, L.B. (1987). Constructing knowledge in school. In L.S. Liben (ed.), *Development and Learning: Conflict or Congruence?* (Hillsdale, NJ: Erlbaum), 19–50.

Roth, W.-M. (1995). *Authentic School Science* (Dordrecht: Kluwer).

Scott, P., Driver, R., Leach, J. and Millar, R. (1993). *Students' Understanding of the Nature of Science* (Leeds: Children's Learning in Science Group).

Shepard, R.N. and Cooper, L.A. (1982). *Mental Images and their Transformation* (Cambridge, MA: MIT Press).

Shepard, R.N. and Mettzler, D. (1971). Mental rotation of three dimensional objects. *Science*, 171, 701–703.

Shepard, R.N. and Mettzler, D. (1988). Mental rotation: effects of dimensionality of objects and type of task. *Journal of Experimental Psychology: Human Perception and Performance*, 14, 3–11.

Sorensen, R.A. (1992). *Thought Experiments* (Oxford: Oxford University Press).

Starkes, J. and Allard, F. (eds) (1993). *Cognitive Issues in Motor Expertise* (Netherlands: North Holland Publishers).

Teng, Y.-J. (1992). *An Enquiry into Thought Experiments*. Unpublished Doctoral Thesis, Southern Illinois University at Carbondale.

Thompson, A.L. and Klatzsky, R.L. (1978). Studies of visual synthesis: integration of fragments into forms. *Journal of Experimental Psychology: Human Perception and Performance*, 4, 244–263.

Ward, T.B. and Scott, J. (1987). Analytic and modes of learning family resemblance concepts. *Memory and Cognition*, 15, 42–54.

Wilkes, K.V. (1989). *Real People: Personal Identity without Thought Experiments* (Oxford: Clarendon Press).

Wise, N.M. (1988). The mutual embrace of electricity and magnetism. In S.B. Brush (ed.), *History of Physics: Selected Reprints* (College Park, MD: AAPT).

CHAPTER 5

MODELLING, TEACHERS' VIEWS ON THE NATURE OF MODELLING AND IMPLICATIONS FOR THE EDUCATION OF MODELLERS

Justi, R. and Gilbert, J.K. 'Modelling, teachers' views on the nature of modelling and implications for the education of modellers'. *International Journal of Science Education*, 2002, 24(4): 369–387

In this paper, the role of modelling in the teaching and learning of science is reviewed. In order to represent what is entailed in modelling, a 'model of modelling' framework is proposed. Five phases in moving towards a full capability in modelling are established by a review of the literature: learning models; learning to use models; learning how to revise models; learning to reconstruct models; learning to construct models de novo. In order to identify the knowledge and skills that science teachers think are needed to produce a model successfully, a semi-structured interview study was conducted with 39 Brazilian serving science teachers: 10 teaching at the 'fundamental' level (6–14 years); 10 teaching at the 'medium'-level (15–17 years); 10 undergraduate pre-service 'medium'-level teachers; 9 university teachers of chemistry. Their responses are used to establish what is entailed in implementing the 'model of modelling' framework. The implications for students, teachers, and for teacher education, of moving through the five phases of capability, are discussed.

Introduction

Many roles in science have been identified for models. At one end of a continuum are apparently straightforward functions such as the representation of entities in simplified depictions of complex phenomena (Ingham and Gilbert 1991; Glynn and Duit 1995). Other functions are to make abstract entities visible (Francoeur 1997), to provide a basis for the interpretation of experimental results (Tomasi 1988), to enable explanations to be developed (Gilbert *et al.* 1998). At the opposite end of the continuum are more complex functions, including the provision of a basis for predictions (Erduran 1999). Overall, modelling – the production and revision of models – has been seen as the essence of the dynamic and non-linear processes involved in the development of scientific knowledge (Leatherdale 1974; Del Re 2000).

Hodson (1992) has suggested three purposes for science education: the learning of science, i.e. to understand the ideas produced by science; learning about science, i.e. to understand important issues in the philosophy, history and methodology of science; and learning to do science, i.e. becoming able to take part in those activities that lead to the acquisition of scientific knowledge. These purposes suggest that there must be a central role for models and modelling in science education. This is so because in order:

1   to *learn science*, students should come to know the natures, scope and limitations, of major scientific models – these are either consensus models, those currently used in research, or historical models, those now superseded for research purposes (Gilbert *et al.* 1998);
2   to *learn about science*, students should be able to appreciate the role of models in the accreditation and dissemination of the outcomes of scientific enquiry; and
3   to *learn how to do science*, students should be able to create, express and test their own models (Greca and Moreira 2000; Harrison and Treagust 2000).

The key to the achievement of all these purposes is therefore the act of scientific modelling, i.e. the formation of appropriate representations. This argument rests on the assertion that all acts of understanding entail the formation and use of mental models (Johnson-Laird 1983). If this is so, then the acts of learning existing scientific models (model learning), of modifying these to accommodate new purposes (model revision), as well as the making of new models (model production), all entail the act of mental modelling. Thus all students of science should, in the broadest possible sense, learn how to make models.

The act of modelling

If modelling is to be taught in science education then, in order for the processes to be authentic (Roth 1995), guidance on how to do so must be sought in the practice of science itself. The search would seem to be in vain, for:

> When we look for accounts of how to construct models in scientific texts we find little on offer. There appear to be no general rules for model construction.... Some might argue that it is because modelling is a tacit skill and has to be learned not taught.... It is, some argue... not susceptible to rules. We find a similar lack of advice in philosophy of science texts.
>
> (Morrison and Morgan 1999: 12)

However the learning of scientific modelling is widely included in curricula, e.g. in DfEE (1999) and NRC (1996), so a way of teaching it has to be found. To facilitate such teaching we have developed a 'model of modelling' framework that is substantially based on Clement (1989) (Figure 5.1).

The elements of the model warrant some expansion. All modelling is undertaken for a *purpose*, whether it be to describe the behaviour of a phenomenon, to establish the entities of which it is thought to consist (together with their spatial and temporal distribution), to ascribe the reasons for – the causes and effects of – that behaviour, to predict how it will behave under other circumstances, or several or all of these (Gilbert *et al.* 1998). The person forming a mental model of an existing consensus model, modifying an existing model, or producing their own model de novo, must be clear as to the purpose being addressed. This process of decision will be enmeshed with some initial, direct or indirect, qualitative or quantitative, experience of the phenomenon being modelled: *making observations* of it. The processes involved in the *selection of the source* from which the model is derived, and the analogical transfer of the relationships between the entities within the

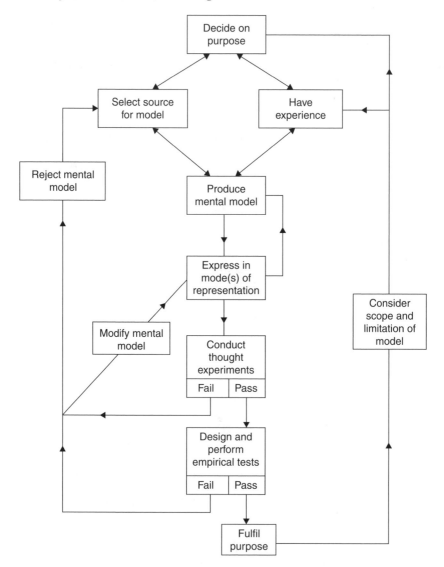

Figure 5.1 A 'model of modelling' framework.

source to those within the target, have been proposed in the 'structure mapping' approach of Gentner and Gentner (1983). Having produced a mental model in this way, the decision has to be taken about the mode of representation in which it is to be *expressed*: material, visual, verbal, mathematical (Boulter and Buckley 2000). This process of expression does seem to be cyclically developmental in respect of the mental model, with the act of expression leading to a modification in the mental model. Having produced a model (or having formed an appreciation of a scientific model), the next step is to explore its implications through '*thought experimentation*' conducted in the mind. As Reiner and Gilbert (2000) have commented, it is likely that scientists always mentally rehearse the design and conduct of empirical experimentation. It is only when the outcomes of this mental activity seem successful that actual empirical testing takes place, where this is possible. If the model fails to produce predictions that are confirmed in the thought experimental testing phase, then an attempt will have to be made to modify it and to re-enter the cycle. However, if it passes the thought experimental phase, it can go on to the '*empirical testing*' phase. This entails the design and conduct of practical work, followed by the collection and analysis of data, and finally by the evaluation of the results produced against the model. If the model fails at this stage, an attempt also has to be made to modify it and to re-enter the cycle. However if it passes the empirical testing phase, the modeller will feel confident that the purpose for which it was constructed has been *fulfilled*. This will be followed by a phase in which attempt will be made to persuade others of its value. During this process of advocacy, the *scope and limitations* of the model will become apparent, leading to a reconsideration of the earliest elements in the model-production cycle. If the sub-cycle of the model modification and thought and/or empirical testing is repeatedly unsuccessful, then the model will have to be rejected. This will lead to a radical reconsideration of the earliest elements in the model-production cycle.

Learning and teaching modelling

Modelling is a complex process, involving many component activities; skill in respect of each of which has to be mastered. It must thus be assumed that it is only gradually acquired with any degree of competence. Lehrer *et al.* (1994) stated that the ability to produce models entails the gradual evolution of a series of epistemological commitments: first, to the notion of a separation between phenomena and noumena: being able to appreciate that a representation may be of, but not identical with, that which is being represented; second, to the possibility of producing a representation through the development and deployment of a system of formal elements; third, to a notion of prediction using simplified representations that enable emergent behaviour to be identified. If the acquisition of these epistemological commitments is gradual, it would seem appropriate to teach students in such a way that their development is supported by an overt capability to implement the 'model of modelling' framework outlined in Figure 5.1. From the literature, we have identified five approaches to learning about models and modelling through which these commitments and capabilities can be acquired.

Whilst these five approaches to acquiring the skills and epistemological commitments entailed in modelling are presented as being distinct, this is to a large extent only a matter of representational convenience. Many of the studies cited above ranged backwards and forward across several or all of them.

Learning consensus/curricular models, if necessary by means of teaching models

Glynn *et al.* (1994) proposed a six-step introduction for any specially constructed teaching model. This could be used to represent a target model, either a scientific model or a curricular model, the latter being a simplified version of the former developed for teaching purposes. The six steps are: the introduction of the target model; the introduction of the teaching model; the identification of relevant and corresponding features between the target and the teaching model; the mapping of the similarities between the two; and the identification of where the analogy breaks down; the drawing of conclusions about the nature of the target as modelled. Treagust *et al.* (1996) showed how this procedure can be used successfully in respect of teaching models introduced or produced by the teacher, whilst Pittman (1999) found that the students can fruitfully construct and use their own teaching models. Idling (1997) summarized good practice in learning through teaching models in general.

The focus of this approach is on the ability to express a teaching model, and subsequently the target model, in a suitable mode of representation. This entails the acquisition of a mental model. All the other elements of the 'model of modelling' framework will be suppressed because the model that is the centre of attention (the target model) is already fully developed and known to the teacher. Of course, if the process by which the target model was originally developed has been 'cognitively reconstructed' (Nersessian 1984), then it may be possible to represent all the elements of the 'model of modelling' framework to the learners in a reasonably authentic manner.

Learning the use of models

This goes one step further than having students just learn the nature of a particular model. It involves having them apply it in contexts where the outcome will be positive, i.e. where it will successfully represent the behaviour of a phenomenon. Arnold and Millar (1996) used a series of allied teaching models to introduce the notions of heat, temperature, thermal equilibrium to 12–13-year olds in the UK and, having done so, required them to apply these ideas to a range of other contexts. Halloun (1996, 1998) had students use scientific models in the solution of a range of paradigmatic problems in physics.

The emphasis in this approach is on the thought experimentation and empirical experimentation elements of the 'model of modelling' within the psychological security provided by seeking only the confirmation of existing ideas.

Learning how to revise models

'Model revision' is the situation where students, having learnt and used a model, have to change it in some way. The intention is that it can represent a phenomenon in contexts other than those initially encountered or so that it can be used for purposes other than those originally envisaged. Stewart *et al.* (1992) taught the skills of model revision by having high school students use a simulation of scientific activity: a phenomenon was observed in groups; that experience was shared between the groups; the groups each designed an explanatory model; the groups defended their models against the critique of other groups; the groups then revised their models until a degree of convergence was achieved. This emergent skill of model revision was then used on standard problems in genetics, using historical papers and computer software. The historical papers showed them how the problem had been solved at the time of the invention of the solution. The computer software enabled them to simulate the process of change and the emergence of an acceptable solution. In an interesting variant of the model revision theme, Frederiksen *et al.* (1999) used computers to support students in progressing through (revising) a sequence of models of higher abstraction/complexity. Students were taken through a sequence of models of electricity (the electricity-particle model, the aggregation model, the algebraic model) such that the entities and interactions of a simpler model provided the emergent properties of the next in the sequence through a process of revision.

Although the thought experimentation and empirical experimental elements are involved, the emphasis in this approach is on the mental activities entailed in the 'modify mental model' element of the cycle.

Learning the reconstruction of a model

This is the situation where students are caused to create a model that they are aware exists but the details of which are unknown to them. Barab *et al.* (2000) had university students re-create the dynamic model for the Solar System by the use of virtual-reality modelling tools. They were presented with a progressive series of questions, the more advanced of which required the construction and running of thought experiments and, later, pseudo-empirical experiments. They successively constructed a static model of the Earth–Sun system, a dynamic model of the Earth–Moon–Sun system, a dynamic model of the Solar System. Groups presented their work to each other, and individuals wrote evaluations of the models produced and of how these related to the 'standard model' when it was finally shown them in detail.

All the elements of the 'model of modelling' framework are present in this approach, but are exercised within the psychological security provided by an awareness that the model is already fully known to and accepted by scientists.

Learning to construct models de novo

The MARS project (Raghaven and Glaser 1995; Raghaven *et al.* 1998a,b) set out to develop model-building skills in grade 6 students in the USA. Working on fundamental topics, e.g. 'mass', 'force', students did practical work, made predictions, inserted these predictions into a model constructed on a computer, and 'ran' the model so produced to see how the outcomes compared with those produced by the consensus model. The long-time scale needed for the development of modelling skills was noted, but an increased use of a modelling approach was found, together with an improvement in the use of the explanation–prediction–evaluation sequence. The construction of a model de novo involves perceiving the emergence of macro-level properties (those of the complete model) from those at the micro-level (those of the components of the model). Recent work by Penner (2000) discussed the issues involved and reported that the relationship of micro to macro cannot be satisfactorily made by grade 6 students.

This approach includes all the elements of the 'model of modelling' framework, without any security derived from the prior knowledge of scientists.

The study

Learning how to model appears, on the basis of the limited literature available, to be a lengthy and somewhat haphazard process, characterized by a high level of uncertainty, partial success, and even failure. The progressive introduction of students to the art of modelling requires that their teachers have, as a necessary condition for success, an appropriate understanding of what modelling as a process entails.

The study reported here formed part of a broader enquiry into teachers' understanding and use of models and modelling. A semi-structured interview methodology was used for the whole study after the questions had been piloted with five high-school science teachers. The interview had three parts. The questions in one part were concerned with teachers' views on the nature of models. The questions in another part were concerned with the use of models and modelling in the science education context. The outcomes are reported in Justi and Gilbert (2001). In respect of the research reported here, the sample of teachers was asked questions around the theme 'What are the knowledge and skills that a person should have in order to produce a scientific model successfully?'

The sample consisted of 39 Brazilian serving science teachers in four sub-samples. The first was of 10 teachers drawn from the 'fundamental' school level (for students aged 6–14 years; referred to below as FT1-10). They had, as professional qualifications, a mixture of primary school teaching certificates and degrees in chemistry or biology. The second was of 10 teachers drawn from the 'medium'-school level (for students aged 15–17 years; referred to below as MT1-10). They each had a degree in one of chemistry, physics or biology. The third was of 10 undergraduate, pre-service, student 'medium'-level teachers (referred to below as ST1-10). The fourth was nine university teachers of chemistry (referred to below as UT1-9). They all had first degrees and doctorates in chemistry.

The interviews, typically lasting 1 hour, were conducted in Portuguese, transcribed in full, translated into English, and analysed with the support of the Q.S.R. NUD*IST Vivo® qualitative data analysis package.

Results

The data are presented here within the framework provided by Figure 5.1. It should be emphasized that the interviewees were *not* aware of the 'model of modelling' framework set out earlier. In respect of 'using a suitable mode of representation' the interviewees seemed to be thinking about the context of teaching. In all other aspects of the discussion, the interviewees seemed to be thinking of modelling as something done primarily by scientists, or by 'other people' who were less effective at this than scientists. The responses that fall within each component of the 'model of modelling' framework are illustrated below using, as far as is possible, quotes from each of the sub-samples:

The purposes for which the model is being constructed

A condition for success was seen to be that the modeller (a scientist or other person) had clear purpose in producing the model, e.g.:

MT8:   In order to produce a model you should have your aims for that model clear, you should know in which context that model is to be used.
ST8:   It is important to keep in mind which ideas you are trying to explain and to express them in a clear way.
UT3:   When a model is being produced it is important to take its aims into account.

For success, the modeller must set out to communicate ideas to a defined audience, to people whose characteristics are known, e.g.:

ST7:   Depending on the level of knowledge of the people to whom the model is to be presented, it is possible to produce different models.
ST8:   The person (producing the model) should think about all the difficulties that other people might have in trying to understand the model.

Another purpose relates to the comprehensiveness of the model. Whilst respondents thought it desirable that any model should be complete as possible, i.e. should represent all facets of the phenomenon, e.g.:

FT5: I think a model has to show all the details (of the phenomenon).

UT9: I think that a model has to show the main aspects of the object or phenomenon in some way.

They did recognize that this was not always possible, e.g.:

UT6: There is no such thing as a complete model. What exist are models that are more complete than others, models that are more adequate to a given level of explanation than others.

In either case, the model must achieve a degree of similarity to the phenomenon that is recognized by the modeller, e.g.:

FT6: It should be as similar (to the phenomenon) as possible in order to provide a better understanding.

MT4: What is going to be different is the complexity of the model. I think that everybody can model something in a way that is consistent with his or her knowledge of the reality.

UT4: Another person may produce a model different from mine. Depending on both the level of information he/she has and the level of complexity required of the model, it could be better or worse than mine.

UT6: It would be possible to think of different models, some of them more complex than others.

The personal experience, knowledge and attributes of the modeller

This is concerned with what the modeller (a scientist or other person) has retained from earlier education or form experience at the time that the modelling takes place. It may either be specific to the field of study in which the modelling is taking place or be more general.

The value of prior experience of the phenomenon being modelled was recognized, e.g.:

ST7: Another person may produce a model different from mine. This depends on the experiences that each person has already had with the reality that is being modelled.

In addition, experiences with the other, perhaps allied, phenomena and models may be useful, e.g.:

FT2: I think another person may produce a model that is different from mine because personal experience influences this a lot. Each person tends to establish comparisons with things s/he has previously seen.

ST4: Scientists have other models with which to establish relationships.

> *ST5:*   I think that a scientist tries to organise his ideas and to establish relationships with things that he already knew in order to be able to represent what he is thinking.

Success in modelling was also seen to be closely linked to the personal attributes of the modeller. There were three facets of character mentioned here. First, a successful modeller has to have an active interest in building the model. Whilst this is seen as a general characteristic of modellers, successful scientists are held to be especially good in this respect because their interest is shaped by specific criteria of success, e.g.:

> *FT3:*   I think that everybody is able to create something if there is engagement with the situation.
>
> *UT2:*   I think that when a scientist produces a model it is different from when an ordinary person does. Scientists tend to be more worried about coherence, accuracy, and validity.
>
> *UT6:*   I think that when an ordinary person produces a model he/she does not think about other issues, whilst scientists are always thinking about additional things that have to be explained.

Second, a successful modeller has, in a broad sense, to be a creative person: again, scientists were believed to be more successful than other people in this respect, e.g.:

> *MT2:*   In order to model something a person needs to have a lot of creativity. Moreover, he/she has to work by intuition.
>
> *MT3:*   It is possible to have more than one model for a given thing because (model building) depends on how a person thinks, on the ideas the person has, on how creative each person is.
>
> *MT8:*   A scientist has to be creative all the time. From some initial observations, he starts to think, to create simple models and to test them at the same time.
>
> *UT1:*   I think it is necessary to have creativity and to dare to deal with new ideas.

Third, the successful modeller has to be persistent, e.g.:

> *FT1:*   The person has to be persistent because s/he should not give up if the first model is no good.

The selection of suitable source for the model

Although the issue of source-selection by the modeller (a scientist or other person) was not touched on directly, the use of analogy, both between phenomena and models and between sources of models, was recognized as being valuable:

> *FT8:*   When I have to model something, I always think of something that I know and which could help me.
>
> *MT7:*   When a scientist has to explain an unknown phenomenon, he tries to make an association with a known phenomenon.

The importance of gauging the strength of the similarity between the source of the analogy and the target for the analogy was recognized, e.g.:

MT2:   In an orrery[1] it is important to use balls and that they can move because these are essential aspects of the phenomenon.

UT8:   When a model is being produced it is important to take account of its aims and the level of accuracy that is needed.

Producing a mental model

Some of those interviewed thought that the process of modelling by a scientist was different from that adopted by other people, e.g.:

FT1:   I think the process of production of a model by a scientist is different from the process of production of a model by another person. This is because a scientist is always testing his model, creating hypotheses, analysing each situation, checking whether the model satisfies a given condition or not. Other people make a model by copying something.

Others thought that the processes were of the same kind but different in quality, e.g.:

FT6:   I think that scientists and other people use the same mental processes when they produce a model. The only differences are that scientists observe more detailed aspects of the phenomenon and are more persistent than other people during the process.

Whilst others made no overt distinction between scientist and non-scientist, e.g.:

MT3:   First of all, the person should have some knowledge about the theme of the model. After this, he/she has to create the model and prove whether it corresponds to reality or not.

Whilst some interviewees saw modelling as an inductive process, e.g.:

FT4:   First of all, a scientist observes something. Then he gets concrete materials in order to communicate what he has observed and his ideas about this to other people.

UT8:   Scientists collect lots of information, sometimes for a long time because such a process is not always simple. Then they try to simply complicate things, to produce models representing their ideas in different ways.

Others saw it as a deductive process, e.g.:

MT4:   Scientists start from hypotheses, check them in different contexts and, dependent on the results of such tests, if the model shows a good reproducibility, it will become a scientific model, a law.

Most seemed think of the inductive and deductive processes involved in model building as being intertwined, e.g.:

> ST5:   I think a scientist tries to organize his ideas and to establish relationships with things he already knows in order to be able to represent what he is thinking. Then he will test his model against reality.

Success would seem to depend on having a fairly clear idea of the approach to the task being adopted.

Using a suitable mode of representation

For the interviewees, the most important issue here in the presentation of a model (by a teacher) was that of expressing it in a mode that offered the greatest support to visualization by its user (students).

Whilst it was thought desirable (for students) actually to see the phenomenon being modelled, where this is possible, i.e.:

> MT9:   When the thing is not abstract, it is also important to observe the object.

It was recognized that this is not always possible, e.g.:

> MT7:   Modelling an object is different from modelling a process. For instance, when modelling a car I just observe it and make a miniature, but when modelling a chemical reaction I have to imagine things I cannot see.

In the latter case, a purpose for representation is to make the abstract or invisible visualizable, e.g.:

> FT7:   I think that a model should bring the thing modelled to a visual level.
> MT7:   Modelling a chemical reaction is difficult. I have ideas about substances and the mechanisms by which substances are changed, but I have to think of ways to make my ideas concrete.

The interviewees showed that they were aware, to some extent at least, of the capacities of each mode of representation, which are linked both to their ability to invoke visualization and to provide the basis for an explanation. These capacities, if known to the modeller, would be a guide in the selection of an appropriate mode of representation.

Thus, concrete models are especially valued in science education for work with younger students. From the range of examples of models, each expressed in a range of modes and presented to the interviewees at the first part of the interview, a grouping of them by mode was readily perceived, e.g.:

> MT10:   I would group the orrery, the concrete model of dissolution, the ball and stick model of water, and the toy car. These models are objects.

Their capacity to represent the three-dimensional aspects of phenomena was identified, e.g.:

> *UT2:* The concrete model of dissolution, the ball and stick model of water and, the orrery are three-dimensional representations.

Visual models were also perceived as a group, e.g.:

> *UT4:* I would group the map, the 'no smoking' sign, the drawing of the car, the formula of water, the drawing of particles during dissolution. They all have a common visual way to represent something.
> *FT4:* Some [of the models] are drawings, they are representations on paper.

Their capacity to represent in two dimensions recognized, e.g.:

> *MT4:* I would group the 'no smoking' sign and the drawing of the car. They are two-dimensional models.

Mathematical models as such were also recognized, e.g.:

> *MT4:* I would group the graph and the mathematical equation. It is a mathematical language for models.

Valued for their ready transferability, e.g.:

> *UT7:* The graph and the equation are mathematical models that might be used to model other things, other phenomena.

The verbal mode was not so readily recognized, perhaps because the mode of presentation used during the interview (on paper, as sentences) distracted interviewees from its essential nature, e.g.:

> *ST6:* I would group the definition of a chemical reaction, the description of the solar system, and the analogy (of the 'car flying'). They are together because ideas are being represented in a written text.

Although there was some success, e.g.:

> *UT9:* It is possible to express a model through words since students understand that an analogy is being used. The analogy would be mental not visual.

There was some indication that the interviewees were flexible in their use of these modes, e.g.:

> *ST5:* It is possible to have different models because it is possible to represent my thoughts about a given reality in different ways.

Conducting a thought experiment

Interviewees' ideas about thought experimentation were not explicitly probed for during the interviews. It is therefore not particularly surprising that no data were collected relating to this aspect of modelling. However, a general lack of awareness in science education of thought experimentation as a component of scientific methodology has been identified by Gilbert and Reiner (2000).

Designing and performing empirical test

Interviewees saw a modeller to need four sets of personal skills that would be invoked during empirical testing. These are presented here in no special order, for none was evident in the data collected.

The successful modeller has to have good manual skills, e.g.:

FT3:   S/he has to be good with her/his hands.
MT7:   The person has to be able to manipulate things in order to carry through ideas for models to completion.

The modeller would need good powers of observation, e.g.:

FT8:   A model results from observations. It is the product of study.
FT10:   The person has to be patient and a good observer.
UT6:   A scientist's modelling processes are different from those of ordinary people in the level of observation of reality.

A capacity for abstract thinking is necessary, e.g.:

FT2:   The person needs to have a good ability to abstract, to build mental images about the process in question.
MT6:   One should be able to think in an abstract way.
ST3:   A scientist's modelling process is different from that of an ordinary person in aspects like the level of abstraction used.

A capacity for logical thought is also necessary, e.g.:

MT5:   The person needs to have some organisation of thought, some logic, in order to think in a series of steps.
ST6:   A scientist's modelling process is different from that of an ordinary person's modelling process in aspects like the level of observation, the analysis of available information, the production of a model. Both scientists and ordinary people think and establish relationship, but scientists think from a scientist's point of view.
UT4:   To produce a model it is necessary to learn how to establish relationships between items of information.

Fulfilling purposes and considering the scope and limitations of a model

These would emerge during the evaluation of the results of the empirical testing of the model and in the process of communicating it to other people, both of which would elicit a critical response. The ability to communicate, the capacity to express ideas in different ways, was seen to be important in modelling, e.g.:

ST8: The person should be able to express his/her ideas in different ways, to keep in mind which ideas you are trying to explain and to express them in a clear way.

UT4: If the model is qualitative, the person should be able to show the relation to the phenomenon and to describe it in some way. If the model is quantitative, the person should also be able to 'translate' what she/he is seeing into a different 'language'.

UT6: It is important to make visible things and ideas that are created.

Rejecting or modifying a mental model

The interviews did not include examples where the teachers were required to reject or to amend a personal mental model. No data on this element of the 'model of modelling' were thus collected.

Discussion

The patterns of responses by the teachers throw interesting light on what they see to be entailed in developing modelling skills.

Purpose for modelling

The teachers were convinced that successful modelling involved a clear perception of why a model was being produced and of the audience that would make use of it. At the same time, the scope of use for the model must be established and as high a degree of similarity as possible to the phenomenon being represented achieved. The issue of purpose for modelling falls within the envelope of the nature of science. What one thinks the processes of science are will govern both the general and specific purposes allocated to the act of modelling. Lederman (1992), in a broad review of the literature, concluded that teachers' understanding of the nature of science was unsatisfactory and that the relationship of that understanding to classroom practice was complex. This implies, but does not state, for modelling was not explicitly discussed in the review, that the purposes that modelling serve in science will be incompletely appreciated by many teachers.

An allied issue will be that of what teachers think a model actually is, i.e. what status is to be attributed to the outcome of modelling activity. Grosslight *et al.* (1991) were of the view that students' notions of the nature of model formed a distinct hierarchy of stages, a finding that can be taken to imply that teachers' understanding might fall within the same pattern. However, teachers do not display such stages, instead showing an understanding made up of positions within a series of seven distinguishable but interlocked *aspects* of the notion of model.

As was identified earlier, one purpose for engaging students in the 'model of modelling' framework would be to facilitate the development of the scientifically important epistemological commitments entailed. That to a separation between phenomenon and noumenon (Lehrer *et al.* 1994) would be reflected in a progression from the use of what Driver *et al.* (1996) call 'phenomenon-based' reasoning, through the use of 'relation-based' reasoning, to the use of 'model-based' reasoning.

Teachers will only be able to support this progression if they themselves have a clear understanding of the role of modelling in science. This will be particularly needed when they are supporting students in the revision or reconstruction of a model, and especially where models are being produced de novo.

Experience of the phenomenon being modelled

The teachers interviewed identified experience of a phenomenon as a prerequisite for the successful modelling of it. Interestingly, they also saw the value of experience with other, allied, phenomena and with models of them. Practical work is established in science education in many countries, although the purposes to be served by it are often unclear (Jenkins 1999). The need to establish the epistemological justification for such work has been demonstrated especially where a distinction is being drawn between the phenomenon under enquiry and the noumenon being established (Tiberghien 1999). These comments suggest that the purposes of such practical experience are distinguishable in cases where a model is being learned or used, and in cases where it is being done as the basis for model revision or model construction.

The selection of the source from which the model is developed

The teachers interviewed saw the value of analogy in the construction of models and also the importance of clarifying the nature and degree of the similarity being proposed. Van Driel *et al.* (1998) pointed to the centrality of this skill of selection of source to the evolution of teachers' pedagogic content knowledge (Shulman 1987) in a given subject area. This would also suggest that teachers should be overtly taught how to draw analogies (Treagust *et al.* 1996).

The construction of a mental model

It is interesting that some of the teachers interviewed saw scientists to have different abilities to other people when it comes to constructing mental models. They saw models to be produced by either induction, or deduction, or a mixture of the two. Successful modellers, they said, must have a personal interest in doing so, be creative in their endeavours, and persistent in their application to the task.

This, the core element in the model-construction process, has been the subject of a diverse but fragmented literature. Holton (1995) pointed to the centrality of 'metaphorical imagination' in scientific creativity, whilst Gentner (1989) explored the operation of analogy within metaphor, pointing to the fact that 'true analogy' was concerned only with the relationships between entities within the source and within the target, not with the nature of those entities. Although Holton (1955) also pointed to the importance of visual imagination in scientific creativity, it was left to Mathewson (1999) to deplore the neglect of visualization in science education. This claim is not entirely justified, even if the literature is scattered. Thus, whilst Tuckey and Selvaratnam (1993) analysed the various forms of visualization and the ways in which they could be tested for and developed in the field of chemistry, Hyerle (1996) tabulated the various ways in which visualization could be supported by the use of diverse schemes of inscription, and Gordin and Pea (1995) proposed ways in which virtual-reality packages on computer could also provide support. A key element to successful modelling must surely be a sense of ownership by students of the task that they are undertaking: this will unleash their creativity and harness their powers of visualization.

The selection of an appropriate mode of representation

The teachers could, over the group of interviewees as a whole, see the virtues of the various modes of representation. Some excellent studies into learning through diagrams have taken place (Kindfield 1993/1994; Lowe 1994). Learning through any mode of representation requires that the students understand the relevant 'code of representation': the set of conventions by means of which the model, the mode and the phenomenon are related to each other. Although the importance of such codes has been recognized (Gilbert *et al.* 1998; Gobert 2000; Harrison and Treagust 2000) they do not seem to have been worked out in any detail for the different modes.

The conduct of thought experimentation

We did not actively seek teachers' views on thought experimentation in modelling and therefore cannot be too surprised to have found no evidence of it. The neglect of thought experimentation generally in science education has been previously noted by Gilbert (Gilbert and Reiner 2000; Reiner and Gilbert 2000). Such activity will be supported by a classroom atmosphere where questioning is actively supported (Costa *et al.* 2000), which will be related to the use of a dialogic mode of discourse (Scott 1998).

The conduct of empirical experimentation

The teachers thought the possession of good manual skills and powers of observation, together with well-developed capacities for abstract and logical thinking, necessary for successful modelling activities. Much of the available literature concentrates on the science-related attributes of performance, e.g. the ability to judge the reliability of data (Lubben and Millar 1996) and the notion of evidence used (Gott and Duggan 1996). Enquiry into the level of manual skills and powers of observation apparently lies outside the envelope of concerns of research in science education. The capacities for abstract and logical thinking and their development within science education have been exhaustively studied, predominately within the template of Piagetian theory (Adey and Shayer 1994).

The persuasion of others over the value of a model

The teachers noted the need for a flexibility of approach in the presentation of a given model to diverse audiences as important. However, they did not generally emphasize either the need for a consideration of the scope and limitations of models during the process of modelling or the importance of the discussion of such matters during the presentation of any model to students. This supports the findings of van Driel and Verloop (1999).

Assuming the relevance of science teachers' comprehensive view of modelling to the learning of modelling skills by students, the aspects discussed earlier, when taken in conjunction within the 'model of modelling' framework, point to specific issues to be addressed within teacher education.

Conclusions

If a student is successfully to learn a scientific/curricular model, that person must have: an understanding of scientists' view of the nature of 'model'; suitable experience of the phenomenon that is being represented; knowledge of why the model was originally constructed and why it has to be learned; an understanding of how analogies operate; knowledge of the source from which the target model and/or teaching model is constructed. A necessary condition for such learning is that the teacher is him/herself competent in all these aspects of modelling. Our work suggested that many teachers in this sample had a less than satisfactory understanding of the nature of 'model' and implies that it should be the subject of in-service (or, better still, pre-service) education. The sample does seem at least aware of the need for competence in all the other above elements of model learning.

If a student is to learn to use a model that has been understood, then all these conditions also apply. To these must be added: the capability to design, conduct and evaluate thought experiments and, if this is successful, empirical experiments. The expectation of successful 'transfer of learning' that 'learning to use a model' implies will reduce the demand levels of these tasks, at least in the initial 'prediction' and final 'evaluation' phases of these activities. As has been discussed by Reiner and Gilbert (2000), the skills of thought experimentation are not widely appreciated by science teachers, a finding tentatively confirmed here. This would suggest that thought experimentation be added to the list of themes for teacher education. As much school empirical experimentation is of the confirmatory variety (Wellington 1998), little specific teacher education would seem to be needed.

One way in which students might learn how to revise and modify a model is by following a 'cognitive reconstruction' (Nersessian 1984) of a historical sequence of models in an area of enquiry. Few attempts at cognitive reconstruction of all the major models in historical sequences exist, e.g. Justi and Gilbert (1999b). Even when they do exist, the frequent use of hybrid models in teaching militates against their effective use (Justi and Gilbert 1999a). If a student follows such a historical sequence, and emphasis that will be novel in respect of prior phases of 'learning to model' will be the acquisition of skills in the evaluation of the scope and limitations of models. As this does not seem to be addressed in teaching at the moment, it might be added to the list of new requirements for the preparation of science teachers.

Assuming that teachers would be well prepared to conduct the teaching of the first three phases of 'learning to model' and that students had become competent in all of them, progression to the last two should be relatively straightforward. Learning to reconstruct a model will engage all the elements of the 'model of modelling' framework (see Figure 5.1), with the psychological safety net of prior knowledge of the outcome in place. That done, 'learning to construct models de novo' will be the last step, with students actually working like scientists: not knowing the outcome beforehand. For the teacher, the remaining major challenge will be how to manage time, for innovation cannot be timetabled!

Acknowledgements

The research described here was partially supported by a grant from the 'Pró-Reitoria de Pesquisa', Universidade Federal de Minas Gerais, Brazil.

Note

1 An orrery is a clockwork model of the Solar System, or just (as in this case) of the Sun, Earth and Moon, named after the 4th Earl of Orrery for whom one was made.

References

Adey, P.S. and Shayer, M. (1994). *Really Raising Standards: Cognitive Acceleration and Academic Achievement* (London: Routledge).

Arnold, M. and Millar, R. (1996). Learning the scientific 'story': a case study in the teaching and learning of elementary thermodynamics. *Science Education*, 80, 249–281.

Barab, S.A., Hay, K.E., Barnett, M. and Keating, T. (2000). Virtual Solar System Project: building understanding through model building. *Journal of Research in Science Teaching*, 37, 719–756.

Boulter, C. and Buckley, B. (2000). Constructing a typology of models for science education. In J.K. Gilbert and C.J. Boulter (eds), *Developing Models in Science Education* (Dordrecht: Kluwer), 41–57.

Clement, J. (1989). Learning via model construction and criticism. In J.A. Glover, R.R. Ronning and C.R. Reynolds (eds), *Handbook of Creativity* (New York: Plenum), 341–381.

Costa, J., Caldeira, H., Gallastegui, J.R. and Otero, J. (2000). An analysis of question asking on scientific texts explaining natural phenomena. *Journal of Research in Science Teaching*, 37, 602–614.

Del Re, G. (2000). Models and analogies in science. *HYLE – An International Journal of the Philosophy of Chemistry*, 6, 3–12.

DfEE (1999). *Science: The National Curriculum for England* (London: Department for Education and Employment).

Driver, R., Leach, J., Millar, R. and Scott, P. (1996). *Young People's Images of Science* (Buckingham: Open University Press).

Erduran, S. (1999). Philosophy of chemistry: an emerging field with implications for chemistry education. Paper presented at the Annual Conference of the American Educational Research Association, San Diego, CA, 13–17 April.

Francoeur, E. (1997). The forgotten tool: the design and use of molecular models. *Social Studies of Science*, 27, 7–40.

Frederiksen, J.R., White, B.Y. and Gutwill, J. (1999). Dynamic mental models in learning science: the importance of constructing derivational links among models. *Journal of Research in Science Teaching*, 36, 806–836.

Gentner, D. (1989). The mechanism of analogical learning. In S. Vosniadou and A. Ortony (eds), *Similarity and Analogical Reasoning* (Cambridge: Cambridge University Press), 199–241.

Gentner, D. and Gentner, D.R. (1983). Flowing waters and teeming crowds: mental models of electricity. In D. Gentner and A.L. Stevens (eds), *Mental Models* (Hillsdale, NJ: Erlbaum), 99–129.

Gilbert, J.K. and Reiner, M. (2000). Thought experiments in science education: potential and current realisation. *International Journal of Science Education*, 22, 265–283.

Gilbert, J.K., Boulter, C. and Rutherford, M. (1998). Models in explanations, Part 1: Horses for courses. *International Journal of Science Education*, 20, 83–97.

Glynn, S.M. and Duit, R. (1995). Learning science meaningfully: constructing conceptual models. In S.M. Glynn and R. Duit (eds), *Learning Science in the Schools* (Mahwah, NJ: Lawrence Erlbaum), 3–33.

Glynn, S.M., Law, M., Gibson, N.M. and Hawkins, C.H. (1994). *Teaching Science with Analogies: A resource for Teachers and Textbook Writers* (Instructional Resource No. 7) (Atlanta, GA: National Reading Research Center, University of Georgia).

Gobert, J.D. (2000). A typology of casual models for plate tectonics: inferential power and barriers to understanding. *International Journal of Science Education*, 22, 937–977.

Gordin, D.N. and Pea, R.D. (1995). Prospects for scientific visualization as an educational technology. *The Journal of the Learning Sciences*, 4, 249–279.

Gott, R. and Duggan, S. (1996). Practical work: its role in the understanding of evidence in science. *International Journal of Science Education*, 18, 791–806.

Greca, I.M. and Moreira, M.A. (2000). Mental models, conceptual models, and modelling. *International Journal of Science Education*, 22, 1–11.

Grosslight, L., Unger, C., Jay, E. and Smith, C.L. (1991). Understanding models and their use in science: conceptions of middle and high school students and experts. *Journal of Research in Science Teaching*, 28, 799–822.

Halloun, I. (1996). Schematic modeling for meaningful learning in physics. *Journal of Research in Science Teaching*, 33, 1019–1041.

Halloun, I. (1998). Schematic concepts for schematic models of the real world: the Newtonian concept of force. *Science Education*, 82, 241–263.

Harrison, A.G. and Treagust, D.F. (2000). A typology of school science models. *International Journal of Science Education*, 22, 1011–1026.

Hodson, D. (1992). In search of a meaningful relationship: an exploration of some issues relating to integration in science and science education. *International Journal of Science Education*, 14, 541–562.

Holton, G. (1995). Imagination in science. In G. Holton (ed.), *Einstein, History, and other Passions* (Woodbury, NJ: American Institute of Physics), 160–184.

Hyerle, D. (1996). *Visual Tools for Constructing Knowledge* (Alexandria, VA: Association for Supervision and Curriculum Development).

Idling, M.E. (1997). How analogies foster learning from science texts. *Instructional Science*, 25, 233–253.

Ingham, A.M. and Gilbert, J.K. (1991). The use of analogue models by students of chemistry at higher education level. *International Journal of Science Education*, 22, 1011–1026.

Jenkins, E.W. (1999). Practical work in school science. In J. Leach and A. Paulsen (eds), *Practical Work in Science Education: Recent Research Studies* (Roskilde: Roskilde University Press), 19–32.

Johnson-Laird, P.N. (1983). *Mental Models* (Cambridge: Cambridge University Press).

Justi, R. and Gilbert, J.K. (1999a). A cause of ahistorical science teaching: the use of hybrid models. *Science Education*, 83, 163–178.

Justi, R. and Gilbert, J.K. (1999b). History and Philosophy of science through models: the case of chemical kinetics. *Science and Education*, 8, 287–307.

Justi, R. and Gilbert, J.K. (2001). *Teachers' Views About Models and Modelling in Science Education*. Paper presented at the Annual Conference of the National Association for Research in Science Teaching, St. Louis, MO, 26–28 March.

Kindfield, A.C.H. (1993/1994). Biology diagrams: tools to think with. *The Journal of the Learning Sciences*, 3, 1–36.

Leatherdale, W.H. (1974). *The Role of Analogy, Model and Metaphor in Science* (Amsterdam: North-Holland).

Lederman, N.G. (1992). Students' and teachers' conceptions of the nature of science: a review of the research. *Journal of Research in Science Teaching*, 29, 331–359.

Lehrer, R., Horvath, J. and Schauble, L. (1994). Developing model-based reasoning. *Interactive Learning Environments*, 4, 218–232.

Lowe, R.K. (1994). Selectivity in diagrams: reading beyond the lines. *Educational Psychology*, 14, 467–491.

Lubben, F. and Millar, R. (1996). Children's ideas about the reliability of experimental data. *International Journal of Science Education*, 18, 955–968.

Mathewson, J.H. (1999). Visual–spatial thinking: an aspect of science overlooked by educators. *Science Education*, 83, 33–54.

Morrison, M. and Morgan, M.S. (1999). Models as mediating instruments. In M.S. Morgan and M. Morrison (eds), *Models as Mediators* (Cambridge: Cambridge University Press), 10–38.

Nersessian, N. (1984). *Faraday to Einstein: Constructing Meaning in Scientific Theories* (Dordrecht: Matinus Nijhoff).

NRC (1996). *National Science Education Standards* (Washington, DC: National Academic Press).

Penner, D.E. (2000). Explaining systems: investigating middle school students' understanding of emergent phenomena. *Journal of Research in Science Teaching*, 37, 784–806.

Pittman, K.M. (1999). Student-generated analogies: another way of knowing? *Journal of Research in Science Teaching*, 36, 1–22.

Raghaven, K. and Glaser, R. (1995). Model-based analysis and reasoning in science: the MARS curriculum. *Science Education*, 79, 37–61.

Raghaven, K., Sartoris, M.L. and Glaser, R. (1998a). Impact of the MARS curriculum: the mass unit. *Science Education*, 82, 53–91.

Raghaven, K., Sartoris, M.L. and Glaser, R. (1998b). Why does it go up? The impact of the MARS curriculum as revealed through changes in student explanations of a helium ballon. *Journal of Research in Science Teaching*, 35, 547–567.

Reiner, M. and Gilbert, J.K. (2000). Epistemological resources for thought experimentation in science learning. *International Journal of Science Education*, 22, 489–506.

Roth, W.M. (1995). *Authentic Science Education: Knowing and Learning in Open-Inquiry Science Laboratories* (Dordrecht: Kluwer).

Scott, P. (1998). Teacher talk and meaning making in science classrooms: a review of studies from a Vygotskian perspective. *Studies in Science Education*, 32, 45–80.

Shulman, L. (1987). Knowledge and teaching: foundations of the new reforms. *Harvard Educational Review*, 57, 1–22.

Stewart, J., Hafner, R., Johnson, S. and Finkel, E. (1992). Science as model building: computers and high-school genetics. *Educational Psychologist*, 27, 317–336.

Tiberghien, A. (1999). Labwork activity and learning science: an approach based on modelling. In J. Leach and A. Paulsen (eds), *Practical Work in Science Education: Recent Research Studies* (Roskilde: Roskilde University Press), 176–194.

Tomasi, J. (1988). Models and modelling in theoretical chemistry. *Journal of Molecular Structure*, 179, 273–292.

Treagust, D.F., Harrison, A.G., Venville, G.J. and Dagher, Z. (1996). Using an analogical teaching approach to engender conceptual change. *International Journal of Science Education*, 18, 213–229.

Tuckey, H. and Selvaratnam, M. (1993). Studies involving three-dimensional visualisation skills in chemistry. *Studies in Science Education*, 21, 99–121.

van Driel, J.H. and Verloop, N. (1999). Teachers' knowledge of models and modelling in science. *International Journal of Science Education*, 21, 1141–1153.

van Driel, J.H., Verloop, N. and de Vos, W. (1998). Developing science teachers' pedagogical content knowledge. *Journal of Research in Science Teaching*, 35, 673–695.

Wellington, J. (1998). Practical work in science: time for a reappraisal. In J. Wellington (ed.), *Practical Work in School Science: Which Way Now?* (London: Routledge), 3–15.

PART 2

RELATING SCIENCE EDUCATION AND TECHNOLOGY EDUCATION

CHAPTER 6

THE INTERFACE BETWEEN SCIENCE EDUCATION AND TECHNOLOGY EDUCATION

International Journal of Science Education, 1992, 14(4), 563–578

This paper reflects on the growing importance of technology education in the school curriculum throughout the world. The nature of technology and its relationship to science are discussed. A series of models for technology education, 'education in technology', 'education about technology' and 'education for technology', are presented. The contribution of science education to these models is explored. A series of broad research questions are identified, an address to which would support a more effective relationship between science education and technology education:

The increased emphasis on technology education

In recent years, many countries have explicitly introduced technology education into the general school curriculum, based on the view that

> technology is the know-how and creative process that may utilise tools, resources, and systems, to solve problems, to enhance control over the natural and man-made environment in an endeavour to improve the human condition.
> (UNESCO 1983)

In passing, it is noted that this definition appears to overlook the interests of the millions of other species which share Earth. However, it is a view which seems to have wide support amongst the human population.

Three sets of reasons seem, to different degrees, to underpin this introduction in northern/western countries (see Wiener 1981; Medway 1989; Jones and Kirk 1990; Layton 1990a).

(a) *Economic*: Technology is the basis of industrial activity of all types and levels of sophistication. One argument here is that, in some countries, the cultural climate is hostile to industry, such that proportionately few of the highest achieving young people take up an industrial career. It is suggested that the inclusion of technology education in the general school curriculum would, at best, improve that image or, at least, provide a basis for a more informed career choice. A second argument is that industries require a highly skilled work-force which must be adaptable in the face of rapidly evolving tasks and technologies: direct experience of existing technologies will lay the foundation for such a future.

(b) *Social*:   Two social arguments, different only in the dimension of time, are put forward. The first is that personal, economic and social decision-taking today requires that the general public, as well as industrialists and law makers, should be aware of the social and environmental consequences of technological acts. Second, solving the problems of tomorrow that today's technology has produced will also require the use of technological processes and products. The knowledge and skills of technological activity, coupled to insight into the consequences of acts, are therefore required.

(c) *Educational*:   The first educational argument rests on the assertion that technology is a, if not the, distinctive human achievement, a point reinforced by Schon's (1983) analysis of the complexity of human judgement in the practice of technologies. Technology, it is argued could be included in the general education curriculum solely for this reason. Moreover, the general over-valuation of an education based on abstractions devalues the achievements of those students with a practical bent, a situation which technology education would remedy. An allied point is that technological products, of one type or another, form part of the homes of everyone (Silverstone 1991): people would acquire a greater sense of control over their immediate environment if they knew more about such products. The second argument is that the inclusion of technology is claimed to support the attainment of general educational goals, e.g., it motivates students to learn because the applicability of knowledge and skills can be readily seen, it provides a context for the integration of knowledge from many school subjects and, by enabling a consideration of technologies from different countries, facilitates multicultural education. The arguments put forward for technology education in southern/ eastern countries differ somewhat in both nature and emphasis from those above: these differences cannot be discussed here.

Despite the diversity and apparent strength of the arguments put forward in its favour, technology education has, so far, not gained a strong foothold in the general school curriculum of many countries (Allsop and Woolnough 1990; Lindblad 1990). The reasons seem various and are interrelated: the dominance of the established subjects in overcrowded timetables; a lack of suitably qualified teachers, books, equipment, rooms; and perhaps, most significantly of all, the absence of a coherent philosophy of technology on which to base technology education (Rapp 1989). This latter point has been compounded by analyses of the complexity of the relation between technology and science, e.g., Brandin and Harrison (1987), Skolimowski (1972) and Russell (1972). As Lewis (1991) has pointed out, the nature of the perceived relation between science and technology will govern the relation between science education and technology education and hence the allocation of resources between the two.

This paper attempts to establish an overview of the concept of technology and to consider the relation between science and technology. The nature of technology education is reviewed and the relationship between science education and technology education analysed. Finally, some questions for research on that relationship are posed.

Technology

Attitudes towards technology, which reflect notions of intrinsic worth, vary widely. Margolis (1984) has observed that there are three conflicting views of the role of technology in society: the 'human' view, which sees it as an evolving response to changing human needs; the 'titanic' view, which sees it as a heroic attempt to subdue unruly nature; the 'satanic' view, in which technology produces destructive power when allied to an instrumental perspective on environment. In a similar vein, there are two, conflicting, views of how technology is pursued and how technological change takes place (Mackay 1991): on the one hand, 'technological determinism' sees technological change to be independent of general human values and goals, yet capable of shaping society; on the other hand, the 'social construction of technology' view sees the development of technologies to reflect human choices such that society shapes what is developed and the purposes to which it is put. Such conflicts have, it seems, led to, or resulted from, a neglect of the study of philosophy of technology. Rapp (1989) has identified four other reasons for the neglect: that Western history of ideas has been dominated by the notion of theory, such that practical activities, typified by technology, have been undervalued; that technology has been viewed as applied science, so that it is subsumed under philosophy of science; that, until the interplay between science and technology became socially visible in the industrial revolution, the traditional, i.e., craft-based, technologies were not perceived to be intellectual entities; and, last, because the phenomenon dealt with by technology were not, of their essence, capable of simplification in the manner achieved by science.

An operational philosophy of technology may be inferred from typologies of technologies, several of which exist (e.g. Mitcham 1978; Kline 1985). The model proposed by Pacey (1983) seems to have especial value. It is based on the concept of 'technology-practice', which is defined as:

> the application of scientific and other knowledge to practical tasks by ordered systems that involve people and organizations, living things and machines...

and which consists of three, simultaneously operational elements: the technical aspect, the organizational aspect and the cultural aspect (see Figure 6.1).

This model can provide a rationale for the interpretative and evaluative tensions outlined earlier. Thus technological innovation will only be far-reaching in scope

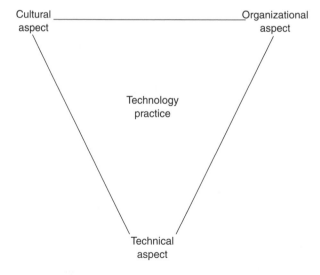

Figure 6.1 Technology practice (Pacey 1983).

and widespread in adoption when the technical, organizational and cultural aspects are in harmony. A linear view of technological progress, which is at the heart of the technological determinism interpretation, is produced when only the technical aspect is considered: a non-linear view, in which all the three aspects are considered, produces the social determinism view.

The work of other authors may be integrated to give a clearer view of the three aspects which comprise Pacey's (1983) model. The technical aspect, i.e.,

> knowledge, skill and technique; tools, machines, chemicals, liveware; resources, products and wastes...

integrates the outcomes of technological activities and the processes by which they are produced. Hodgkin (1990) has produced a fivefold typology for the artefacts which are the outcomes of technological activity, i.e.:

(i) objects or contrivances for attaining a defined and recognizably useful purpose, e.g., train, domestic cooker. This is the commonly understood meaning of technological outcome;
(ii) implements with which to seek understanding of the universe. This subsumes all instruments used in scientific enquiry, e.g., telescopes, microscopes;
(iii) toys, e.g., construction kits, dolls;
(iv) weapons, e.g., guns, rockets;
(v) art. Here art is seen as a way of extending meaning by alerting, focusing and developing feelings and awareness which are then held in tension, e.g., in museums, in legal codes, which are themselves forms of art.

The other outcomes of technology are systems, which may be seen as sets of objects and/or activities which collectively perform tasks, and environments, which may be seen as modifications to natural environments to varying degrees.

The processes of technology are too complex to explore in any depth here. The most simplistic linear model involves four elements: the identification of needs, the design of one or more possible solutions, the construction of the most promising solution, its evaluation against the original need. However, as Jeffrey (1990) has observed, the linear model has given way to a cyclic variant, where the same components are iteratively revisited. The tasks will be complex, for according to Powell (1987), the problems of real-world design cannot be solved linearly because they: involve identifying what the actual problems are; cannot be completely defined before a solution is formulated; are only solved within a given context, so that vital information has to be collected during the process itself.

The organizational aspect of technology, in Pacey's (1983) model, refers to 'economic and industrial activity, professional activity, users and consumers, trade unions'. As such it draws on the insights found within: economics, particularly the economics of investment and of production systems; sociology, particularly the sociology of professional organizations and trade unions, and of consumer decision-taking; psychology, particularly the psychology of risk assessment and decision-taking in industrial contexts.

Pacey's (1983) cultural aspects of technology is the most complex to represent, for it is seen as:

> goals, values and ethical codes, belief in progress, awareness and creativity.

Layton (1988) makes use of Longino's (1983) notions of constitutive and contextual values to discuss technology. Constitutive values refer to the framework of expectations which contribute to the determination of 'good practice', here in technology. These make a major contribution to a philosophy of technology. Contextual values here relate to those commonly held in the social and cultural setting within which technology is being undertaken. For Layton (1988), they will be revealed: by the form that a technology takes; when it is transferred to another culture; and by the criteria by which a technology is deemed to be out of date. It would seem that constitutive and context values are inter-dependent: the latter will influence the former. The range of possible values is large, with some achieving dominance: as this comment:

> we ought to recognize that the culture of technology comprehends at least two overlapping sets of values, the ones based on rational, materialistic, and economic, goals, and the other concerned with the adventure of exploiting the frontiers of capability and pursuing virtuosity for its own sake.
>
> (Pacey 1983: 89)

The range may even exist within one person: a professional technologist may vigorously pursue a chosen field, irrespective of its economic consequences, when at work, yet take a strongly economic view of technology in private life.

Science and technology

The relationship between science and technology is undoubtedly close, for most definitions of technology refer directly to science. However, the nature of the link is much disputed. Standenmaier's (1985) book *Technology's Storytellers: Reweaving the Human Fabric*, which discusses the nature of technology, is based on an analysis of articles, mostly written by historians of technology, in the journal *Technology and Culture*. Standenmaier (1985) identifies seven distinctive types of statement which were made about the relationship in that journal: several are mutually contradictory whilst others contain internal inconsistencies. Mayr (1982) argues that the failure to arrive at a clear definitional relationship between science and technology, which is applicable to a range of historical examples, arises because, first such attempts begin by postulating models and then only try to confirm them and, second, because the general meanings of the words involved have changed in passages across time and linguistic barriers.

The final word is probably best left, for the time being, with Barnes (1982). For Barnes, the classical model of the relation was hierarchical, with science in the superior position. The relationship between the two was mediated by words on paper. The outcomes of the model for the development of knowledge were predictable: technology deduced the implications of science and gave them physical representation, but without feedback. Some of the artefacts of technology were used by science as instruments. Whilst science evaluates discoveries in a context-independent way, technology is evaluated in terms of its ability to infer the implications of science. Successful technology makes proper and component use of science. According to Barnes (1982), the currently used model is egalitarian and interactive, and is mediated by direct personal contact between the people involved. The outcomes of the model for the development of technique and knowledge are unpredictable, with each occasionally making use of the other. Both science and technology are inventive, being evaluated in terms of the ends achieved (see also Layton 1990b).

Forms of technology education

It might be argued that it is possible to derive a rationale for technology education without clear recourse to a philosophy of technology. After all, this argument would proceed, science education, as actually practised in schools, does not bear any close relation to science as practised by researchers or as postulated by philosophers of science and could not, for the intellectual circumstances of science are different to those of science education: why should technology education be any different? This argument is firmly resisted on two grounds: first, one of the main purposes of science education is to introduce students to the practice of science, which an aphilosophical approach must inherently fail to do; second, because the failure of science education to sustain the interest of many students can be attributed to a lack of clear relation to science as practised. These counter-arguments are strengthened in the case of technology education because of the practical skill and judgemental elements which are at the heart of technology itself.

A number of perspectives may be taken on the scope of 'technology education' which derive from views held on the nature of technology itself and which reflect the different sets of purposes which may be addressed. If all aspects of Pacey's (1983) model of technology are held to be equally important, then the most closely related intention of technology education might be 'to provide initial preparation to be a citizen practising as a technologist': this process might be described as 'education in technology' if more limited view was adopted, where the technical aspect of Pacey's (1983) model was emphasized, then the purposes might be described as 'education for technology', with the student being prepared for work in an overtly technological industry. However, such an approach might lead to a reinforcement of technological determinist views in students. If an alternative, equally limited, view was taken, one which emphasized Pacey's (1983) organizational and cultural aspects, then this set of purposes might be called 'education about technology': students would learn how technology is organized and about its consequences, yet would be unable to acquire a destructive attitude to technology: Margolis's (1984) 'satanic' view might be reinforced.

Whatever perspective on the nature of technology is adopted, a prerequisite for its introduction must be clear identification of the entity involved. Here Schwab's (1964) idea of 'disciplines' is of use. Schwab (1964) views a discipline as any coherent body of Knowledge and enquiry. A discipline may be defined through the possession of both a substantive structure and a syntactical structure. The substantive structure of a discipline is a conceptual structure which enables: telling questions to be posed; data to be collected by experimentation; the significance of that data to be evaluated. The syntactical structure of a discipline is a structure for the conduct of enquiry which enables: appropriate methods of enquiry to be selected; the quality of data obtained to be evaluated; canons of evidence to be erected; routes to conclusions to be established. Shulman (1987) identified four major sources of the knowledge base required for teaching: scholarship in the discipline to be taught; materials and settings which define the institutional settings of the education to be provided; research into the circumstances and processes of teaching and learning; the wisdom acquired by the practice of teaching the discipline involved. In the conduct of teaching, the teacher integrates aspects of these sources into what Shulman (1987) terms 'pedagogical reasoning and action' which has six elements: comprehension (of the discipline as perceived); transformation (of specific aspects of the discipline, by identifying ways and means by which those aspects can be presented to students in the light of what is known about their characteristics as learners); instruction (by a range of methods within a managed environment); evaluation (of teaching performance and student learning); reflection (on what was attempted and achieved); new comprehension (of purposes, subject-matter, students, teaching and self). For technology education, however, conceived of in terms of Pacey's (1983) aspects, the discipline has not been clearly identified, the sources of Shulman's (1987) knowledge base are not fully established, and there is inadequate documentation of case studies of pedagogical reasoning and action.

It may be helpful to review the extent to which the three approaches to technology education are currently being provided and to what effect.

Education for technology: the technical aspect

This may be divided into 'education involving technological outcomes' and 'the processes by which those outcomes are produced'.

Technological outcomes

Rennie (1987), using an on-the-spot open-ended questionnaire with 94 heads of science from secondary schools in Western Australia, found that the consensus view of technology was as the application of science in the modern world to produce artefacts, with the intention of improving the quality of human life, or as equipment, the use of which led to an increase in knowledge. Working with primary school teachers, Symington (1987) found that the majority view of technology was as modern technological artefacts. Only a limited range of artefacts are considered (Medway 1989) which fall within Hodgkin's (1990) first two types (see p. 104). Moreover, systems and environments seem to have been generally neglected.

This limited view of technology arises, in the UK at least (Allsop and Woolnough 1990), from a preoccupation with science processes in the 1960s, with the integration of the separate sciences in the 1970s, and with the emergence of a highly conservative National Curriculum for Science in the 1980s and 1990s. Indeed, Chapman (1991) regrets that there has been a preoccupation with science education at all, seeing technology education as more important for most citizens. He points out that, whilst the practice of advanced science requires an understanding of abstract concepts, which are inaccessible to many, advanced technology can be engaged in at a wide range of levels of abstraction. The exploitation of technology does not require an understanding of the underlying science, even where this is known. To quote a memorable sentence:

> A condom worn by, or inserted in, a material scientist is not made more effective because of his/her knowledge of the elastic properties of polymers.
>
> (Chapman 1991: 55)

Much of the teaching in schools which focuses on technological outcomes has taken place in science classes. Five approaches can be identified.

(a) *Minimal reference to technological outcomes during science lessons*: This is what Fensham (1988) refers to as a 'motivational' use and the use of 'factual references to randomly chosen examples'. The relationship between the science concepts and the technological outcomes remains unexplored.

(b) *The teaching of science followed by a discussion of how the science is relevant to technological outcomes*:   In some cases, the science and technology are presented within a framework provided by the technology. Thus Raat and de Vries (1986) have produced integrated module guides, e.g., on 'making music instruments' in which the relevant physics concepts are explained, issues of design discussed, case studies of the design of specific technological outcomes are presented and social issues discussed. In other cases (Gadd 1991) basic science is taught within compulsory modules and then options, each relating to one technological outcome, may be chosen. The potential weakness of such approaches is that the understanding of the technology will be limited to that provided by the selection and depth of treatment of the antecedent science concepts.

(c) *Teaching starts from either the associated science or from particular technological outcomes*:   Wells (1990) suggests that the capacity of micro-technology (computer) based devices to produce, store and modify sounds, provides a potential link between physics and music. It would seem possible to start from either the physics or the music theory and to proceed to the technological outcome, or vice versa. Such an approach offers an excellent forum within which to explore the relation between the forms of concepts used in science and in technology (Layton 1988).

(d) *Teaching starts by considering a technology outcome, the science concepts are then taught, and then other technological outcomes are discussed*:   This approach, which has been described by Fensham (1988) as 'thematic or tropical', has been subject to systematic development. Examples include:

- the initial use of a single technological outcome. Kahn (1989) had trainee science teachers in Botswana build a working radio from plans, but without any prior theoretical instruction; disassemble the radio into its component part and then learn the physics of each part; progressively reassemble the radio, as a generic example from those parts, using sub-assemblies whose theory was understood. Jones and Kirk (1990) developed a scheme for 17-year-old New Zealand physics students based on a learning model in five stages: focusing (where a single technological outcome, e.g., a camera flash, was examined to identify its component parts and the underlying scientific principles); exploration (where students investigated the component elements and principles involved); reporting (where the outcomes of investigations were pooled by the class); formalization (where the physics involved was clarified and recorded); applications (where further examples of technological outcomes were considered). The technological outcomes used were selected against the following criteria: relevance to human relationships and everyday life; relevance to students' personal interests and existing knowledge; suitability for introduction at the beginning of a lesson; the availability of a range of developed laboratory experiments; the use of important concepts which also inform a range of other technological outcomes. The impact of this approach was investigated using a combination of questionnaires, interviews and classroom observation. Teachers reported: strong and sustained student interest in the material dealt with; higher levels of student involvement and achievement; the use of more interactive teaching styles; improved learning from practical work. Students reported: seeing more links between scientific theory, technological outcomes and everyday life; a valuation of the experimental approach; better understanding and higher achievement;
- the use of a broad-based technological system. Some of the more interesting examples of this group have been drawn from the so-called 'key technologies' (Meeke 1989) which are: 'newly emerged topics in science and engineering

which are likely to have a major revolutionary impact on an existing product or process or may lead to a revolutionary new product or process'. In general, ACARD (1986) saw key technologies as activities in which information is: acquired (by sensors); organized (using concepts drawn from artificial intelligence; processes (using parallel processing techniques); transmitted (e.g. by fibre optics or satellites); used (particularly in robotic systems). A contrasting classificational system was proposed by Meeke (1989) as: materials (using building blocks for products, e.g., biological material, semiconductors, ceramics); components (sub-assemblies of products, e.g., optoelectronics, mechatronics); manufacturing and processing (techniques for assembling systems from materials and components, e.g., image processing, expert systems). Although there has been no systematic consideration of the possible role of high technology in schools, development work has taken place on the contribution of satellite-based remote sensing (Gilbert 1992) and biotechnology (Henderson and Knutton 1990) to school science education. Medway (1989) noted that the range of key technologies being addressed in school was small;

- the use of native technologies. George (1988) has proposed that 'native', or indigenous technologies should be used as a starting point for school science in developing countries. In her view, this would be valuable because they: have a similarity in function and evolution to modern technologies; enable a greater understanding of the immediate environment to be achieved by students; enable problem-solving teaching techniques to be used; most importantly, enable a cultural continuity in society to be maintained. George (1988) has developed materials built around these premises. The approach seems capable of wide extension, for all countries make use of technologies which have been long established, e.g., in horticulture, in domestic governance.

(e) *The use of technological outcomes in the teaching of science*:   The use of technology outcomes in the actual processes of teaching science, an example of what Kerr (1989) refers to as 'teaching-with-technology', provides an unusual perspective on the technology–science education relationship. Focusing on the computer, Kerr (1989) has pointed to the need to accept teachers' views on what is feasible, to develop high-quality software which reduces the load of routine work for teachers and which strengthens the teachers' tutorial role. One approach is to use the computer as a way of teaching science concepts, as what Salomon *et al.* (1991) call a 'technology of the mind', i.e., one which supports the development of higher-order thinking skills. Other uses for computers, under the general heading of information technology, include the provision of data base, data-logging and spreadsheet facilities (Sparkes 1989; Rogers 1990). The computer can also be used to control other technology outcomes of use in science education, e.g., to produce interactive video (Royal Society 1990).

Processes by which technology outcomes are produced

The subject of Technology within the UK National curriculum (DES 1990a) is a recent and major example of a curriculum which places a heavy emphasis on the design cycle within the technical aspect. In summary, it has the following features:

- an address to four Attainment Targets (identifying needs and opportunities; generating a design; planning and making; evaluation);

- the establishment of ten Levels of Attainment within each Attainment Target, to be sought during the four Key Stages of compulsory education (5–7, 7–11, 11–14, 14–16 years);
- the design and making of artefacts (people-made objects), systems (sets of objects and activities which together perform a task) and environments (surroundings made, or developed by people);
- work within a range of contexts: home, school, recreation, community, business and industry;
- work with a range of materials, e.g., textiles, graphic media, construction materials, food;
- a drawing upon knowledge and skills from other subject areas, e.g., science, mathematics.

Although the new subject is still being established in schools, albeit by a review of existing subjects, i.e., Home Economics, Craft Design and Technology, Art, Business Studies, Information Technology, it has already been possible for Medway (1992) to observe that the new subject:

- represents an uneasy compromise between the wish to establish a new discipline and a concern to disseminate more wider aspects of selected practical activities;
- assumes that the student will master the knowledge and skills of a wide diversity of specialists in the everyday world, e.g., market researcher, clothes designer, engineer;
- presumes the separateness of the elements of the design cycle, whilst in technological practice they are integrated;
- emphasizes the concerns of business in respect of technology, at the expense of broader societal concerns;
- draws on only a limited set of spheres of practical activity, e.g., those of 'caring' are omitted.

Design cycle technology education is still an optical element in the school curriculum in many countries. Research into students' views on this approach to the subject has been sparse. The main focus of research has apparently been on gender issues, and particularly on why girls seem less inclined to opt for the subject. Smail and Kelly (1984a,b) found that 11-year-old boys in the UK: have better spatial abilities and skills of mechanical reasoning (thought to be important underlying factors in successful learning); were Keener on the physical sciences (which have some relationship to design-based technology); and had more experience of tinkering activities (seen as allied to design work). Kelly (1988) found that in the UK: boys enjoyed the subject more; were more likely to continue with it if given the choice; and received greater support in their study of the subject from teachers, parents and friends. Rennie (1987) found that 13-year-old Australian boys were significantly more certain than girls that: technology is important in life; technology has a history; they knew what the word meant; they were interested in a technical job. They also claimed that they did not hear much about technology at school and would welcome more information. This 'knowledge of the scope of the subject' factor is undoubtedly important, for Nash *et al.* (1984) found that more students, of both sexes, would have opted for the subject if they had known more about it. Where technology courses become compulsory this problem will be overcome. However, the main issue, how to make technology more interesting to girls, will remain. Burns (1990) believes that progress in this direction will be made if: 'girl-friendly' topics are studied; a co-operative learning environment prevails; teachers use a non-sexist vocabulary; teachers raise the expectations that they make of girls; and if a wide range of skills, including verbal skills, can be used in assessment.

Education about technology: the organizational and cultural aspects

Such an approach focuses on: the organization and economics of industry; consumer choice in respect of technology outcomes; good practice in the conduct of technology processes; the contextual values which inform the choice of technologies which are developed, the nature of the outcomes produced and the use to which the outcomes are put. Such courses would come within Fensham's (1988) category of 'Science and Society' and would fall within the province, in many school systems, of Social Studies teachers: certainly science teachers would have to acquire new knowledge and skills in order to contribute directly to such causes.

However, science could indirectly contribute, in respect of good practice in the conduct of technology processes at least, through the growing incidence of industrial experience in science courses. Marsden (1989) has pointed out the scope of such arrangements: single visits by groups of students to industrial plants; extended placements by individual students, often in connection with a project; extended placements by teachers; the involvement of industrialists in the school curriculum, e.g., as advisors to projects. The extent of all such schemes is growing in the UK (DES 1990b), whilst Davies *et al.* (1991) have pointed to elements of good practice in the management of industrial experience: central to these is the need to establish a clear curricular role for the experience.

Education in technology: all Pacey's aspects

The provision of this form of technology education is by far the most challenging. All the challenges associated with 'education for technology' together with those associated with 'education about technology' have to be met. It would seem likely that, in many countries, the focus in preservice technology teacher education on the design cycle facet to the technical aspect, and the equivalent focus in science teacher preparation on 'pure' science, would make a cross-curricular provision of 'education in technology' a necessity. The strong subject boundary maintenance in many schools would make this difficult to take effect in practice.

Nevertheless, the fact that Fensham (1988) identified courses on 'social and scientific concepts related to a technology' and on 'scientific aspects of technology or of socio-technical topics', suggests that efforts in this direction are being made. Certainly the most exciting proposals for technology in Project 2061 in the USA suggest that the major challenges involved will be more widely addressed (AAAS 1989). As Solomon (1988) points out, the technology, the science and the society elements, will each have to be treated with an up-to-date knowledge and sophisticated insight.

The contribution of science education to technology education

Many of the concepts used in science, together with scientific methodology itself, have a major role to play in technology, and hence in technology education.

Layton (1988, 1990a,b) has pointed to some of the things that might be done to knowledge derived from science education before it can be used in technology education:

(a) The level of abstraction can be reduced. Whilst science tends to value the most inclusive, abstract, mathematical models, technology is often satisfied with less inclusive, more concrete and iconic models. For example, whilst a science teacher will focus on the kinetic-molecular theory of heat, the technology teacher will find a fluid flow model entirely adequate.

(b) The knowledge can be repackaged. Technology, being more overtly problem-centred, makes use of knowledge drawn from several fields of science. This knowledge must be brought together and integrated before it can be used. Aikenhead (1989) gives the example of the breathalyzer (a device to detect alcohol levels in the blood by analysing exhaled air): it uses ideas drawn from respiratory mechanism, chemical change and equilibrium, the laws on the solubility of gases in liquids, and photometry.

(c) The knowledge can be reconstructed. Science concepts must be reframed to meet the apparent demands of a particular technological problem. Thus whilst organic chemists may discuss a collection of molecules in terms of their functional groups, pharmacists talk about the same molecules in terms of their use, e.g., as antidepressants, decongestants.

(d) The knowledge can be contextualized. Scientific knowledge is produced within definitional frameworks which are not usually met with in reality, e.g., frictionless motion, pure compounds, perfect health. The problems addressed by technology, being based more closely on reality, use concepts in which these constraints have been relaxed, e.g., in motion when friction is present to different and varying degrees, of compounds of a less-than-perfect purity, of organs functioning when partially degenerated.

Given that technology sets out to provide answers to practical problems, it is inevitable that the range of science concepts used will be large. In order to identify the science concepts needed for a particular technology, Layton (1990b) has suggested that specific questions be asked, i.e., what science is needed to:

- decide on the adoption of a technology;
- use a technology effectively;
- maintain a technology in an operational form;
- adapt a technology to a local situation;
- improve a technology to enhance its performance;
- develop an existing technology to extend the range of its use and application;
- create a new technology;
- critically evaluate competing/alternative technologies.

If either education in technology or education for technology is sought, then the concepts required from science education will arise from this analysis.

Layton *et al.* (1989) found that four strategies were used in UK schools to make science concepts available to students of technology:

- the general teaching of science concepts before a particular class of technology problem was to be addressed. This was particularly used in respect of structures, mechanics and electronics;

- bespoke teaching in response to individual students' needs. This was expensive;
- student experimentation. This seems allied to discovery learning and was evidently often a source of heavy time use and student frustration;
- the 'eclectic pillaging' of ideas, including those of science, by students from whatever source could be identified.

In addition to 'reworked' knowledge, Layton (1988) has observed that technology also makes use of scientific methodology to collect what Standenmaier (1985) calls 'problematic data' (information which is specific to the context in which technological processes are being applied). It would seem likely that students of technology would have to adopt procedures more closely allied to those of practising scientists than to those taught in the contemporary reductionist approach to 'the processes of science'.

Some research questions

The main conclusion of this paper, which has drawn on the available empirical (as opposed to exhortatory) literature, suggests that, whilst some attention has been given to the relationship between technology and science, the relationship between technology education and science education is, at school level at least, fairly uncharted waters. Most of the relevant research questions have not been addressed, certainly not thoroughly enough for general advice to be provided to teachers. A few of the key questions are:

(i) How comprehensive a description of technology is that provided by Pacey (1983)?

(ii) What is 'good practice' in technology and how does that relate to the cultural and organizational milieux in which it is embedded?

(iii) What relationship is there between design, as practised in technology, and the design cycle, as taught in schools?

(iv) How are the concepts of science reworked (Layton 1988) when used in technology?

(v) What range of conceptions of the relationship of science and technology are held by technologists, science teachers and school students?

(vi) What is the composition of the 'discipline' (Schwab 1964) of technology education and how does it relate to the forms of provision currently being made in schools?

(vii) What conceptions do students have of the ideas used in technology education and how might these be developed?

(viii) What knowledge base (Shulman 1987) is needed by science teachers if they are to contribute fully to technology education?

(ix) What 'pedagogical reasoning and action' (Shulman 1987) is undertaken by science teachers in contributing fully to technology education?

(x) What range of technological outcomes might contribute to science education?

(xi) Which concepts of science underpin these technological outcomes?

(xii) How might a consideration of technological outcomes and the concepts and processes of science be most effectively related to each other in science education and in technology education?

(xiii) What contribution might science education make to 'education in technology', 'education for technology' and 'education about technology'?

(xiv) How might teachers be educated to provide an 'education in technology'?
(xv) What impact does technology education have on attitudes to, and participation in, technology-based industry?
(xvi) What impact does technology education have on student's general attitudes to, and achievements in, education as a whole?

Anning *et al.* (1992) have also produced a list of relevant questions.

Technology education research currently receives little funding in many countries. The preceding agenda suggests that, if broadly supported social goals involving technology education are to be achieved, this deficiency must be remedied.

Acknowledgements

I am most grateful to Patrick Dillon, Alister Jones, David Layton for incisive comments on an earlier draft.

References

AAAS (American Association for the Advancement of Science) (1989). *A Project 2061 Panel Report: Technology*. Washington, DC, AAAS.
ACARD (Advisory Council for Applied Research and Development) (1986). *Exploitable Areas of Science*. London, HMSO.
Aikenhead, G. (1989). *Logical Reasoning in Science and Technology*. Ontario, Wiley.
Allsop, T. and Woolnough, B. (1990). The relationship of technology to science in English schools. *Journal of Curriculum Studies*, 22(2), 126–136.
Anning, A., Driver, R., Jenkins, E., Kent, D., Layton, D. and Medway, P. (1992). *Towards an Agenda for Research in Technology Education*. Leeds, School of Education, University of Leeds.
Barnes, B. (1982). The science–technology relationship: a model and a query. *Social Studies in Science*, 12, 166–172.
Brandin, D. and Harrison, M. (1987). *The Technology War: A Case for Competitiveness*. New York, Wiley.
Burns, J. (1990). *Students' Attitudes Towards, and Concepts of, Technology: A Report to the Ministry of Education*. Wellington, New Zealand, Ministry of Education.
Chapman, B. (1991). The over-selling of science education. *School Science Review*, 72, 53–55.
Davies, T., Gilbert, J. and Dillon, T. (1991). *Six Counties Technology: Management Folder*. London, Longmans.
DES (Department of Education and Science) (1990a). *Technology in the National Curriculum*. London, Longmans.
DES (Department of Education and Science) (1990b). *Survey of Schools/Industry Links*. London, DES.
Fensham, P. (1988). Approaches to the teaching of STS in science education. *International Journal of Science Education*, 10(4), 346–356.
FEU (Further Education Unit) (1988). *The Key Technologies*. London, FEU.
Gadd, K. (1991). Wessex Advanced Level Chemistry. *Steam*, 14, 15–16.
George, J. (1988). The role of native technology in science education in developing countries: a Caribbean perspective. *School Science Review*, 69, 815–820.
Gilbert, J. (ed.) (1992). *Remote Sensing in Science Education*. Hatfield, ASE.
Henderson, J. and Knutton, S. (1990). *Biotechnology in Schools: A Handbook for Teachers*. London, Open University Press.
Hodgkin, R. (1990). Techne, technology and inventiveness. *Oxford Review of Education*, 16(2), 207–217.
Jeffrey, J. (1990). Design methods in CDT. *Journal of Art & Design Education*, 9(1), 57–70.
Jones, A. and Kirk, C. (1990). Introducing technological applications into the physics class room: help or hindrance for learning? *International Journal of Science Education*, 12(5), 481–490.

Kahn, M. (1989). Physics in a technological context. *School Science Review*, 71, 9–13.

Kelly, A. (1988). Option choice for girls and boys. *Research in Science and Technology Education*, 6(1), 5–23.

Kerr, S. (1989). Technology teachers and the search for school reform. *Educational Technology Research and Development*, 37(4), 5–17.

Kline, S. (1985). What is technology? *Bulletin of the Scientific and Technical Society*, 1, 215–218.

Layton, D. (1988). Revaluing the T in STS. *International Journal of Science Education*, 10(4), 367–378.

Layton, D. (1990a). Science education and the new vocationalism. In E. Jenkins (ed.), *Policy Issues and School Science* (pp. 52–62). Leeds, Centre for Studies in Science and Mathematics Education.

Layton, D. (1990b). *Inarticulate Science*. Liverpool, School of Education, University of Liverpool.

Layton, D. (1991). Science education and praxis: the relationship of school science to practical action. *Studies in Science Education*, 19, 43–79.

Layton, D., Medway, P. and Yeomans, D. (1989). *Technology in TVEI 14–18: The Range of Practice*. Sheffield, The Training Agency.

Lewis, T. (1991). Introducing technology into school curricula. *Journal of Curriculum Studies*, 23(2), 141–154.

Lindblad, S. (1990). From technology to craft: on teachers' experimental adoption of technology as a new subject in the Swedish primary school. *Journal of Curriculum Studies*, 22(2), 165–175.

Longino, H. (1983). Beyond bad science: sceptical reflections on the value-freedom of scientific enquiry. *Technology and Human Values*, 1, 7–17.

Mackay, H. (1991). Technology as an educational issue: social and political perspectives. In H. Mackay, M. Young and J. Beynon (eds), *Understanding Technology in Education* (pp. 1–12). London, Falmer Press.

Margolis, J. (1984). Three conceptions of technology: satanic, titanic, human. In P. Durbin (ed.), *Research in Philosophy and Technology*, Vol. 7 (pp. 145–176). Richmond, VA, Virginia Polytechnic Institute.

Marsden, C. (1989). Why *Business Should Work with Education*. London, PB Education Unit.

Mayr, O. (1982). The science–technology relationship. In B. Barnes and D. Edge (eds), *Science in Context: Readings in the Sociology of Science* (pp. 155–163). Milton Keynes, Open University Press.

Medway, P. (1989). Issues in the theory and practice of technology education. *Studies in Science Education*, 16, 1–24.

Medway, P. (1992). Constructions of technology: reflections on a new subject. In J. Beynon and H. Mackay (eds), *Technological Literacy and the Curriculum* (pp. 65–83). London, Falmer press.

Meeke, N. (1989). *The Concept of key Technologies*. London Further Education Unit.

Mitcham, C. (1978). Types of technology. In P. Durbin (ed.), *Research in Philosophy and Technology*, Vol. 1 (pp. 229–294). London, JAI.

Nash, M., Allsop, T. and Woolnough, T. (1984). Factors affecting pupil uptake of technology at 14+. *Research in Science and Technology Education*, 2(1), 5–19.

Pacey, A. (1983). *The Culture of Technology*. Oxford, Blackwell.

Powell, A. (1987). Is architectural design a trivial pursuit? *Design Studies*, 8(4), 191–194.

Raat, J. and De Vries, M. (1986). The Physics and Technology project. *Physics Education*, 21, 333–336.

Rapp, F. (1989). Introduction: general perspectives on the complexity of philosophy of technology. In P. Durbin (ed.), *Philosophy of Technology: Practical Historical and Other Dimensions* (pp. ix–xxiv). Dordrecht, Kluwer.

Rennie, L. (1987). Teachers' and pupils' perception of technology and the implications for the curriculum. *Research in Science and Technology Education*, 5(2), 121–133.

Rogers, L. (1990). I.T. in science in the National Curriculum. *Journal of Computer Assisted Learning*, 6, 246–254.

Royal Society (1990). *Interactive Video and the Teaching of Science*. London, Education Committee, Royal Society.

Russell, B. (1972). *A History of Western Philosophy*. New York, Simon and Schuster.

Salomon, G., Perkins, D. and Globerson, T. (1991). Partners in cognition: extending human intelligence with intelligent technologies. *Educational Researcher*, April, 2–9.

Schon, D. (1983) *The Reflective Practitioner*. London, Temple Smith.

Schwab, J. (1964). Structure of the disciplines. In D. Ford and L. Pugno (eds), *The Structure of Knowledge and Curriculum* (pp. 6–30). Chicago, IL, Rand McNally.

Shulman, L. (1987). Knowledge and teaching: foundations of the new reform. *Harvard Educational Review*, 57(1), 1–22.

Silverstone, R. (1991). The body electric. *The Times Higher Education Supplement*, 2 August, 13–15.

Skolimowski, H. (1972). The structure of thinking in technology. In C. Mitcham and R. Mackey (eds), *Philosophy and Technology: Reading in the Philosophical Problems of Technology* (pp. 42–49). New York, Free Press.

Smail, B. and Kelly A. (1984a). Sex differences in science and technology among 11 year old school children, I: Cognitive. *Research in Science and Technology Education*, 2(1), 61–76.

Smail, B. and Kelly, A. (1984b). Sex differences in science and technology among 11 year old school children, II: Affective. *Research in Science and Technology Education*, 2(2), 87–106.

Solomon, J. (1988). Science technology and society courses: tools for thinking about social issues. *International Journal of Science Education*, 10(4), 379–387.

Sparkes, B. (1989). Information technology in science education. *School Science Review*, 71, 25–31.

Standenmaier, J. (1985). *Technology's Storytellers: Rewarding the Human Fabric*. Cambridge, MA, Society for the History of Technology and MIT.

Symington, D. (1987). Technology in the primary school curriculum: teachers' ideas. *Reasearch in Science and Technology Education*, 5(2), 167–183.

UNESCO (1983). *Technology Education as Part of General Educational*. Paris, UNESCO.

Wells, C. (1990). Microtechnology: a link between music education and physics education? *Physics Education*, 25, 10–14.

Wiener, M. (1981). *English Culture and the Decline of the Industrial Spirit*. Cambridge, Cambridge University Press.

CHAPTER 7

POSITIONING MODELS IN SCIENCE EDUCATION AND IN DESIGN AND TECHNOLOGY EDUCATION

Gilbert, J.K., Boulter, C. and Elmer, R. 'Positioning models in science education and in design and technology education'. In J.K. Gilbert and C. Boulter (eds), *Developing Models in Science Education*. Dordrecht: Kluwer, 2000, pp. 3–17

Introduction

The purpose of this chapter is to establish the place of modelling and models in science education and in technology education (the UK terminology of 'design and technology education' is introduced and used in the chapter). It is argued that both the processes and outcomes of science and of technology *per se* have a great deal in common. 'Authentic' educations in science and in technology must reflect the natures of the parent disciplines as far as is practicable. Modelling and models are common to both, thus providing a potential bridge between science education and technology education. The basic terminology of modelling and models is presented.

The role of modelling in scientific enquiry

The central roles that modelling plays in the processes of scientific enquiry and that models play as the outcomes of that enquiry are well established (e.g. Giere 1988). As a consequence, modelling and models should make major contributions to 'authentic' (Roth 1995) science education. However, there is a secondary purpose. Barnes (1982) has argued that there are considerable similarities between the processes and outcomes of science and of technology. This suggests that some commonalities ought to exist between science education and technology education. Modelling and models should be capable of forming a bridge between the two. This is a first step in constructing such a bridge. Whilst the emphasis is on the role of modelling and models in science education, because much of the relevant research and development work so far has been done there, this chapter makes the case for such a bridge.

The essence of much of the thinking that underlies this chapter is reflected in the report *Beyond 2000: Science Education for the Future* (Millar and Osborne 1999). A recommendation made is that: 'The science curriculum from 5 to 16 (years) should be seen primarily as a course to enhance general "scientific literacy"' (para. 4.2). It is suggested that one structural element of such a curriculum should be 'explanatory stories', which are:

> The heart of the cultural contribution of science...a set of major ideas about the material world and how it behaves...(presented in) one of the world's most powerful and persuasive ways of communicating ideas...narrative form. It is these accounts...which interest and engage pupils.
>
> (para. 5.2.1)

It is also proposed that 'Work should be undertaken to explore how aspects of technology and the applications of science currently omitted could be incorporated within a science curriculum designed to enhance "scientific literacy"' (para. 5.2.3).

We intend to establish that the theme of 'modelling and models' is both a highly suitable basis for the construction for many 'explanatory stories' and that it can provide a valuable link between science and technology in education.

The conduct of science and of technology

Educational provision under the labels of 'science' and 'technology' should be as 'authentic' as possible (Roth 1995), that is they should be as faithful to the intellectual structures of the parent disciplines as possible. Syllabuses should reflect three things. First, the processes by which science and technology are conducted (their epistemologies). Second, the value systems underlying such activities, the situations in which and the purposes for which they take place (their contexts). Third, the entities with which they deal and which are their outcomes (their ontologies).

This reflection of epistemologies, contexts and outcomes should be as accurate as is possible under the circumstances within which education is conducted. For 'authenticity' to be possible, there must be a reasonable prior understanding of them by both practitioners and educators. The natures of these processes and outcomes are discussed in the following sections. These are complicated matters: for example, only simplified versions of 'processes' (given here) are even partially acceptable to their practitioners.

The nature of technology as process and as outcome

Pacey (1983) has defined 'technology-practice' as: 'the application of scientific and other knowledge to practical tasks by ordered systems that involve people and organisations, living things and machines'. The 'practical tasks' most commonly addressed focus on the improvement of the physical conditions of human life (UNESCO 1983). Pacey's (1983) 'technology-practice' consists of three, simultaneously operational, elements: the technical aspect, the organisational aspect and the cultural aspect. The technical aspect consists of: 'knowledge, skill and technique; tools, machines, chemicals, liveware; resources, products, wastes…'. In short, it is the aggregate of human resources brought to bear on these 'practical tasks', the means by which these are deployed, and the material focus and outcomes of this deployment. The organisational aspect is: 'economic and industrial activity, professional activity, users and consumers, trade unions'. These are the social organisations in which technology as an activity takes place, together with those which support, in one way or another, the conduct of that activity. The cultural aspect consists of relevant: 'goals, values and ethical codes, belief in progress, awareness and creativity' within which solutions to practical problems are both framed and evaluated. In respect of the core idea of 'values', Pacey (1983) notes that:

> the culture of technology comprehends at least two overlapping sets of values, the ones based on rational, materialistic, and economic goals, and the other concerned with the adventure of exploiting the frontiers of capability and pursuing virtuosity for its own sake.

(p. 89)

Striking a balance between the influence of these two sets of values in technology education is very difficult. It will be manifest in the outcomes of technological activity, the technologies that are produced, the solutions to the practical tasks arrived at: objects (products, e.g. cars, clothes) and systems (processes, e.g. ways of making cars, clothes). What emerges from Pacey's (1983) ideas is that technological process consists of thoughtful actions by individuals taken within social contexts to produce solutions to problems which it is intended will be valued.

The nature of science as process

Science is about finding explanations for natural phenomena in the world-as-experienced. The document *Science for All Americans* (Rutherford and Ahlgren 1990) states that: 'Science presumes that the things and events in the universe occur in consistent patterns that are comprehensible through careful, systematic, study' (p. 3). Matthews (1994) has identified ten philosophical theses which inform the view of science-as-a-process in *Science for All Americans*. These may be summarised as follows:

1 *Realism.* The material world exists independent of human experience and knowledge.
2 *Fallibilism.* Although human knowledge of the world is imperfect, it is possible to make reliable comparisons between competing theories about the nature of the world.
3 *Durability.* Science modifies the ideas that are produced about the world, rather than abandoning them if they are found to be inadequate.
4 *Rationalism.* The validity of scientific arguments is tested, sooner or later, against the criteria of inference, demonstration and common sense.
5 *Antimethodism.* There is no fixed set of steps in a scientific enquiry, for knowledge involves an element of human creativity rather than emerging directly from experiment.
6 *Demarcation.* Although there is no fixed method for scientific enquiry, it does involve a series of features which enable it to be distinguished from other, non-scientific, endeavours.
7 *Predictability.* Successful science predicts observations which are then made.
8 *Objectivity.* Although science is a human activity, it attempts to rise above subjective interests in the pursuit of truth.
9 *Moderate externalism.* The direction of scientific research is influenced by prevailing views on what questions are worth addressing, and what methods will prove productive.
10 *Ethics.* Ethical considerations determine what topics are researched and arise in the actual conduct of research.

The outcomes of science are the broadly conceived notion of 'scientific methodology', together with descriptions of how the material world behaves, ideas about the entities of which the world is believed to consist or with which it can be reliably analysed (concepts), proposals for how these entities are physically and temporarily related to each other in the material world (models), and general sets of reasons why these behaviours, concepts and models can be thought to occur (theories). Science then consists of thoughtful actions by individuals within social contexts producing explanations of the natural world which it is hoped will be valued. The similarity of these overviews of science and technology suggests that there is a relationship between them.

The relationship between science and technology

The ways that science and technology relate, which cover both the processes involved and the outcomes achieved are undoubtedly complicated. It is possible to argue that the processes of technology first provide solutions to problems. Science afterwards explains the reasons for the success of these solutions. For example, steel was initially developed empirically as a way of producing harder iron, whilst the consequences for the structure of iron of the addition of small amounts of other elements, for example cobalt, were only explained long afterwards. It is possible also to argue that science precedes technology in time, such that technology is the application of science. For example, that enquiries into the sequences of amino acids within genetic material are leading to the rapid development of the industry of biotechnology.

A third interpretation is that the two are bound together in a synergistic relationship, as Barnes (1982) argues, a view which justifies the similarities between their definitions. Both involve invention, being creative, constructive activities conducted within social contexts which draw extensively on prior achievements and which are subject to no one major constraint on their success. People constitute the link between the two, with an individual often moving between scientific and technological activity. There is a traffic in knowledge and skills between the two, whilst they are both concerned to achieve definite outcomes. In short, science and technology are interdependent. We argue that this interdependency should inform 'authentic' science education and technology education. At the moment they generally do not, hence the call by Millar and Osborne (1998).

The nature of education in science and in technology

In most countries the curricula in science and in technology are currently organisationally separate.

The nature of science education

Science has, in one form or another, been a 'subject' at school level in many countries for well over a century. As a consequence, there has been an extensive sharing of experience, so that the structure and substance of that provision has become fairly homogenous at world level. Beyond transitory fashions (e.g. 'learning by discovery') the differences, such as they are, tend to be couched in terms of the 'applications' of science to 'practical tasks' of local (often national) importance, for example water purification in rural areas, wind-powered electrical generation. In general terms, there are three major components to all curricula which are drawn from diluted versions of the academic subjects of biology, chemistry and physics. These are either taught independently of each other, or with the curricula co-ordinated so as to avoid repetition and the teaching of different interpretations of the same ideas, or in an integrated fashion, with the material built around some common theme or topic.

Hodson (1993) has identified three purposes for science education which cut across the structure and content of whatever provision is made. One major purpose is to 'learn science', that is, to come to understand the major achievements of science, the concepts, the models, the theories. A second major purpose is to 'learn about science', that is, to develop an understanding of the nature and methods of science, how it is conducted. The third purpose is to 'learn to do science', that is to become able to engage in and develop expertise in the practice of scientific enquiry.

The UK National Curriculum for Science (DfEE 1995a), for example, is organised into four threads ('Attainment Targets') which run across the full age spread of compulsory science education in state schools (5–16 years). These are concerned with the processes of science ('Experimental and Investigative Science'), biology ('Life Processes and Living Things'), chemistry ('Materials and their Properties') and physics ('Physical Processes'). The emphasis is on 'learning science' and to some extent on 'learning how to do science', with a relative neglect of 'learning about science'.

The nature of technology education

In many countries, 'technology education' is evolving, at an uneven pace, from 'craft education' (McCormick 1991). Craft education, the physical making of things, was traditionally reserved for students of lower academic achievement, for example in Canada (Ontario Ministry of Education and Training 1993), and of lower social status, for example in South Africa (Department of National Education 1991). There are global trends towards the automation of industrial production and increased competition in the innovation of products. These are leading to an increased instrumental valuation of the design element of technology education. At the same time, the intrinsic valuation of this element is also increasing as the contribution that it can make to the development of creativity is appreciated. The design element is concerned with deciding on the optimum fit (a value-laden term) between the problem for which a solution is sought, the structure of that solution and the materials from which it is to be made.

The trend towards the emphasis on the design element is reflected, for example, in the UK National Curriculum for the significantly named school subject of 'Design and Technology' (DfEE 1995b). This consists of the two threads ('Attainment Targets') of 'Designing' and 'Making'. We adopt the UK nomenclature of 'Design and Technology' (hereafter D&T) for the school subject of 'technology education'. This is done for two reasons. First, to differentiate it from the USA use of the word 'technology' as meaning 'anything to do with computers'. Second, as a reminder of the curricular tension between 'making' and 'design' which is reflected in the titles of the two Attainment Targets.

Exploring links

There are a number of strong reasons why an exploration of possible links between science education and D&T education is desirable. Living in the everyday world entails solving a continuous series of practical problems. Decisions have to be taken about personal matters, for example, what type of diet to follow in order to remain healthy. Decisions have to be taken socially about matters of collective importance, for example, whether a waste incinerator should be built in the neighbourhood. Decisions have to be taken about economic matters, for example, whether an individual should seek work in an emerging field of employment. Taking informed decisions about an ever-increasing number of examples of all three of these types requires a substantial knowledge and understanding of the processes, contexts and outcomes of both science and of technology. The perspective brought to bear on such problems often involves an integration, or at least co-ordination, of the insights drawn from the two fields of endeavour.

It was argued earlier that science and technology, as spheres of intellectual activity and as practical pursuits, are interdependent. However, a review of the curricular interactions between these two domains, within timetabled 'science', shows a wide and fragmentary range of provision (Gilbert 1992): minimal reference to technological outcomes during science lessons; the teaching of science followed by a discussion of its 'application' in technology; teaching which starts from science and which then leads on to technological outcomes; teaching which starts from technological outcomes and which leads on to the associated science; and, of course, the use of technological outcomes in the teaching of science. To these should be added the design and making of technological outcomes within 'design and technology education' with very varying degrees of reference to the processes and ideas of science. 'Authentic' education in both, whether as a preparation for everyday decision-taking or for disciplinary activity *per se*, must draw them closer together by focusing attention on those aspects which they share or where considerable similarity exists between them. If students are to learn about these shared aspects, it would seem more desirable to use them as a partial basis for structuring the overall curriculum pursued rather than persisting with the present *ad hoc* arrangements.

The theme of 'modelling and models' can be manifest in three ways in the school subjects of science and of D&T. First, modelling as a process and the models that are an outcome of that process can be part of the substance of curricula. The issue is one of the values of modelling and models *in* curricula. Second, science education and D&T education can be represented for analysis with the aid of models of the curriculum. The issue is one of the value of modelling and models of curricula. Third, modelling and models in and of science curricula and D&T education can form an important way of exploring the relationship between them. The emphasis here is, because of the circumstances created by the emphasis of research to date, on the first of these three, and on science education in particular. However, we attempt to establish the conditions of a future treatment of the other two possibilities.

On modelling and models

Modelling and models in science education

A model in science is a representation of a phenomenon initially produced for a specific purpose. As a 'phenomenon' is any intellectually interesting way of segregating a part of the world-as-experienced for further study, models are ubiquitous. The specific purpose for which any model is originally produced in science (or in scientific research, to be precise) is as a simplification of the phenomenon to be used in enquiries to develop explanations of it. Many models are composed of entities which are concrete, objects viewed as if they have a separate existence (e.g. a wheel) or as if they are part of a system (e.g. a wheel on a car). A model of an object can be either smaller than the phenomenon which it represents (e.g. of a train), or the same size as it (e.g. of the human body) or bigger than it (e.g. of a virus). Other models are composed of abstractions, entities which are treated as if they are objects, for example forces, energy. A model can thus be of an idea. A model can consist of a mixture of entities which are concrete (e.g. masses) and of entities which are treated as if they are concrete (e.g. forces acting on masses). A model can be of a system, a series of entities in a fixed relation to each other (e.g. of the stations and the connections between them in a metro railway). A model can be of an event, a time-limited segment of the behaviour of one or more entities in a system (e.g. a model of an athletics race). A model can be of a process, one or more events within a system which have a distinctive outcome (e.g. of the Bosch-Haber method of making ammonia from nitrogen and hydrogen). A Thought Experiment (see chapter 8) is a model of that group of processes known as a 'scientific experiment' carried out entirely within the mind as an idea, a mental model.

A classification of the ontological status of models is possible:

- A *mental model* is private and personal cognitive representation. It is formed by an individual either on their own or whilst within a group.
- An *expressed model* is placed in the public domain by an individual or group, usually for others to interact with, through the use of one or more modes of representation (see next paragraph). The relation between any one mental model and the apparently corresponding expressed model is complex. Any reflective person who has set out to express a mental model will be aware that the act of expression has an effect on a mental model: expressing it changes it.
- Different social groups, after discussion and experimentation, can come to an agreement that an expressed model is of value, thereby producing a *consensus model*. In particular, scientists produce a wealth of expressed models of the phenomena which they are investigating. An expressed model which has gained acceptance by a community of scientists following formal experimental testing, as manifest by its publication in a refereed journal, becomes a *scientific model*. It then plays a central role in the conduct of scientific research for a length of time which is governed by its utility in producing predictions which are empirically supported.
- Those consensus models produced in specific historical contexts and later superseded for many research purposes are known as *historical models*.
- That version of a historical or scientific model which is included in a formal curriculum, often after some further simplification, is a *curricular model*.
- As the understanding of consensus, historical and curricular models (as well as the phenomena that they represent) is often difficult, *teaching models* are developed to assist in that process. Teaching models can be developed by either a teacher or by a student.
- A *hybrid model* is formed by merging some characteristics of each of several distinct scientific, historical or curricular models in a field of enquiry. It is used for curriculum and classroom teaching purposes as if it were a coherent whole.
- A *model of pedagogy* is used by teachers during the planning, practical management and reflection on, classroom activity, and is concerned with the nature of science, the nature of science teaching and the nature of science learning.

One or more of five *modes of representation* are significant in expressed models of any phenomenon and can be used to construct typologies:

- The *concrete mode* consists of the use of materials, for example a metal model of a railway engine, a polystyrene model of a molecule.
- The *verbal mode* consists of the use of metaphors and analogies in speech, for example when talking about teaching models and in written form, for example in textbook descriptions.
- The *mathematical mode* consists of mathematical expressions, including equations, for example the universal gas equation.
- The *visual mode* makes use of graphical pictorial forms in graphs and in diagrams.
- The phrase '*symbolic model*' includes the visual, verbal and mathematical modes.
- The *gestural mode* consists of actions, for example movements of the hand.

Modelling and models make three major contributions to science education. First, it is believed that the formation of mental models and the public presentation of expressed models are central to the development of an understanding of any phenomenon or body of information. Mental modelling is thus as important in achieving all three of Hodson's (1993) purposes for science education as it is in the learning of any other subject. Second, the production and experimental testing of expressed models plays a central role in the processes of science (see points 2, 3, 5–7 in Matthews (1994) list on p. 119). Hodson's (1993) purposes of 'learning about science' and 'learning to do science' thus involve modelling and model testing. Third, historical and scientific models are major outcomes of science (see points 1–4, 8, 9 in Matthew's (1993) list). Hodson's (1993) purpose for science education of 'learning science' must involve the development of an understanding of major historical and scientific models, if only through curriculum models.

Modelling and models in design and technology education

Modelling and models are also used for specific purposes in D&T education. Harrison (1992) puts it thus: 'The critical question to ask about a model is... for what purpose is it intended? Indeed this intention will determine the nature of the model, against which the usefulness will be evaluated' (p. 32).

Several systems for classifying these purposes have been put forward. Harrison (1992) divides them into: helping with thinking; communicating form or detail; evaluating a design or features of it. Whilst others, for example, Liddament (1993), have longer lists, a telling division is into 'communicating with oneself' and 'communicating with others' put forward by Kelly *et al.* (1987). This latter division reflects the situation in design *per se*, where the designer uses modelling to reflexively develop personal ideas, to work with other members of the design team by facilitating communication, as well as in communicating intentions to and in negotiation with clients (Baynes 1992).

Design and technology education, in contrast to science education, is often substantially 'authentic' in nature, that is it involves a student acting as a 'designer' and then going on to be a 'maker'. However, there are three major problems inherent with such an expectation. First, the educational system is badly capitalised, so that the range of resources available to the student with which to construct concrete models is bound to be poorer than that available in an industrial context. Second, although industry has to produce solutions to problems within a given period, the time allocated to a task is usually fairly negotiable at the outset. Schools, on the other hand, divide time into blocks of rigid size, perhaps leading to conservative designs and poor 'making'. Indeed, the act of producing a concrete model may act as a substitute for full product realisation. Third, the matter of assessment becomes complicated, for:

> There is some confusion between industrial and educational perspectives on the activity (of D&T Education). In education, the concern is to expose pupils to designing technological experiences in order that they may develop understanding and capability. In industry, that design and technological capability is directed towards a manufacture of a product.
>
> (Kelly *et al.* 1987: 7)

Within the educational perspective on student work, what Downey and Kelly (1986) call the 'intrinsic' aim, the teacher-as-assessor has to evaluate the quality of the processes undertaken during the activity and their personal significance for the student, including the capability to work appropriately with others. This latter tension will be manifest in terms of the models presented: good work within the industrial perspective, for example, within the visual mode, may lead students to use professional means of presentation, for example computer graphics, whilst good work within the educational perspective may lead students to use more personally expressive means of presentation, for example a series of rough sketches on paper. Liddament (1993) has explored the conflict between models used to teach concepts of design and technology and models used to assist and advance design activity *per se*. They include issues of the 'ethics of representation' (Baynes 1992): modelling-for-the-public-domain will be in tension with modelling-for-the-private-domain.

Modelling and models is not a well-developed theme in the literature of design and technology education. As is typical of any discipline in what Kuhn (1970) calls a pre-paradigmatic phase, a wide range of words is used to cover ill-defined, perhaps similar, meanings. For 'modelling', these include 'imaging', 'cognitive modelling', 'concrete modelling', 'making'. For 'model', these include 'mock-up', 'lash-up', 'prototype'. One aspect of an exploration of the scope of modelling and models in forming links between science education and design and technology education must be the extent to which a common terminology can be used in the two fields.

That which goes on in a designer's head has been described by Kimbell *et al.* (1996: 114) as 'creative concrete thinking'. However, the processes involved and the consequences achieved seem to be very close to what were called 'mental models' earlier in this chapter. The next step in the design process is the generation of an expressed model in an appropriate mode of representation. A visual mode, for example a sketch, or a concrete mode, for example the use of modelling clay, are commonly used. This is then subjected to a cycle of development, testing, further development and so on, until the designer is convinced that the outcome can be presented to the client (or, in an educational context, the teacher-as-surrogate client) in the form of a *prototype*. This prototype will be subsequently altered in response to the client's reaction and, perhaps more significantly, in the light of the materials used in fabrication when the product is manufactured.

This developmental process parallels, in many ways, the changes that take place as a consensus model is produced in a D&T classroom. Common to both is:

- The notion of a developmental cycle, with changes taking place to the nature of the testing imposed and to the model itself, leading to an outcome through a rolling programme.
- The notion of 'fitness for purpose' in respect of a 'design specification' being used as a judgemental criterion at the end of the developmental programme.
- The notion of that evaluation being conducted on behalf of both the immediate social group (the class) and an external reference agency (the client).

However, perhaps there is greater variability in the design and technology education context, as compared with a science education context, in respect of:

- *The modes of representation used.* Thus, whilst both make use of visual modes, for example diagrams, D&T education makes more use of the concrete mode.
- *The range of materials used within the concrete mode in D&T education.* Card, plastic, modelling clay, are all commonly used. At each stage in the developmental process the 'convenience of use' of a material is balanced against its 'analogical capability' (the range of ideas that it can express).

- *The range of perspective adopted.* Whilst models are all too often only presented in 2D in science education, not only is much greater use of 3D made in D&T education but the perspective adopted can vary during the developmental cycle.
- *The range of scales used.* Whilst modelling in the science education context often sticks to one scale (that which can be contained within one page of a school exercise book!), scale in D&T education can either remain small (where the product would be large, e.g. a bridge) or be gradually increased to that of the prototype (where the product would be of 'human scale', e.g. a kettle).

The notion of 'teaching model' has some relevance to D&T education. This is where an 'exemplary solution to a problem' is used to teach the principles of D&T. There is no direct equivalence to the idea of a scientific or historical model, in that there are, axiomatically, no generic solutions to design problems. However, the existence of a 'school' of design, where general principles of colour, form and composition are employed. For example, the Bauhaus (Pevsner 1960; Wingler 1969) comes moderately close through the notion that a recognisable approach to problem solution is widely adopted.

Modelling and models as a bridge

It would seem that modelling and models do constitute a possible bridge between science education and D&T education. It is a defining feature in the conduct and outcomes of both science and D&T. It will be possible to speculate in the future on the existence of possible parallel models of pedagogy for D&T education.

The purpose of modelling in both fields is to facilitate communication through a visualisation of the relation between the intention and the outcome of the activity. In the case of science education, the intention is to provide an explanation, which can be defined as an answer to a question about the nature of the world-as-experienced. The quality of an explanation produced can be evaluated by consideration of the predictive value of the model produced. In the case of D&T education it is usual to define the problem with some precision before the design process begins. A model enables the suitability of any proposed solution(s) (the fitness-for-purpose) to be evaluated against that problem before fabrication of the final outcome takes place.

The first stage of creative activity in both fields is the construction of a mental model. In both cases this mental model is subjected to a developmental process, through the medium of a series of expressed models, towards a version which is socially accepted (a scientific or historical or curriculum model and prototype respectively). The range of modes of representation used in both fields is somewhat similar, although the incidence of their use probably varies between the two. Whilst a consensus model, a scientific/historical/curriculum model, is the 'final' outcome of science education, the prototype model in D&T education should be followed by another stage (the manufacture of the product outcome) but this happens all too infrequently. Exemplary designs in D&T education have a role similar to that of historical models in science education: they represent solutions to problems which, whilst they have now been overtaken by events, were valued in their day.

This chapter has outlined the basic terminology which has developed and through looking at the nature of authentic education in science and D&T has suggested that modelling and models should be taught across both fields as a way of linking them.

References

Barnes, B. (1982). 'The science–technology relationship: a model and a query'. *Social Studies in Science*, 12, 166–172.

Baynes, K. (1992). 'The ethics of representation'. In *Modelling: The Language of Designing, Design: Occasional Paper No. 1*, B. Archer, P. Roberts, K. Baynes (eds), Loughborough: Department of Design and Technology, University of Loughborough.

Department of National Education (1991). *A Curriculum Model for Education in South Africa*. Pretoria: Department of National Education.

DfEE (1995a). *Science in the National Curriculum*. London: HMSO.

DfEE (1995b). *Design and Technology in the National Curriculum*. London: Department for Education.

Downey, M.E. and Kelly, A.V. (1986). *Theory and Practice of Education: An Introduction* (3rd edn). London: Harper and Row.

Giere, R. (1988). *Explaining Science*. Chicago, IL: The University of Chicago Press.

Gilbert, J.K. (1992). 'The interface between science education and technology education'. *International Journal of Science Education*, 14(5), 563–578.

Harrison, P. (1992). 'Modelling in Key Stages 1 and 2'. Paper presented at the IDATER 92 Conference, Loughborough.

Hodson, D. (1993). 'Re-thinking old ways: towards a more critical approach to practical work in school science'. *Studies in Science Education*, 22, 85–142.

Kelly, A.V., Kimbell, R.A., Patterson, V.J. and Stables, K. (1987). *Design and Technological Activity: A Framework for Assessment*. London: Assessment of Performance Unit/HMSO.

Kimbell, R.A., Stables, K. and Green, R. (1996). *Understanding Practice in Design and Technology*. Milton Keynes: Open University.

Kuhn, T.S. (1970). *The Structure of Scientific Revolutions (second edition)*. Chicago, IL: The University of Chicago Press.

Liddament, T. (1993). 'Using models in design and technology education; some conceptual and pedagogic issues'. Paper presented at the IDATER 93 Conference, Loughborough.

McCormick, R. (1991). 'Issues in the Current Practice of Technology Education'. Paper presented at the ITEA Conference, Salt Lake City, Utah.

Matthews, M. (1994). *Science Teaching: The Role of the History and Philosophy of Science*. New York: Routledge.

Millar, R. and Osborne, J. (eds) (1999). *Beyond 2000: Science Education for the Future*. London: School of Education, King's College London.

Ontario Ministry of Education and Training (1993). *The Common Curriculum Grades 1–9*. Ontario.

Pacey, A. (1983). *The Culture of Technology*. Oxford: Blackwell.

Pevsner, N. (1960). *Pioneers of Modern Design*. Harmondsworth: Penguin.

Roth, W.-M. (1995). *Authentic School Science*. Dordrecht: Kluwer.

Rutherford, F. and Ahlgren, A. (1990). *Science for All Americans*. New York: Oxford University Press.

UNESCO (1983). *Technology Education as Part of General Education*. Paris: UNESCO.

Wingler, K. (1969). *The Bauhaus*. Cambridge, MA: MIT Press.

CHAPTER 8

MODELLING

Promoting creativity while forging links
between science education and design and
technology education

Davies, T. and Gilbert, J.K. 'Modelling: promoting creativity while forging links
between science education and design and technology education'. *Canadian
Journal of Science, Mathematics and Technology Education*, 2003, 3(1): 67–82

Educational reforms in many countries currently call for the development of
knowledge-based societies. In particular, emphasis is placed on the promotion
of creativity, especially in the areas of science education and of design and
technology education. In this paper, perceptions of the nature of creativity and
of the conditions for its realization are discussed. The notion of modelling as a
creative act is outlined and the scope for using modelling as a bridge between
science education and design and technology education explored. A model for
the *creative act of modelling* is proposed and its major aspects elaborated
upon. Finally, strategies for forging links between the two subjects are
outlined.

Building a knowledge-based society

For the first time in history, knowledge is becoming both a primary outcome of
economic production and a core resource for commercial and non-commercial
organizations. As a result, possessing knowledge is a mark of employability for the
individual and the output of knowledge is explicitly sought through research-
policy development – in Europe, for example (European Union (EU), 2002). Thus
Seltzer and Bentley (1999) have identified four clear trends that are being driven
by two ongoing developments: the introduction of information and communication technologies and the emergence of economic globalization. These four trends
are that the *weightless economy*, based on human resources, information and
networks, has become a very influential source of productivity and competitiveness; that workers are developing the skills to manage themselves in increasingly
unstable organizational environments; that more *horizontal* organizational structures within and between organizations are taking the place of *vertical* structures;
and that new patterns of exclusion are emerging among those who are unwilling
or unable to develop marketable knowledge. These changes set the pace and
drive associated patterns of social, economic and cultural development that
are reflected in personal and social values, in attitudes, and in behaviour.
Creativity in organizational, social and intellectual contexts is the key to a
prosperous future.

Perceptions of the nature of creativity

Piaget (1962), one of the first psychologists to seek to explain creativity, started from the premise that individuals exist in a structured, stable world, and not, as is the case today, in one dominated by unpredictable change. There certainly is a strong element of individuality in creativity, whatever the social circumstances in which the individual acts. Thus Feldman *et al.* (1995) recognize that, to be creative, individuals have to believe that they can change the world and add to its knowledge themselves. Similarly, McKeller (1957) believes that the essence of creativity consists in individuals striving to do better than did their predecessors. There is an element of deliberation: Poïncaré found, following self-reflection, that 'a period of preliminary conscious work always preceded fruitful unconscious work' (as cited in McKeller 1957: 116). However, individuals must act within supportive social frameworks if there are to be opportunities for people to realize their creative potentials.

> Socialization not only shapes behaviour, it also moulds consciousness to the expectations and aspirations of the culture, so that we feel shame when others observe our failings, and guilt when we feel that we have let others down.
> (Csikzentmihalyi 1997: 77)

It comes as no surprise, then, to learn that Csikzentmihalyi believes that 'focusing on the individual alone when studying creativity is like studying how an apple tree produces its fruit by only looking at the tree and ignoring the sun and the soil' (as cited in Feldman *et al.* 1995: 147).

The social and economic cultures in which individuals work, in addition to the psychological factors operating, play an important role in the recognition of creative contributions in all fields. Judgements about the outcomes of creativity cannot be separated from the more general value norms of a culture. Creative individuals show a desire to create a new order by breaking down the existing order. This takes place through constructing and testing new knowledge, believing in a changeable reality, and working with detail and complexity within a domain. New individual and social realities are constructed and reconstructed in the process of relentless change that is itself construed as *culture*. It is through culture that we judge the qualities associated with creativity.

Overall, evidence for creativity in individuals can be sought by:

- identifying the frequency and nature of the apparently creative acts for which they are responsible;
- identifying their distinctive personal characteristics (abilities, patterns of conduct, declared needs, motivations observed, etc.);
- collecting the opinions of those who make decisions, choices and judgements about the merits of the work of individuals within a domain, whether working alone, in groups or in institutions.

How, then, is creativity manifested in educational settings? This is a matter that has been the object of much careful scrutiny (e.g. Fryer 1996; Beetlestone 1998). Although recognizing that creativity 'belies simple definition and measurement' (Davies 2000: 28), the Department for Education and Employment has suggested that there are four main features of creative acts:

- using imagination, often to make unusual connections or see unusual relationship between objects, ideas or situations;

- having targets and reasons for working which are capable of resulting in new purposes being discovered;
- being comparatively original in relation to the work of a small closed community, such as peers or family, or uniquely original in comparison with those working historically or currently in a field or discipline;
- judging value, which demands critical evaluation and reflection, standing back and gaining an overview position.

(Department for Education and Employment (DfEE)
1999a, as cited in Davies 2000: 28)

This formulation conforms in general to that of both Koestler (1964) and Feldman *et al.* (1995). A creative act, of its very nature, must involve doing something that leads to a novel outcome. We suggest that *modelling*, as a generic activity, entails the exercise of creativity. Modelling is a particular feature of design and technology education and, at least in theory, of science education.

Modelling as a creative act

Modelling may be defined as the process of producing a representation of an object, an idea, a system, an event or a process. A language with which to discuss models and modelling is well developed in science education. A *mental model* is the personal and private representation that is the outcome of the modelling process. A version of a mental model that is expressed in the public domain, for example through speech, may be called an *expressed model*. An expressed model that gains acceptance by any community may be called a *consensus model*. Where that community comprises scientists, the consensus model may be called a *scientific model* if it is used at the cutting edge of research and a *historical model* if it has been superseded. Expressed models of all types may be shared with others through the use of one or more *modes of representation*: concrete (or material), visual, verbal, mathematical or gestural (Gilbert *et al.* 1998). The equivalent language in design and technology education seems less extensively developed. *Cognitive model* (Archer 1980) is a readily recognizable term, whilst the list, *iconic model* (e.g. sketch, card model), *analogue model* (e.g. computer simulation) and *symbolic model* (e.g. mathematical model) (Baynes 1992), places more emphasis as a scheme of classification on the methods of expression than on the ontological status of the various types.

In science education, the learning of a consensus model – which falls within what Hodson (1992) calls 'learning science' – does entail the formation of a mental model through the process of modelling. The major context for modelling in science education should, however, be the formation of a mental model of a phenomenon for which none is available to the student – this falls within what Hodson calls 'learning to do science'. The development and use of the skills of modelling in science education are widely, if often obliquely, advocated in the UK (DfEE 1999b), the US (Rutherford and Ahlgren 1990), Australia (Curriculum Corporation 1994) and New Zealand (Ministry of Education 1993). This is a recognition of the core commitment of science to produce causal explanations of the world-as-experienced. A model in science or science education, once produced, is a simplification of a phenomenon produced for the purpose of making predictions about its behaviour under different circumstances, predictions that are then experimentally tested.

The core commitment of design and technology is to produce solutions to human problems. For this purpose, two interactive roles for modelling were identified by Kimbell *et al.* (1991). They are modelling ideas in the mind (the role of communicating with oneself) and modelling ideas in a material form (the role of communicating with others). Two main modes of representation seem to dominate the activity of modelling in material form. The first of these is the visual mode, making especial use of sketching, technical drawing and pseudo-3D images – for example, through computer-based formats. The second is the material (or concrete) mode. The 3D representations are typically made in modelling clay, shaped in polystyrene foam or constructed from softwood. In design and technology *per se*, material models (often called *mock-ups* or *prototypes*) are translated into the final outcome (or product), a process that often involves a change of scale and a change of medium. Because of the special circumstances of schools, particularly the lack of resources, of skills in making and of the traditional values of the field, together with the tyranny of an inflexible timetable, the material model is all too often the only product. The purpose of a model in both the industrial and educational contexts is to permit efficient and economical evaluation of the *fitness for purpose* of the proposed solution as a response to a design brief. In many cases, the function of the model is to permit predictions about the behaviour of the final product, a function that has close parallels in the science education context.

Modelling: a creative link between science and design and technology in schools

Barlex and Pitt (2000) advance five arguments in favour of stronger links between science and design and technology (D&T) education:

- Both require the pupil to reflect on his/her practice. This is difficult in science education: since, as a rule, experiments are conducted only once, there is often little in the macro domain (concrete evidence) to reflect upon. D&T education does provide this concrete evidence, often in abundance. Pupils have to conduct fair tests on their D&T education products, a skill extensively developed in science education.
- A consistent approach to the promotion of metacognition (the ability to think about one's thinking) could be developed across the two subjects because of the high degree of epistemological congruence between them.
- Both subjects require that pupils develop the capacity to visualize – that is, to see 'in the mind's eye' what cannot be, or has not yet been, seen.
- The explanatory concepts of science education, such as force and energy, are often used in producing the solution to D&T education design briefs.
- The use of technological contexts can be motivational in the teaching of science, for such contexts provide utilitarian reasons why abstract, often difficult, science concepts should be mastered by students.

We believe that the development and use of modelling skills can be a route to forging such links because the development of modelling skills is a major objective of both subjects; because modelling in the two domains makes use of the same modes of representation, especially of the concrete and visual modes; because both subjects require mental/cognitive modelling, which is a major tool in thinking itself (Johnson-Laird 1983); because both make extensive use of and provide a context for the development of visualization skills; and because the explanatory modelling of science education complements the product modelling of D&T education. The actual knowledge and skills entailed in the act of modelling must be explored with this high degree of commonality in mind.

The creative act of modelling

Lakoff and Johnson (1980) have argued persuasively that all our ideas are developed by analogy with an original set of ideas associated with our bodily postures. This idea has been adapted to the context of science education by Glynn *et al.* (1994), who state that a model is developed when a target (that which is to be understood) is viewed by the analogical consideration of a source (that which is already understood). In the light of what has been argued earlier in this paper, we believe that the process of modelling – the cognitive acts by means of which a model is produced – is, to a large extent, common to both science education and design and technology education. This process may be thought of as five interactive elements, as detailed in the following sections.

Having suitable experience of the phenomenon/problem

In science education, this usually consists of practical work activities, conducted either directly in a laboratory or indirectly through video. The general pattern of behaviour in an exemplary case of the phenomenon is established either qualitatively or quantitatively. For example, the phenomenon of electrostatics is investigated with the aid of various soft materials, rods and small pieces of paper. In the case of D&T education, the activities consist in establishing the nature of the problem that has to be solved, together with gathering any clues as to what might constitute an efficient and effective solution – for example, what might be entailed in producing a lift to be attached to a staircase in order to carry a physically disabled person to and from an upper storey in a house? The nature of a suitable seat, its attachment to the wall, the power requirements and safety issues, might be separately considered.

Identifying a suitable metaphor and drawing the associated analogies

A metaphor is, by definition, 'a figure of speech in which a word or phrase is applied to something to which it is not literally applicable' (*The Concise Oxford Dictionary*, s.v. 'metaphor'). The nature of metaphor – how it works – has been and continues to be hotly disputed by linguists. However, a view that seems widely accepted is the *interactive* (Black 1979), according to which the two juxtaposed words or phrases are temporarily equated and each is subjected to a mutual modification of meaning. For example: '*The sun is a furnace*'. Here, the sun is seen to take on the attributes of a furnace: the burning of fuel, the production of by-products, the idea of a finite time of operation, the production of a large quantity of heat and the serving of human purposes. At the same time, a furnace is seen to take on some of the characteristics of the sun: the apparently inexhaustible production of heat; the absence of evidence of fuel use; and the serving of vital human purposes. However, this identity is very easily challenged, there being little evidence to support the attribution of some of the furnace-like characteristics to the sun (e.g. the idea of burning, the evidence of gaseous by-products). It is at this point, where specific attributes are considered, that the drawing of an analogy is triggered: 'a comparison between one thing and another made for the purpose of explanation or clarification' (*The Concise Oxford Dictionary*, s.v. 'analogy').

Hesse (1966) states that any metaphor can be interpreted in terms of a series of comparisons, of analogies. Where a comparison can be traced from the source to the target, a *positive analogy* can be drawn. For example, the sun produces a continuous supply of heat. Conversely, where no comparison can be justified, a *negative analogy* has been drawn. For example, the sun produces no gaseous by-products. There are often attributes where the status of the comparison is unclear. This is the case of the *neutral analogy* – for example, whether or not the sun's supply of fuel is inexhaustible. Neutral analogies can be the source of creative thought: Are they really unrecognized positive or negative analogies?

But what is actually entailed in drawing an analogy? Gentner's (1983) widely employed 'structure mapping theory of analogy' rests on the predicate that any object (or idea treated as an object) consists of ontological entities and relationships (temporal, spatial, causal) between those objects. To take two examples: a crystal of an ionically bonded substance can be viewed as a regular array of a few types of ions, held 'in place' by the electrostatic forces operating; a bridge can be viewed as a regular array of girders held in place by the forces generated by their masses. Gentner has identified three types of comparison (see Table 8.1).

Of the three types, *mere appearance* comparison provides the least insight into the nature of the target, for only the entities of which the source is composed are transferred to the model. The fact that no relationships are transferred means that no causal explanations can then be inferred. *Literal similarity* is quite a powerful form of comparison, for it assumes that both the entities and the relationships among them can be transferred. However, the *tightness* of the transfer does mean that the source and the target have to be conceptually very close. This inhibits the exercise of radical creativity: the seeing of comparisons where none had been seen before. *True analogy* offers the greatest scope for imagination: no assumptions are made that the entities of which the source is comprised can be transferred to the target via the model, but it *is* assumed that the relationships among them can. Perhaps the most famous example in physics is that of the *mechanical wave* (e.g. ripple on still water), where the relationships operating were used to represent first light and then sound.

These ideas have been used in science itself (Nersessian 1984) and in science education (Gentner and Gentner 1983). However, they do not seem to have been used to analyse D&T education, although there seems no a priori reason why they should not provide fruitful insights. There are two much bigger issues before metaphor and analogy can be used to bring science education and D&T education closer together. First, teachers and students must understand how metaphor and analogy 'work': there is evidence that the use of analogy develops gradually with age (Goswami 1992). Second, those using metaphor must have a broad range of sources on which to draw. This assumes both a broad education base and a conscious awareness that knowledge acquired in one context may be of use, later, in another.

Table 8.1 Gentner's classification of similarity

Type of similarity	Entities transferred from source to model	Relationships transferred from source to model	Example
Mere appearance	Yes	No	Table top gleamed like a pool of water
Literal similarity	Yes	Yes	Milk is like water
True analogy	No	Yes	Heat is like water

Visualizing the outcome of the modelling process

There does seem to be general agreement that visualization is an important component in both scientific achievement (Miller 1987) and in the design and production of technological artefacts (Baxter 1995). Major advances often seem to be made through the use of the *metaphorical imagination*, the linking of metaphor use with visualization (Holton 1995).

Christopherson (1997) has advocated the development of *visual literacy skills* during the process of formal education. Such skills are seen to be of central importance in science education generally and in chemistry education in particular (Tuckey and Selvaratnam 1993) and in D&T education (Archer 1980). There is a close link between skills of visualization and achievement in science education (Coleman and Gotch 1998) and in D&T education (Kimbell *et al.* 1991). These skills do improve with age during childhood and adolescence, with relevant experience playing a major role in that development (Tuckey and Selvaratnam 1993). Studies of gender and visualization seem inconclusive: any possible initial advantage boys possess can readily be nullified by suitable experience for girls (Tuckey and Selvaratnam 1993).

The terminology of the field lacks uniformity, with the term *visualization* often being used to cover a range of spatial abilities. Barnea (2000) has analysed the terms *spatial ability* and *visualization ability*, often used interchangeably, into three skills in ascending order of importance:

- *spatial visualization*: the ability to understand 3D objects from their 2D representations;
- *spatial orientation*: the ability to imagine what a representation would look like from a different perspective;
- *spatial relations*: the ability to visualize the effect of operations such as rotation, reflection and inversion, or to manipulate objects mentally.

These skills are used in understanding the model in all modes of representation – especially in the concrete and visual modes. For example, the National Science Foundation (NSF) (2001) has pointed to two major types of problem that students may have in visualizing models of molecules, however represented. First, *visual subtlety* – the values of angles in non-orthogonal relationships and the values of distances are both difficult to perceive; the limited scope of an individual's visual attention makes the spatial relationships among the parts of a complex object (e.g. molecule or artefact) difficult to perceive simultaneously; and, while symmetries can readily be perceived in 2D representations, those in 3D are found more difficult. Second, *complexity* – in molecules or objects of any complexity, the amount of information to be processed visually is large; and, while the range of conventions available within any one mode of representation can be of help in simplifying this complexity, the conventions attached to each have to be learned.

Given the importance of visualization skills in understanding models, it is surprising that the explicit teaching of such skills has been so neglected, an issue noted for science education by Mathewson (1999) and for D&T education by Kimbell *et al.* (1991).

Producing a representation of the model

Each of the *modes of representation* (concrete, visual, verbal, mathematical, gestural) has a *code of representation* that governs what can be represented within it.

Each mode enables focusing on a particular aspect of a model. For example, the concrete mode enables focusing on the relative spatial and temporal relationships among entities within the model, using the tactile sense. The visual mode enables summarizing all the entities and the relationships among them in the model, so that a personal overview of the whole can be produced. The choice of mode thus depends on what is to be the focus of attention. The concrete and visual modes are the two most commonly used in science education and D&T education. However, the extent and manner of use of each mode differs.

The concrete mode is used in science education for a variety of purposes. An array of laboratory equipment can be viewed as a way of modelling the behaviour of a phenomenon. Concrete models are used in biology to show the structure of a system, such as the network of veins and arteries in the body. The concrete mode is used in chemistry to make the sub-microscopic visible and to show how causal mechanisms operate – for example, space-filling models (Francoeur 1997). However, its use is much more predominant in D&T education, where concrete models in a variety of media (clay, wood, plastic) are often the main output of activity. Visual representations are prominent in both subjects. There is a whole literature on the complex sub-mode conventions used in respect of diagrams in science education (Kress and Van Leewen 1996; Unsworth 2001). Sketches play a major role in the early stages of D&T educational activities, with technical drawing techniques often being used to show the final product, especially where that is a system (e.g. computer circuitry).

There is a great deal of work to be done before teachers can select a mode, or sub-mode, with which to express a model, confident that it will serve the purposes they desire.

Evaluating the scope and limitations of the model produced

In science education, models are either simplifications of phenomena or idealizations created to show the applicability of theory. In these processes of simplification or idealization, perceived features of reality-as-experienced are emphasized, suppressed or removed. All models are developed in science and science education to provide the basis for explanations that are adequate for the purposes intended (Gilbert *et al.* 1998). In D&T education, models are produced to show what the final product might/will look like/feel/behave. They are also a major vehicle for assessment, although here the process of producing the model is being taken as indicated by the final product (Gilbert *et al.* 2000).

In both cases, the *scope and limitations* of the model produced are statements of its adequacy. In science education, the *scope* involves the aspects of reality the model is intended to represent, those purpose(s) for which it was developed and the acceptability of the explanations that it provides. In D&T, the scope is an estimate of the degree to which the model responds to a design brief. The *limitations* of a model in science education comprise statements of those aspects of reality that it does not represent, of purpose(s) that it cannot address and of a level of explanation above which it cannot aspire. The limitations of a model in D&T education involve those aspects of the design brief that it does not meet, together with less tangible concerns about the potential acceptability of the product to the client and about its aesthetic appeal.

Students must be taught to evaluate the scope and limitations of any given model that they produce. In science education, this knowledge will prevent their concluding that the model is a *copy* of reality, and they will see their model as being something capable of being changed or replaced at some time in the future. In D&T education, evaluating the scope and limitations of their model will enable them to develop their judgement of the overall success of their solutions to problems.

Promoting creativity in general

There is a real need to promote creativity in schools and in society at large. As Rogers (1959) comments:

> Many of the serious criticisms of our culture and its trends may best be formulated in terms of a dearth of creativity. These can be stated briefly:
>
> - in education we tend to turn out conformists; stereotypes, individuals whose education is 'completed', rather than freely creative and original thinkers;
> - in our leisure-time activities, passive entertainment and regimented group action are overwhelmingly predominant, whereas creative activities are much less in evidence;
> - in the sciences, there is an ample supply of technicians, but the number who can creatively formulate fruitful hypotheses and theories is small indeed;
> - in industry, creation is reserved for the few.
>
> (p. 72)

However, specific conditions are needed if creativity is to flourish. Seltzer and Bentley (1999) identify the following six key characteristics of learning environments that encourage creativity in an individual or group:

> *Trust*: secure trusting relationships are essential. People must feel able to take risks and to learn from failure.
>
> *Freedom of action*: creative application of knowledge is only possible when people are able to make real choices over what they do and how they do it.
>
> *Variation of context*: learners need experience in applying their skills within a range of contexts in order to make connections between them.
>
> *The right balance between skills and challenge*: creativity emerges in environments where people are engaged in challenging activities and have the right level and mix of skills to meet them.
>
> *Interactive exchange of knowledge and ideas*: creativity is fostered in environments where ideas, feedback and evaluation are constantly exchanged and where learners can draw on diverse sources of information and expertise.
>
> *Real world outcomes*: creative ability and motivation are reinforced by the experiences of making an impact – achieving concrete outcomes, changing the ways that things are done.
>
> (p. 10)

If the six conditions are met, the exercise of creativity is readily perceived by practitioners as having resulted in novelty value and originality. In an established field such as science, where sophisticated rules and regimes must be interpreted and understood before an original statement can be made, it is necessary to take the total sum of prior work in the domain into account. It is more difficult to make creative contributions in science, where there are already complex sets of rules, prescriptions and procedures. It is also more difficult to convince practitioners in the field of science of the originality of a contribution. Where there are no pre-existing rules, traditions or domain knowledge – for example, in the case of certain types of computer games – there is evidence of abundant creativity in the responses of young people.

The six conditions are only fully met within curricular frameworks designed with that in mind. However, doing so is not easy. Schools are communities with many roles at many levels: promoting social and cultural norms; providing for individual care and improvement; and equipping learners with bedrock academic skills. Judgements about the work of learners and teachers are made by many agencies – for example, educationalists, parents, government, industry and commerce – who interpret the activities and outcomes of schooling in ways that serve their interests and accord with their value systems. Each make judgements based on what they perceive to be in the best interests of the learners, in relation to their future roles as citizens and participants in complex communities and society as a whole. Handy (1995) posits the view that schools, in response to the multiple pressures resulting from these diverse judgements, prioritize the maintenance of cycles and structures rather than moving towards the 'new age' and are resistant to change. Consequently, although there are well-recognized exceptions, schools can be uncomfortable places for creativity and innovation. In the post-1988 Education Reform Act period in the UK, the teachers' role has become increasingly subject to compliance with conservative pressures to conform to politically mediated targets and approaches, with little room for discovery or innovation. *Assessment-led* approaches to learning and teaching have been particularly constricting in classrooms. We cannot expect teachers to act creatively or imaginatively unless they work in an environment that supports and values creative, innovative approaches to teaching and learning.

The world trend to *national curricula* prescribes the content of school subjects in varying degrees of detail. Often overlooked are the strategic contexts within which creativity will flourish. Thus the presence or absence of demands for links between science education and D&T education is simply a matter of the politics of educational bureaucracies. As Barlex and Pitt (2000) point out, the detailed prescriptions for each of the two subjects in England and Wales make no mention of each other. At the opposite extreme, Ontario is implementing a combined science–technology curriculum (Gardner and Hill 1999a,b; Van Oostveen *et al.* 2000). So, given political will, demands for links can be put in place and creativity fostered.

A number of authors have described the process of promoting creativity. LeBoeuf (1994) discusses the promotion of creative thinking in everyday life and De Bono (1992: 239) argues convincingly and methodically that creativity can be taught if attitudes are appropriate: 'There is a creative and constructive attitude. There is a willingness to look for new ideas and to consider the new ideas that are turned up by others.' He suggests the following as a set of 'basic attitudes' to underpin the development of techniques to promote everyday creativity:

- the creative pause – a willingness to stop and think;
- challenge – not being critical but believing that there are better ways of solving a problem;
- green hat – looking for alternatives;
- simple focus – a deliberate search for a particular solution;
- alternatives – searching for a broad range of alternatives;
- provocation – when the culture of creativity is established, a willingness to consider strange or unlikely ideas;

- listening – gaining 'tuned judgement', so helping others to realize their creative potential;
- sensitization – paying attention to possible instances of creativity;
- training – teaching formal techniques of lateral thinking;
- programmes and structures – developing the organizational features required to promote and encourage creativity.

Advocacy of such techniques is certainly easy to find. Seltzer and Bentley (1999), for example, state that 'creative learners need a wider array of contexts within which to apply their skills and knowledge. They also need "teachers" or guides who can expose them to the strategies for thinking about the connections between their experiences' (p. 29). Such techniques will only be successful if applied to definite tasks. We suggest that these might be modelling tasks.

Strategies for linking science and design and technology education

Barlex and Pitt (2000) reject the *integrationalist* view, that the curricula for science and for D&T can be unified, on the grounds that 'science and design and technology are so significantly different from each other that to subsume them under a "science and technology" label is both illogical and highly dangerous to the education of pupils' (p. 42). It will be interesting to see how this statement relates to the outcomes of the long-term evaluation of the new curriculum in Ontario. Barlex and Pitt also point to the use of *coordination* strategies between the two subjects, so that the timing of the treatment of topics of mutual interest is orchestrated and common vocabularies are developed. Successful coordination could, they suggest, be followed by the use of *collaboration* strategies so that some activities between the two subjects are taken in common. Work on *the nature of models and modelling* could readily fall within both these frameworks.

There are, as far as we know, no extensive materials yet available for the compulsory-education sector that take a modelling approach to the teaching and learning of either science or design and technology, although some preliminary materials did appear at one time (Gilbert 1983). However, materials do exist that seek to forge general links between science education and D&T education, some starting from science education (Bath University 1992; Salters Science Teaching Project 1999) and others from D&T education (Nuffield 1995).

Specific strategies for promoting modelling

There is evidence that both students (Grosslight *et al.* 1991) and their teachers have an inadequate understanding of the nature of 'model'. Without an appreciation of the consensus view on 'model' held by scientists and by design technologists, little can be accomplished. The development of an understanding of the consensus view can be seen to involve progress in a number of *aspects*, the extreme positions in each aspect being:

- *from* a model as a copy of reality *to* a model as a mental image;
- *from* a model can only exist in the concrete model *to* a model may exist in a range of modes;
- *from* the entities of which model are composed are derived from specific objects *to* the entities of which models are composed are concepts;

- *from* a model can only be of an object *to* a model can be of an object, idea, system, event or process;
- *from* a model is made as a representation of reality *to* a model is made in order to make predictions or to present a solution to a problem;
- *from* the valuation of a model is conferred by an individual *to* the valuation of a model is conferred by a community of scientists or design technologists;
- *from* a model cannot be modified *to* a model can be modified when explanatory or representational inadequacies emerge;
- *from* a model exists in isolation *to* a model often exists within a historical sequence, either within a field of science or within a family of solutions to a problem in design and technology.

Although there currently exists no body of work on how to promote an understanding of the nature of models we do have some clues about how to go about it. Abd-El-Khalick and Lederman's (2000) review of the literature into the development of the concept of the *nature of science*, albeit by teachers, suggests that a mixture of *explicit* methods, where the ideas are directly taught, and *implicit* methods, where the ideas are encountered during other tasks, is the best way forward.

There seem to be no accounts of the direct teaching of the *model of model*. Dagher (1985) reported on changes found when the word *model* was used by students as they were introduced to teaching models – in her terminology *teaching analogies*. Providing students with extended experience in the use of *modelling kits* in chemistry, as advocated by Laszlo (2000), may have the desired effect. It was found that, as one consequence of a model-centred, computer-supported, semester-long course for a Grade 6 class in the US, many of the students developed the consensus view (Raghaven and Glaser 1995; Raghaven *et al.* 1998a,b).

There is a need for the development of direct teaching sequences, of indirect approaches, and of the orchestration of the two. Such approaches will need to include the *model of analogy* used (Goswami 1992). An evaluation of the effectiveness of such attempts will need the development of a research-based, effective and efficient means of establishing students' *model of the nature of model*. It is only when such approaches and techniques are in place that theoretically important issues in the developmental trajectory of understanding can be addressed.

Mathewson (1999) has suggested a series of categories of *master images* used in science education – for example, boundaries, branching, chirality, circuits, containers, coils, colour and 14 others. He has also suggested a series of *visualization techniques* used in science education: data display, data manipulation, encoding; gestalt, location, ordering, perceptual extension, reference point and signs. Both he and others (Tuckey and Selvaratnam 1993; Kelsey 1997; Piburn *et al.* 2002) suggest the direct teaching of such master images and visualization techniques. This would include activities that provide experience of the physiological experiences of vision (focus, resolution, peripheral vision, colour) and visualization techniques (shadows, mental rotations, reflection). It has been suggested that frequent (e.g. three times a week) and short (e.g. 10–30 minutes) sessions of such activities are the most effective (NSF 2001).

There is a need for these activities to be collected together from their origins, often in university departments of psychology and orchestrated for this important task. Only then will it be possible to look at the developmental trajectory of key mental skills, such as rotation and reflection, across both science education and D&T education.

There has been at least one study of science teachers' views on the teaching of modelling (Justi and Gilbert 2002). It is to be hoped that its findings, coupled with the ideas presented in this paper, will lead to classroom-based, efficient and effective approaches. It is only when these are available that the ambitions set out in this paper will be capable of realization.

Conclusion

In this paper we have argued that the process of modelling can serve both to forge desirable links between science education and D&T education and to provide a medium through which creativity can be fostered. Bringing about these links will not be easy. Both science education and D&T education are steeped in history, but in accordance with changes in *universal culture*, both are, themselves, subject to change. Thus, while Barlex and Pitt (2000), in their analysis of the relationship between the two, consider that 'science is essentially explanatory in nature whereas design and technology is aspirational' (p. 42), Nichol (1998) proposes that 'creativity concerns a sensitivity to similarities and differences'. Very significantly, the success of bringing the subjects together in educational settings demands the application of creative minds: here lies the potential for excitement and radical developments.

D&T, as such, is served by many specialists with different interests and episte-mological positions. Much scientific research is undertaken to fulfil the demands of society for new products and systems or to solve the problems in the implemen-tation of new technological solutions to, for example, environmental pollution. For *education systems* to serve their learners well, ways must be derived to serve the similar and different aspects of both subjects.

'We believe that the importance of context in the learner's experience is para-mount. First, this must be achieved through offering *authentic learning*: education that is coherent, personally meaningful, and purposeful, within a social frame-work' (Hennesy and Murphy, cited in Turnbull 1998: 30). This will allow learners to be sensitized to the essential relationships between scientific and D&T knowledge content, embedded as they are in the processes of application of that knowledge to the definition and solution of problems. Turnbull recognizes how school activity is usually different from the activity of practitioners, and the tasks worked on by learners become part of a classroom culture significantly removed from their origins. Second, there are similarities and differences between creativity in science and in D&T that reflect the aims and purposes of the subjects. In science, creativity is concerned with 'creative science experiments, creative problem finding and solv-ing and creative science activity and must depend upon scientific knowledge and skills' (Hu and Adey 2002: 392). In D&T, creative work can be found applied to the derivation and solution of ill-defined problems. Solutions are never right or wrong, only better or worse, and rely often on having a feel for the market place (Davies 2000). We believe that success in promoting effective learning in both science education and D&T education rests on an increasing emphasis on the understanding and use of the role of modelling.

References

Abd-El-Khalick, F.F. and Lederman, N. (2000). Improving science teachers' conceptions of the nature of science: a critical review of the literature. *International Journal of Science Education*, 22, 665–702.
Archer, B. (1980). The mind's eye: not so much seeing as thinking. *Designer*, 8(9), 8–9.

Barlex, D. and Pitt, J. (2000). *Interaction: The Relationship between Science and Design and Technology in the Secondary School Curriculum*. London: Engineering Council and Engineering Employers' Federation.

Barnea, N. (2000). Teaching and learning about chemistry and modelling with a computer-managed modelling system. In J.K. Gilbert and C. Boulter (eds), *Developing Models in Science Education* (pp. 307–324). Dordrecht: Kluwer.

Bath, University of. (1992). *Bath Science 5–16 Materials*. Walton-on-Thames: Nelson.

Baxter, M. (1995). *Product Design*. London: Chapman and Hall.

Baynes, K. (1992). The ethics of representation. In P. Roberts, B. Archer and K. Baynes (eds), *Modelling: The Language of Design*. Loughborough: Department of Design and Technology, Loughborough University of Technology.

Beetlestone, F. (1998). *Creative Children, Imaginative Teaching*. Buckingham: Open University.

Black, M. (1979). More about metaphor. In A. Ortony (ed.), *Metaphor and Thought* (1st edn, pp. 19–43). Cambridge: Cambridge University Press.

Christopherson, J.T. (1997). *The Growing Need for Visual literacy at the University*. Paper presented at Visionquest: Journeys Towards Visual Literacy. 28th Annual Conference of the International Visual Literacy Association, Cheyenne, Wyoming.

Coleman, S.L. and Gotch, A.J. (1998). Spatial perception skills of chemistry students. *Journal of Chemical Education*, 75, 206–209.

Csikzentmihayli, M. (1997). *Living Well*. London: Phoenix.

Curriculum Corporation. (1994). *Science: A Curriculum Profile for Australian Schools*. Carlton, Victoria: Author.

Dagher, Z. (1985). Review of studies of the effectiveness of instructional analogies in science teaching. *Science Education*, 79, 295–312.

Davies, T. (2000). Confidence! Its role in the creative teaching and learning of design and technology. *Journal of Technology Education*, 12(1), 19–32.

De Bono, E. (1992). *Serious Creativity*. London: Harper Collins Business.

Department for Education and Employment. (1999a). *All our Futures: Creativity, Culture and Education*. London: Author.

Department for Education and Employment. (1999b). *Science: The National Curriculum for England*. London: Author.

European Union. (2002). *Learning in the Information Society: Action Plan for a European Education Initiative*. Brussels: Author.

Feldman, D., Csikzenmihayli, M. and Gardner, H. (1995). *Changing the World: A Framework for the Study of Creativity*. New York: Praeger.

Francoeur, E. (1997). The forgotten tool: the design and use of molecular models. *Social Studies of Science*, 27, 7–40.

Fryer, M. (1996). *Creative Teaching and Learning*. London: Paul Chapman.

Gardner, P.L. and Hill, A.M. (1999a). Technology education in Ontario: evolution, achievements, critiques and challenges. Part 1: the context. *International Journal of Technology and Design Education*, 9(2), 103–136.

Gardner, P.L. and Hill, A.M. (1999b). Technology education in Ontario: evolution, achievements, critiques and challenges. Part 2: implementation and evaluation. *International Journal of Technology and Design Education*, 9(3), 201–239.

Gentner, D. (1983). Structure-mapping: a theoretical framework for analogy. *Cognitive Science*, 7, 155–170.

Gentner, D. and Gentner, D.R. (1983). Flowing waters and teeming crowds: mental models of electricity. In D. Gentner and A.L. Stevens (eds), *Mental Models* (pp. 99–129). Hillsdale, NJ: Erlbaum.

Gilbert, J.K. (1983). *Models and Modelling in Science Education*. Hatfield: The Association for Science Education.

Gilbert, J.K., Boulter, C. and Rutherford, M. (1998). Models in explanations, Part 1: horses for courses. *International Journal of Science Education*, 20(1), 83–97.

Gilbert, J.K., Boulter, C.J. and Elmer, R. (2000). Positioning models in science education and in design and technology education. In J.K. Gilbert and C.J. Boulter (eds), *Developing Models in Science Education* (pp. 3–18). Dordrecht: Kluwer.

Glynn, S.M., Law, M., Gibson, N.M. and Hawkins, C.H. (1994). *Teaching Science with Analogies: A Resource for Teachers and Textbook Writers* (Instructional Resource No. 7). Atlanta, GA: National Reading Research Center, University of Georgia.

Goswami, U. (1992). *Analogical Reasoning in Children*. Hillsdale, NJ: Lawrence Erlbaum.
Grosslight, L., Unger, C., Jay, E. and Smith, C.L. (1991). Understanding models and their use in science. *Conceptions*, 28(8), 799–822.
Handy, C. (1995). *The Empty Raincoat*. Reading: Arrow Books.
Hesse, M. (1966). *Models and Analogies in Science*. London: Sheen and Ward.
Hodson, D. (1992). In search of a meaningful relationship: an exploration of some issues relating to integration in science and science education. *International Journal of Science Education*, 14(5), 541–562.
Holton, G. (1995). Imagination in science. In G. Holton (ed.), *Einstein, History, and other Passions* (pp. 160–184). Woodbury, NJ: American Institute of Physics.
Hu, W. and Adey, P. (2002). A scientific creativity test for secondary school students. *International Journal of Science Education*, 24, 389–403.
Johnson-Laird, P.N. (1983). *Mental Models*. Cambridge: Cambridge University Press.
Justi, R. and Gilbert, J.K. (2002). Modelling, teachers' views on the nature of modelling, and implications for the education of modellers. *International Journal of Science Education*, 24(4), 369–387.
Kelsey, C.A. (1997). Detection of visual information. In W.R. Hendee and P.N.T. Wells (eds), *The Perception of Visual Information* (2nd edn, pp. 33–55). New York: Springer.
Kimbell, R., Stables, K., Wheeler, T., Wosniak, A. and Kelly, V. (1991). *The Assessment of Performance in Design and Technology: Final Report*. London: Schools Examination and Assessment Council.
Koestler, A. (1964). *The Act of Creation*. London: Pan.
Kress, G. and Van Leewen, T. (1996). *Reading Images: The Grammar of Visual Design*. London: Routledge.
Lakoff, G. and Johnson, M. (1980). *Metaphors we Live by*. Chicago, IL: University of Chicago Press.
Laszlo, P. (2000). Playing with molecular models. *Hyle*, 6(1), 1–11.
LeBoeuf, M. (1994). *Creative Thinking*. London: Judy Piatkus.
McKeller, P. (1957). *Imagination and Thinking: A Psychological Analysis*. London: Cohen and West.
Mathewson, J.H. (1999). Visual–spatial thinking: an aspect of science overlooked by educators. *Science Education*, 83(1), 33–54.
Miller, A.I. (1987). *Imagery and Scientific Thought*. Cambridge, MA: MIT Press.
Ministry of Education. (1993). *Science in the New Zealand Curriculum*. Wellington: Learning Media.
National Science Foundation (NSF). (2001). *Molecular Visualization in Science Education*. Washington, DC: Author.
Nersessian, N. (1984). *Faraday to Einstein: Constructing Meaning in Scientific Theories*. Dordrecht: Matinus Nijhoff.
Nichol, L. (ed.). (1998). *On Creativity*. London: Routledge.
Nuffield. (1995). *Nuffield Design and Technology Project Materials*. Harlow: Longmans.
Piaget, J. (1962). *Play, Dreams, and Imagination*. New York: Norton.
Piburn, M.D., Reynolds, S.J., Leedy, D.E., McAuliffe, C.M., Birk, J.P. and Johnson, J.K. (2002). *The Hidden Earth: Visualization of Geologic Features and their Subsurface Geometry*. Paper presented at the Annual Meeting of National Association for Research in Science Teaching, New Orleans.
Raghaven, K. and Glaser, R. (1995). Model-based analysis and reasoning in science: the MARS curriculum. *Science Education*, 79(1), 37–61.
Raghaven, K., Sartoris, M.L. and Glaser, R. (1998a). Impact of the MARS curriculum: the mass unit. *Science Education*, 82(1), 53–91.
Raghaven, K., Sartoris, M.L. and Glaser, R. (1998b). Why does it go up? The impact of the MARS curriculum as revealed through changes in student explanations of a helium balloon. *Journal of Research in Science Teaching*, 35(5), 547–567.
Rogers, C. (1959). Towards a theory of creativity. In H. Anderson (ed.), *Creativity and its Cultivation* (pp. 69–82). New York: Harper and Rowe.
Rutherford, F. and Ahlgren, A. (1990). *Science for all Americans*. New York: Oxford University Press.
Salters Science Teaching Project. (1999). *Salters' Science Focus Materials*. Oxford: Heinemann.

Seltzer, K. and Bentley, T. (1999). *The Creative Age: Knowledge and Skills for the New Economy*. London: Demos.

Tuckey, H. and Selvaratnam, M. (1993). Studies involving three-dimensional visualisation skills in chemistry. *Studies in Science Education*, 21, 99–121.

Turnbull, W. (1998). The place of authenticity in technology in the New Zealand curriculum. *International Journal of Technology and Design Education*, 12(1), 23–40.

Unsworth, L. (2001). *Teaching Multiliteracies across the Curriculum*. Buckingham: Open University Press.

Van Oostveen, R., Corry, A., Bencze, L. and Ayyavoo, G. (2000). Teaching a combined science–technology curriculum. *OISE Papers in STSE Education*, 1, 145–159.

CHAPTER 9

RESEARCH AND DEVELOPMENT ON SATELLITES IN EDUCATION

Gilbert, J.K., Temple, A. and Underwood, C. (eds), *Satellite Technology and Education*. London: Routledge, 1991, pp. 205–218

The current situation

From the pioneering work at Kettering Grammar school in the 1950s and 1960s, activity in the field of satellites in education grew slowly. Enthusiasts built their own ground-stations and collected data, sometimes just to show that the equipment actually worked. However, the pace of new initiatives and the general level of activity has risen sharply since the early 1980s. This can be traced to a number of influences. The evolution of a truly global system of telecommunications has focused attention there, and within the UK, British Telecom has developed a series of well-produced materials to explain the basis of its enterprise. Public news broadcasts have made increasing use of cloud-pattern photographs in weather forecasting and of pictures of the earth's surface to show ecological events, both of which are produced by remote sensing satellites.

These and other highly visible public events have encouraged equipment manufacturers to produce user-friendly prefabricated ground-stations at relatively low cost. Individuals have raised the money to purchase such equipment, are introducing the equipment and the data so collected into the curriculum, and are gradually establishing networks with those who are like-minded. So far it is all *ad hoc*. The wheel is re-invented daily. Equipment is purchased and then not used for the benefit of students. Opportunities are lost. Good practice goes unapplauded and without emulation. Bad practice continues, unnoticed or unremarked. There is, in short, a grave shortage of appropriate research and development.

Research on 'satellites in education' might ask such questions as: Who is doing what, how, with whom, to what effect? What consequences has this activity for the curriculum for teaching, for learning generally? Development, which might proceed concurrently or consecutively with research might produce and test resource packs for different groups of teachers, so that they can exploit the potential of satellites when using varying teaching styles, e.g. project work or supported self-study, within their subjects, produce computer software with which to analyse and present data in differing ways, and produce in-service education materials and courses with which to disseminate good practice.

Whilst the need for research and development (R&D) would be unquestioned in an industrial context, the lack of a tradition of systematic R&D in education requires that it be justified in terms of the interests of participating groups. For teachers and their students, in whatever part of the educational system, this should lead to more appropriate and effective teaching and learning. For those charged with curriculum development, the place, value and methods of conduct of satellite-related work can be demonstrated. The legion of informal educators, in museums, zoos and the media, might see a value for satellites within their activities which do so much to influence public attitudes to science and technology. The hidden army of radio amateurs might see where their contribution may best be made, for they have in-depth relevant experience. Satellite engineers may be inspired to render their creations more educationally user friendly. Equipment manufacturers may see what is really required to support worthwhile education. Why, then, is so relatively little R&D currently taking place?

To some extent, the reason must include the nature of the satellite engineering industry. It is necessarily hardware driven, with the emphasis placed on the optimum design and production of electronic systems to undertake specific tasks. It often seems to overlook, or ignore, the software or human aspects of its artefacts. The introduction of any new piece of technology, e.g. the car cell-phone, does carry with it implications for changes in how people manage their lives, e.g. for safe and effective use. Anything to do with social science, including education, is neglected, perhaps because it is seen as soft, i.e. incapable of multiple reproduction in identical form. This is to be very much regretted, not least because the abrasive encounters that citizens have with new technology does colour their inclination to support the funding of new ventures of the same kind. Another major contribution to the reason must stem from a lack of awareness of how satellites, and their data, can contribute to the main educational trends of our times. Educationalists either do not propose appropriate features to satellite designers or, where these are included, do not exploit the potential thus made available. I shall attempt to rectify these issues in what follows, drawing both on contemporary documents and on the implications of earlier chapters in this book.

Satellites and the new education

Throughout the world, and certainly most markedly in the UK, the recent trend in educational reform has been to attempt to make it directly supportive of industrial development or regeneration. These reforms are sometimes referred to as the 'new education' (Tomlinson 1986) and have a number of loosely associated attributes. The first of these is a move to provide an 'education in technology' which will have the following three aims:

1   the development of an awareness of technology and of its implications as a resource for the achievement of human purposes and its dependence on human involvement in judgements and decision-making;
2   to develop in pupils, through personal experience, the practical capability to engage in the central task actions of the processes of technology;
3   to help pupils learn to acquire those resources of knowledge, of physical and intellectual skills, of personal qualities and of experience which need to be available for calling upon when engaged in the task actions of technology.

(Manpower Services Commission 1987)

A second attribute of the new education is a greater reliance on experiential learning, especially that which involves out-of-institution activity, including work within the general community. This links directly to the third attribute, a greater emphasis on vocationalism, meaning either a preparation for the world of work or as viewing industry as an opportunity for study. A fourth attribute involves a shift in teaching methods away from didactic exposition (chalk and talk) towards participative techniques with an emphasis on active learning, including the use of group work. The valuation placed on problem-solving techniques of teaching and learning raise it to the status of a fifth attribute. The placing of these trends within a modular design of the curriculum so that students can perceive a clear relationship between goals accepted and their realization (the sixth attribute), the reliance on criterion-referenced forms of assessment (the seventh attribute) and a greater involvement of the students themselves, their parents and the community generally in the organizing of opportunity and the negotiation of curricular choice (the eighth attribute) complete the profile of the new education.

Satellites, ground-stations, and the reception, storage, processing, display and interpretation of data collectively offer a valuable route to the realization of the new education. I now consider each attribute in turn.

The provision of education in technology

Satellites as such are the product of technology. Handling the data that they produce offers opportunities to students, likely to be seen as very relevant to today's world, to engage in the processes of technology. Whilst the younger among them may begin by constructing models of satellites, older students may construct ground-stations from circuit drawings or bring commercially available kits into use. Data produced or merely transmitted by satellites offer a window into modern computer-managed information technology. In short, a broad awareness of space, space science and the development of space technology can be readily provided.

The use of more experiential learning

The general use of telecommunications within society, whether the well-established international network of telephones or the rapidly emerging continent span of television reception, as well as the use of weather maps, e.g. by farmers and fishermen, offer a wide diversity for experiential learning involving satellites in commerce and industry. In addition to multi-national and major national companies, many small enterprises now exist within the service sector of industry. Many are approachable for educational support: the problem is to avoid overloading those that are willing to provide opportunities for teachers and students.

Greater emphasis on vocationalism

The decision by the UK Government in 1987 not to increase its financial commitment to the space industry may lead it to a relative or absolute decline. It may be that the financial institutions of the UK will provide substitute investment. In any event, other European countries, the USSR and the USA are increasing their investment, and this will lead to the creation of more jobs in satellite engineering, communications and remote sensing. These are strong enough pointers to justify recommending to young people that they consider entering the relevant fields of employment, and of giving them a flavour of what is entailed during their formal education. Satellite engineering is, of course, an excellent advertisement for modern engineering practice generally; engineers of all types are constantly in demand.

The use of more active methods of learning

The data transmitted or generated by satellites offer a wealth of material which can be stored, processed, displayed, analysed and interpreted by students working alone or in groups with a teacher as a manager of resources and as a consultant. Certainly students must not be allowed to drown in the sea of highly encoded data that satellites generate or the undifferentiated flood of television programmes or opportunities for telephone conversations that they make accessible. A graduated and structured introduction will be needed. However, once basic skills have been acquired, there is much to say for a progressive introduction to the wild excess of informational possibilities that exist. After all, skills of differentiation, selectivity and evaluation are at the core of life in today's communication-rich environment.

The development of problem-solving skills

As the Third Industrial Revolution has produced information at an exponential rate, much of which may make a claim for inclusion in educational provision, the search has begun in earnest for generative skills – ways of locating, structuring and deploying information of all types in order to answer questions posed or needs identified. This has crystallized into a call for the teaching of 'problem-solving skills', and generalized algorithms for this activity set have appeared. Whilst I am far from convinced that humans use only one way of solving problems, or indeed that the concept of 'problem' is capable of a single definition, such ambitions are worthwhile, if only because factual information is treated as a route to an answer and not as the answer in itself. Satellite-related data, as we have already seen, are capable of being utilized to address a wide variety of questions in analysing a plethora of contexts and situations and thus are capable of making a valuable contribution to an exploration and development of problem-solving skills.

Inclusion of technology in the curriculum

The Secondary Science Curriculum Review (1987) observed that technology can be included in the curriculum in four ways: as a single subject; as an enrichment and extension offered to traditional subjects; within an interdisciplinary topic approach; within open-ended project work. Modularization, i.e. the cutting up of a theme or subject into short periods of instruction, e.g. 1 hour per week for 10 weeks, framed between aims for achievement and the assessment of learning, is compatible with any of these approaches. Indeed, until the potential of satellites is more fully realized their contribution may be confined, in many subject areas, to a few modules. However, there are plentiful opportunities for cross-curricular activities, and satellite-related work may make its greatest contribution in this way.

The introduction of criterion-referenced forms of assessment

The only unusual contribution that satellites can make in the evolution of assessment procedures from being norm referenced, where an individual's performance is graded within an anticipated structure of response from all students, to being criterion referenced, where an individual effectively competes with him/herself, is that they can produce or communicate large quantities of unique data. Thus the question set can always be based on novel information, so as to preclude the deployment of unwanted memorization skills by students.

A broader participation in the curriculum

As many parents now work in the communications or other satellite-related industries, it can be anticipated that there would be an increased willingness to support satellite-related activity when local control over the implementation of the curriculum is exercised. Indeed, the possibility of satellite-related work offers many parents and industrialists the opportunity for contribution to decision-taking in education on the basis of their own expertise.

If satellites have a broad potential for contributions to contemporary views of the desirable curriculum, how far is this being realized?

Current research on satellites in education

In a study conducted in 1987 by Gilbert *et al.* (1987) and undertaken by questionnaire and interview, to explore the needs of local education authorities (LEAs) with respect to in-service training (INSET) related to satellite work and to identify ways in which this INSET might be most advantageously provided, a 38 per cent response was obtained to a questionnaire sent to 380 science and humanities advisers and Technical and Vocational Education Initiative (TVEI/TVE) Coordinators. The picture that emerged was one of satellite-related work just getting under way in many LEAs with some pockets of relatively advanced activity being identified, a view substantiated by interviews and correspondence with twenty-three enthusiasts in the field. An auxiliary enquiry into the use of a television receive only (TVRO) system in ten schools with a project mounted jointly by the British National Space Centre and the Royal Signals and Radar Establishment showed a similarly uneven pattern of response.

The main enquiry, where data were collected between September and December 1987, showed that 218 schools were active in 98 LEAs, although the distribution of activity was very uneven: Dyfed had 12 active schools (perhaps due to the presence there of Annette Temple, co-editor of this book) whilst 34 LEAs had no activity recorded. Much of the work was reported to be in its infancy. The curriculum areas of geography, the sciences, modern languages and computer studies were those where satellite work was being most frequently pursued. However, over forty subject areas/themes were cited in replies. Those mentioned by 10 per cent and over are listed in Table 9.1. Whilst cross-curricular initiatives were mentioned, e.g. within a 'technology for all' course aimed at lower secondary pupils, the pattern of satellite systems from which data was received is revealing (Table 9.2). It would seem that many schools are still only using data from one type of satellite. The early stages of development of satellite-related work is revealed in the 106 replies to a question on the type of activity being pursued (Table 9.3). Whilst the level of project work is encouraging and interschool communication networks are being set up, many schools are still bringing equipment into use and trying to make sense of data received; the issues of curriculum exploitation have not yet been extensively addressed.

The amount of support needed for the potential of satellites to be realized is shown by the figures on INSET activity: only 32 per cent of replies reported that INSET was underway. In order of decreasing contribution, the providers of INSET were LEA advisory staff, school teachers, higher-education based, British Telecom sponsored, provided by data base, conferences, commercial demonstrations and through teacher secondment. In a similar order the preferred methods of delivery for such INSET were distance learning materials, consultancy to local working groups, day-length courses and weekend courses.

Table 9.1 Subject areas in which satellite work was pursued

Area	Mentioned, as percentage of all replies
Geography	53
Information technology	48
Physics	47
General sciences	41
Computer studies	32
French	22
General studies	21
German	19
Mathematics	13
Modern languages (overall)	10

Table 9.2 Satellite systems from which data was received by schools

Satellite system	Mentioned, as percentage of all replies
Weather (NOAA, Meteosat)	89
Scientific/educational (UoSAT)	42
TVRO (Eutelsat)	32
Imaging (Landsat, SPOT)	25
Amateur communications (JAS)	15
Weather, scientific, TVRO	15
Other (Cosmos)	1

Table 9.3 Types of activity being pursued in satellite-related work

Activity type	Mentioned, as percentage of all replies
Building systems from commercial equipment	71
Collecting and interpreting weather/surface images	64
Collecting and interpreting scientific data	38
Project work on satellites	38
Reception of TV broadcasts	35
Construction of equipment from plans	33
Collecting and interpreting all types of data	13
Communication with other schools	9

The TVRO enquiry produced mixed results. It was evident that the equipment provided was not appropriate given the levels of technical support available in schools or the uses to which it was put. Nevertheless, edited versions of newscasts and other items were used in modern language classes to good effect, providing authentic examples of contemporary pronunciation within the context of current social and political issues. The editing was thought necessary by the teachers not only to make best use of class time, but also to allow for appropriate language levels and to remove material thought undesirable on other grounds.

Future development work

It is evident that three types of development work, taking that phrase in its broadest meaning, are needed.

1   Plentiful supplies of data of an appropriate type should be made available when and where required.

The commercial equipment manufacturers are now producing ground-stations which are both relatively cheap and user friendly with which to collect data from scientific research satellites, e.g. UoSAT, and from weather satellites, e.g. NOAA. Appropriate TVRO equipment is still under development. There is a clear market for cheap high-resolution earth-observation images, e.g. from Landsat, which are related to curriculum needs, as opposed to what is available as a by-product of scientific and commercial research. The management of resources will present problems if opportunities are to be fully grasped. The TVEI-related in-service training (TRIST) project 'Satellites in Education' at the University of Surrey in 1987 showed that one controlling factor in satellites work was access to micro-computers and VDUs: all too often these are held, with excessive security precautions, in one place, e.g. the 'resources room', or by one special interest group, e.g. the mathematicians. It will be necessary either to locate satellite ground-station equipment where the microcomputers are stored, which will have the disadvantage that curriculum integration of satellites work may be hindered, or vice versa, which is only possible if schools have a plentiful supply of microcomputers so that they are readily available to all subject departments.

The use of real-time data, i.e. that directly collected from a satellite at the time of use, is attractive because the relationship between transmission and reception is reinforced, and because large quantities of data are readily available. However, this does present difficulties for schools either because satellites do not transmit during the school day, e.g. the French station TV5, or because the school time-table will not allow students and signal come together. In these circumstances, schools may wish to rely on recorded signals, which will imply editing, or even, *in extremis*, that video/audio cassettes are carried from a central reception point, e.g. at a Teacher's Centre.

2   The analysis of examination syllabuses must be undertaken to show how satellites can contribute to education provided.

At school level, Cooper and Underwood (1985) have analysed public examination syllabuses for a mention of satellites or satellite-derived data. Whilst some examination requirements, particularly at A level, e.g. physics, geography and electronics, do mention satellites, many current syllabuses do not where they might. Certainly, many GCSE syllabuses refer to aspects of data handling, and there is unpublished evidence that training and vocational education (TVE) schemes, e.g. in Surrey, and joint support activity (JSA) schemes, e.g. in Dyfed, are including work that is satellite related. However, the full range of possibilities has not been fully identified.

3   Teachers should be aware of the potential contribution of satellites to education and their skills developed in order to realize that potential.

How do teachers, in general, find out about new ideas and gauge the relevance and practicability of innovations to their professional circumstances? The evidence is that the two main mechanisms are being involved in the primary development of ideas or of having the opportunity to see how innovations have worked out in practice in other institutions. How this might be done is discussed in the last section of this chapter. There is a clear need for materials, both printed and in the form of videos, to support satellite work across the curriculum. Given the universal use in schools of worksheets as the main medium of support for individual and small group work, there will be a ready market for such an output. However, worksheets are usually very highly structured, and there is also a need for a more open-ended type of material, e.g. a collection of satellite data sheets, articles on the theme from popular journals and magazines and outlines of projects suitable for school realization. The pack might include some industry-related materials, i.e. those showing the background to industrial satellite activities, and also videos of actual satellites and their support agencies, for these will not normally be accessible to students.

Future research work

Many interesting questions for future research are embedded within the previous section. In some ways this is as it should be: innovation and evaluation proceeding hand-in-hand within a model derived from engineering practice. In other ways this is not so healthy: the current climate, both within the UK and elsewhere, which presents enquiry as a luxury, is not favourable for long-term development. Strategic research is necessary to ensure that appropriate opportunities are identified and addressed in an efficient and effective manner. From a plethora of possibilities, I will set out an agenda based loosely on the five features of satellite use.

1   (a)   What population might usefully be involved with 'satellite education'?
        Secondary school pupils and higher and continuing education students. Yet one could equally well ask: What might primary school pupils obtain from satellite education? How might technical college students benefit? What might that vast majority of the population not involved in any formal education or training derive from it?
    (b)   What is actually learnt during satellite-related activities?
        If, as has been advocated, students have direct access to satellite-derived data, what psychomotor skills are developed during such work? As much activity will be based around the computer, the whole field of 'research into the processes of information technology education' is opened up.
    (c)   What satellite-related work is currently taking place?
        Given the rapid pace of developments in the field, a continuous monitoring procedure is needed.
    (d)   What similarities and differences exist between the satellite-related education and training provided in different countries?
        The moves to integrate fully the economic systems of the EEC countries and the underlying trend towards the global village point to a need for enquiries into Stevenson's idea of trans-border educational flow.

2   *The production of educational resources*   The questions here flow into each other: What resources currently exist? How can existing resources be made more widely available? What additional resources are needed? How can new resources best be produced? Systematic enquiry is needed.

3   *Technical matters*   Assuming that signal transmission is in the hands of satellite engineers, the questions to be asked thus include: How can ground-station equipment systems be evaluated? What technical support is needed for equipment to be fully exploited over a long period of time? What computer software is needed?

4   *Legal and regulatory arrangements*   This key set of issues must be addressed, at national and international levels, if extensive progress is to be made. Moreover, the outcomes of such deliberations must be presented to those active in the field in simple language. The question here is: How might this be done effectively?

5   *Finance*   The questions surrounding the financing of satellite-related work are really a subset of those involved in the broader context of education and training generally. However, as satellites are a prime example of high technology, and therefore problematic in many countries, the question may be: How can satellite-related educational activity be so presented as to relate to the legitimate values and aspirations of influential citizens who have knowledge bases other than in the sciences?

The conduct and organization of research and development

Enquiry into the educational implications of satellites falls squarely into the area of educational or applied social science research. Controversy has raged for many years about the most appropriate view of knowledge on which to base such research. On the one hand, it is argued that, in order to be researched in a valid and reliable way, social phenomena must be defined such that they reflect an assumption of an invariant occurrence throughout humanity, e.g. the notion of intelligence quotient (or IQ). This leads to enquiry procedures being adopted which mirror those thought appropriate for the study of the physical world as conceived by Newton, i.e. the isolation of variables and their manipulation against each other in an experimental manner. Whilst there are some social phenomena which have an uncontested physical reality, e.g. numbers of individual people, experience of conducting social research in this manner has led to many doubts about whether the results obtained have a predictive value, i.e. can be used as a basis for guiding future action. The alternative view of knowledge recognizes the infinite plasticity of the human mind, and accepts that the most worthwhile perceptions are those of the perceivers, and that these perceptions are almost certain to change over time, if not rapidly. This set of assumptions leads to the study of 'cases', whether of individuals or naturally occurring groups, by means of observation and interview. This type of research has been found to influence the future actions of teachers, not by direct emulation, but my metaphorical transfer to new circumstances.

From the point of view of enquiry into satellites in education, questionnaires are an appropriate way to establish obvious facts, e.g. the amount and type of equipment held by an institution and the quality and type of data captured by ground-stations. The case study approach, based on observation and interview, is appropriate for context-dependent phenomena, e.g. the contribution of satellites to the curriculum, the teaching styles which most appropriately present learning opportunities and the learning styles that students adopt. Notions of 'good practice' approaches, which seem to yield an educationally valuable process or outcome, emerge from all this.

The skills of educational research, like those in any field of enquiry, take time and the concentrated application of effort if they are to be mastered such that high-quality outcomes are to be obtained. This certainly means that enquiries into satellites in education must be associated with individuals who are skilled educational researchers – a relatively rare breed, nearing extinction in the current intellectual climate. However, it would be entirely appropriate for practising teachers, perhaps on full- or part-time secondment, to carry out some, if not all, of the research. Their own experience will yield a ready capacity to identify important features of an educational situation, although safeguards against prejudice will be needed.

Traditionally, most education development work has been conducted on a 'centre-to-periphery' model. A few experts, gathered in one place, constructed a development of the basis of their perceptions of what was needed by teachers. This was then tested by teachers, modified by the experts and disseminated to teachers by means of highly structured courses, the intention being that the innovation was adopted and implemented *in toto*. Alas, painful experience has shown that this model is flawed: the experts misjudge the requirements of the educational system, teachers do not see the underlying assumptions of the innovation and introduce them in such a way that the novelty is nullified, and schools vary enormously in their circumstances and requirements. The alternative, or 'periphery-to-centre', model allows for a loose guidance by experts, fully negotiated with teachers, with the latter actually undertaking the development. The Secondary Science Curriculum Review in the UK was an example of this approach, where both process and product had considerable influence on the development of science education. It is argued that this 'periphery-to-centre' model is the most appropriate for 'satellites in education' work.

Turning, finally, to the financing and management of work on 'satellites in education', it is obvious that the immediate outlook is dismal. Within the UK there are relatively few government agencies that fund R&D work in educational matters. Of late, these have been priority driven, i.e. they will only fund work in areas of their own initiative and choice. Given the plethora of other demands on a miniscule R&D budget, satellites are unlikely to achieve a high priority. However, other alternatives do exist. Industrial companies may be brought to see that such educational work encourages the emergence of a constituency of support for satellite engineering, telecommunications and remote sensing. Additionally, pan-European initiatives, e.g. the Olympus satellite, will produce opportunities for education-related funding. Perhaps the overriding need is to ensure, by normal professional networking augmented by conferences which draw together diverse groups, that funding agencies are mutually aware and willing to work cooperatively in support of broad ventures.

What seems to be needed is a few centres specializing in satellites in education. These would be lightly staffed, but over a sufficient period to ensure stability for sustained research and development, and augmented by a core of consultant experts. These centres could form a focus for teacher secondments, with the outcomes of work being disseminated within a cascade model driven by the use of data bases, e.g. National Education Resource Information Service (NERIS), distance learning materials, mobile practical work units, and access to practitioner – consultants. Several such centres exist within the UK: the National Resource Centre on Satellites in Education, the Dyfed Satellite Education Project Centre and the Lancashire Satellites in Education Centre. Based on the energy of enthusiasts, they need a small, but continuous, dose of financial fertilizer to realize the enormous potential of the educational opportunities afforded by the exciting technological advances which seem likely to be sustained well into the twenty-first century.

References

Cooper, A. and Underwood, C. (1987). *Satellites and Current Examination Syllabuses.* National Resource Centre on Satellites in Education, University of Surrey, Guildford.

Gilbert, J., Underwood, C. and Sweeting, M. (1987). *The Report of a Feasibility Study into the Development of Models of INSET Concerning Earth-Orbiting Satellites in Education and Training.* University of Surrey, Guildford, p. 30.

Manpower Services Commission. (1987). *Technology for TVEI.* Manpower Services Commission, London.

Secondary Science Curriculum Review. (1987). *Technology and Science in the Curriculum.* Secondary Science Curriculum Review, London, p. 5.

Tomlinson, J. (1986). Changes in education. In I. Jamieson and D. Blandford (eds), *Education and Change.* CRAC, Cambridge.

PART 3

INFORMAL EDUCATION IN SCIENCE AND TECHNOLOGY

CHAPTER 10

INFORMAL CHEMICAL EDUCATION

Stocklmayer, S. and Gilbert, J.K. 'Informal chemical education'. In J.K. Gilbert, O. De Jong, R. Justi, D.F. Treagust and J.H. Van Driel (eds), *Chemical Education: Towards Research-Based Practice*. Dordrecht: Kluwer, 2002, pp. 143–164.

Introduction

We will define 'informal chemical education' as taking place entirely outside the purview of the examination orientated, highly structured, system of schools, colleges and universities. It is acquired, for example, through watching television, reading books, magazines and newspapers, visiting museums and science centres. Its characteristics are that the learner has control over what is learned, as well as why, when, how and where, learning takes place.

As we came to write this chapter we were dismayed to find that chemistry, of the major sciences, has the lowest profile in the provision of opportunities for informal science education. Moreover, little systematic research has been done into what is currently available, how the public responds to that provision, and what else is needed. Whilst we have collected together the little research evidence that does exist, much of what follows was created by our enquiries and through talking to interested individuals. This being the case, it is inevitable that we have mainly drawn evidence from our own countries – Australia and the United Kingdom – although the situation does seem roughly similar throughout the world.

Attitudes to chemistry

It is evident that many individuals have poor images of chemistry and negative attitudes towards it. To some considerable extent this image has its origins in the school system. For example, in 2000, the UK House of Lords Select Committee on Science and Technology investigated public attitudes to chemistry through the conduct of small 'focus groups' in four UK cities. They concluded that: 'Most people's perceptions of chemistry were derived from school, and most recalled school chemistry as boring' (House of Lords 2000, Section 2.22). Even teachers were found to regard chemistry in a negative light:

> Most of the primary teachers saw chemistry as a difficult and boring subject, pursued by intelligent but unimaginative people. Of the secondary teachers, the women tended to be more negative than the men. Many of the teachers displayed 'green' sympathies.
>
> (House of Lords 2000: 16)

The treatment of chemistry within the formal educational system is not addressed in this chapter. Until the solutions proposed there are widely implemented, the provision of opportunities for informal chemical education will be the main way of changing negative attitudes. Such attempts are necessary because of the major significance of chemical ideas for personal, social and cultural life: we take up the issue of justification at the end of this chapter.

Some efforts have been made to combat this negative image of chemistry. For example, in 1995–1996 the UK Royal Society of Chemistry (RSC) mounted a three-month long public relations campaign in the UK town of Huddersfield to: 'raise the awareness of the people of Huddersfield of the central role played by science and the profession of chemistry in enhancing the well-being and prosperity of society at large' (Webster 1996: 1).

The evidence from the RSC campaign suggests that people were more positive about chemistry in general after those initiatives (Webster 1996). But, as Dr Tom Inch, the RSC General Secretary, told the House of Lords Select Committee: 'In Huddersfield we were still looking for a short-term fix. We have to go for more fundamental solutions to the problem'.

In seeking a basis for those solutions we have to look to experience elsewhere, that is, to successful informal education through themes from biology and (especially) physics.

Narrative, context and situation: keys to success

The thesis that we advance in this chapter rests on a series of definitions. These are that:

- Successful informal science education involves contact with *situations* that are found to be interesting and for the understanding of which scientific ideas are central.
- These situations are transformed into *contexts* in which meaning is created – in which learning can take place – by personal mental activity as experiences take place.
- Learning takes place when a context is linked to a *narrative* in the life of an individual or in the lives of others. A physical situation that is most readily converted into a context has a close relationship to at least one existing narrative.

In the broadest sense, narrative is a way of portraying the lives of individuals, the evolution of events and the evolution of ideas. For any individual, there is a natural tendency to seek continuity across their experiences in any aspect of their life by constructing a series of narratives that link these experiences together. These narratives provide continuity, whilst simultaneously accommodating significant changes in the nature of those experiences by incorporating them into evolving *stories* of which the individual is often aware. Particularly strong and enduring stories become reified into *metaphors* in terms of which a person may describe some aspect(s) of their life (Clandinin and Connelly 2000). Because of their structural analogy to the narratives of personal experience, stories from the lives of others, for example from the history of science, can readily vicariously engage the attention of an audience.

Our analysis of the work of successful authors who take a scientific or technological theme suggests that the notion of narrative is a key to that success. They tell stories that engage the reader by portraying the interweaving nature of the evolution of people, events and ideas. Dava Sobel has told the tale of the solution of the 'longitude' problem by John Harrison, involving the interplay of events in his life, the political events of his day, and his extraordinary work on metallurgy and the technology of watches (Sobel 1995). She has also told the tale of Suor Maria Celeste, Galileo's daughter, who played a major role in the history of science by emotionally, physically, and perhaps intellectually, supporting her father as his ideas evolved (Sobel 2000). Both Steven Jay Gould and Richard Dawkins have written books of elegant essays in which Darwin's Theory of Evolution is used to interpret occurrences, events or historical trends in the natural world (Dawkins 1996; Gould 1985, 1991). Paul Davies (1995) has brought together the diverse contributions of theoretical and experimental physicists to explain the cosmos and its evolution. Heroic tales of imagination, tenacity and (often) triumph, are told in such books.

How do current opportunities for informal chemical education relate to the notions of situation, context and narrative? Introducing more narrative into informal chemical education presents two important challenges. The first of these concerns the notion of narrative itself. The simplest form of narrative is that of an individual person who is fixed in time, place and in relationships to other people, for example anything about Charles Darwin. This fixation does not readily enable complex trajectories of change in participants, time, place and relationships to be portrayed within an evolving story, for example, in the development of the idea of 'the periodic table'. The recently published *Mendeleyev's Dream* (Strathern 2000) is a scholarly and unusual exposition of this theme. Such a fixation is sometimes itself not completely justified, in that the person producing the narrative, the narrator (usually not the subject of the narrative), may, perhaps inadvertently, be selective in respect of data and of its significance. This uncertainty is magnified when the narrative covers the lives of many people. So creating narratives across people, events and places, is a difficult task, on themes from chemistry as elsewhere.

An additional problem is that in chemistry we are concerned with the forging of links between specific types of narratives, situations and the chemical substances that contribute to their explanation. The latter can be progressively considered at the macro level (e.g. as the white solid commonly called 'salt' and, in its chemical context, 'sodium chloride'), at the sub-micro level (e.g. as a lattice of sodium and chloride ions) and at the corresponding symbolic representational level (e.g. $(Na^+Cl^-)_s$). The depth of the understanding acquired will be related to the level of the explanation reached.

The second challenge lies in the traditional view of science as

> the one true source of authorised knowledge. In the service of the Enlightenment vision of reason, science has developed canons of evidence, patterns of discourse and forms of reasoning which are argued to be superior to the knowledge held in common sense and narrative accounts of the world.
> (Wallace and Louden 2000: 4)

It will thus be impossible to reconcile the traditional view of chemistry, usually portrayed as a series of chemical changes represented as chemical equations, with a narrative view of the subject that would seek to portray the lives and achievements of individuals within the times and contexts of events. Even the 'textbook' chemical equations would have to be represented as the culmination of many failed and partially successful efforts at analysis and synthesis. Yet the effort to move on from the traditional view would be worthwhile, for chemistry would then, more faithfully, be portrayed as a human endeavour set within temporal and social circumstances. This portrayal would, we argue, enable the subject to be seen as set within the general warp and weft of human creativity. We will show that the notions of situation, context and narrative are currently often poorly represented in opportunities for chemical education. To do this, and then to see what might be done to increase their contribution to future provision, we look at the various ways in which opportunities for informal chemical education is provided, starting with books.

Chemistry in popular books

There is a genre of successful books with scientific themes for children based on practical activities in the home. Here, specific situations are used instead of narratives as an organising principle. Activities involving the ideas of physics link readily to the narratives of children's experience. For example, the everyday situations of skateboards, of walking, of music, are easy to connect both to the ideas of physics and to the joys of children's lives. Activities built on chemical ideas fall less frequently into the category of everyday situations: examples are chemistry in the garden and chemistry in the kitchen. Such situations are often, however, highly contrived. Certainly it is fun to add vinegar to bicarbonate of soda, or to make red cabbage indicator, but their scope is limited. What connections to narratives can children make from these activities? In the kitchen, nearly everything is acidic rather than basic – and one never needs to test pH anywhere else but in a swimming pool. One of the best books for making connections to everyday narratives that we have seen is *The Science Chef* (D'Amico and Drummond 1995). This book is a series of simple recipes within the scope of children's ability that, along with the recipes, explains interesting things like the chemistry of why popcorn pops and how sauces thicken. Such books, which include the chemistry almost in passing and within situations from which contexts are readily constructed and linked to intrinsically interesting narratives, are rare.

There is an, as yet unmet, demand for books and perhaps activities that go beyond the home into the broader lives of young people. It was identified in an enquiry by the Royal Society of Chemistry (RSC 2001), which, although not overtly focused on chemistry, was conducted to learn how young people find out about science-related issues in their daily lives. A number of single-sex 'focus groups', each of 6–8 people aged 11–18 years, were conducted, each for 90 minutes, throughout the United Kingdom. The young people were found to be highly critical about the lack of information specifically intended for them concerning those science-related issues that they perceive to affect their lives. They would welcome materials on such issues (both printed and on the web) that were readily physically accessible and which had been produced using good principles of communication. They had narratives waiting to be used and extended by contact with suitable situations.

At an adult level, kitchen science has given rise to a number of books ranging in depth of treatment from the eminently accessible *The Inquisitive Cook* (Gardiner and Wilson 1998), which is a publication of the San Francisco Exploratorium, through *Kitchen Science* (Hillman 1989), which is a more extensive treatment, to *The Science of Cooking* (Barham 2000), which includes deep explanations. All three books organise their information under chapters about the kitchen such as 'Meats', 'Eggs' and so on. The genre of kitchen science is amusing, interesting and informative, while the books resemble many other cookery books in layout and appearance.

There is one genre of successful chemistry book, aimed at adult readers, where situation rather than narrative is a vital ingredient. Three outstanding examples are Ben Selinger's *Why the Watermelon won't Ripen in your Armpit* (2000), Joe Schwarcz's *Radar, Hula Hoops and Playful Pigs* (1999) and John Emsley's *Molecules at an Exhibition* (1998). Emsley has also written (amongst others) *The Consumer's Good Chemical Guide* (1994) and Selinger has produced *Chemistry in the Marketplace* (1998). The evident fascination of these authors with the chemical world is transparent in their books. Who could fail to be intrigued by 'Why damp clothes iron better' (Selinger 2000: 110–111); 'The worst smell in the world', a description of methyl mercaptan which Emsley (1998: 18–20) shows us in 'Gallery 1 – an exhibition of some curious molecules in the foods we eat'; or the '67 digestible commentaries in the fascinating chemistry of everyday life' presented by Schwarcz (1999). These books are similar in their intent: they take particular, often everyday situations, and show how chemistry can explain them. The authors have approached the problem of making chemistry accessible in similar ways and have arranged the content in small, bite-sized pieces with catchy titles. The books are, in their way, delightful, and may well make a contribution to informal chemical education. But they are not a good *read*, in our view because they do not have distinct narratives to provide continuity between the situations provided. Lacking these, it would be difficult to sustain concentration, to absorb them from cover to cover, and their authors clearly did not expect this to happen.

To illustrate the point that popular books on chemistry are rare, we analysed the *Guide to Popular Science Books* (Rennison 2000), which consists of titles and summaries of content of popular science books as sold by Waterstones, the major UK bookshop chain. We found that the number of books that have an overtly chemical theme was under 4% of the total. The balance of chemistry to other disciplines reflected in this analysis is echoed by displays on the shelves of 'Popular Science' in most bookshops. A quick check at any shop will confirm that chemistry is under-represented. Even in Singapore, that most technological of societies, a major bookshop features 27 shelves of popular physics, another 29 of astronomy and just 1 of chemistry.

In looking at the titles of the 13 books from Waterstones' list that had an obvious input of chemistry, it was clear that a basis of narrative, linking the background to a great chemical discovery/invention, the biography of the discoverers/inventors, the chemistry itself and the consequences of the discovery/invention, seems to be a key to the popularising of chemistry through these books.

Although this approach is evidently successful, there is another issue to be considered. For many of the books in the Waterstones' list, chemical ideas are often subsumed under a presentation of the book as being about biology. One of the problems faced by those who attempt to popularise chemistry is thus that other disciplines often subsume important chemical discoveries, so the perception that such books are 'about chemistry' is dulled if not lost. It would seem that chemists do need to reassert the place of their subject in scientific advances. Genetics is, after all, chemistry: the deep interpretation of specific biological phenomena in terms of genetics is only just beginning in earnest.

There are some excellent narrative accounts already available of major achievements in chemistry, for example, *Sexual Chemistry: A History of the Contraceptive Pill* (Marks 2000). This narrative links the political circumstances in which the female contraceptive pill was produced, the people and chemistry involved, and the social circumstances of the clinical trials and its later widespread use. The political circumstances were provided by the awareness in the 1960s of the rapidly growing world population and its consequences. The development took place in the United States and was funded by a charity, rather than by the state, because of the long history of opposition to artificial means of contraception by the Roman Catholic Church. The chemistry involves a number of individuals identifying naturally occurring substances that inhibited ovulation in animal models and culminating in the development and industrial implementation, by Pincus, of a way to synthesise the most promising chemical (progesterone). The clinical trials were surrounded with a series of ethical issues, whilst the consequent use of the pill by many women has had a major impact on their lives.

Rich and highly contextualised accounts of this nature are, however, regrettably few.

Newspapers and popular magazines

All articles that appear in newspapers and magazines do so as a result of selection by the editor, usually based on a notion of 'what the readers will be interested in'. What does this mean for the incidence and nature of articles on chemical themes? We conducted three analyses of articles with a broadly scientific content in mainstream newspapers and magazines.

In the first analysis, Masters degree students at the Australian National University's National Centre for Public Awareness of Science were asked, as part of their unit on science journalism, to take turns to clip such articles from the major Australian east-coast dailies. Every day, from 5 papers, 8–10 articles were netted and then classified according to content. A quick inspection revealed that the categories of 'physics' and 'chemistry', as such, had few articles in the file. Each of these disciplines, however, had sub-classifications that were treated separately. For physics, these included astronomy, space science, cosmology, nuclear science and radiation: categories which were obviously related to physics and for which large numbers of articles were found in the files. For chemistry, two sub-groups were chemical industry and chemical technology, but little was written about these subjects, for there were few articles. However, when we looked at forensic science, therapeutic drugs or recreational drugs, we found many articles. Equally, a look at the atmosphere, or pollution, or food and nutrition, or agriculture, or even mining, found chemistry revealing its many faces to the public. Here, finally, we could see a true reflection of media interest in chemistry – but it was so well disguised that it could only be seen by those trained to look. The situations chosen were such that the underlying chemical ideas were obscured.

In the second analysis, we looked at the 'science in the news' section of the Royal Society of London's website (http://www.royalsoc.ac.uk), a daily digest of articles on scientific themes in UK newspapers. The great majority of articles was overtly on biological themes, for example: animal experiments; biological weapons; climate change; gene research; biotechnology; foot-and-mouth disease. Again, many of these articles had an intellectual core that was drawn from chemistry, yet the connection was so deeply embedded as to be nearly invisible.

The third analysis was of the items from the regular column *The Last Word* in the weekly magazine *The New Scientist*. Here, readers can write with burning questions and have other readers offer answers. In a book of items from the column, published in 1998, questions are placed in categories such as 'Your Body', 'The Physical World' and so on (O'Hare 1998). An analysis indicates that of the 103 questions asked, 53 were about physics, 26 about zoology and 21 about chemistry. Of the last group, however, it is the way they are categorised that is most revealing. Nine are in 'Household Science' and range from smelly dustbins to making tea. Six are in 'Bubbles, Liquids and Ice' and are strictly 'physical chemistry' – organic chemistry is not featured here. The remaining six questions are scattered through other, not obviously 'chemical' areas, such as why bananas brown faster in the fridge (classified under 'Plants and Animals'). There is no category called 'The Chemical World'. Again, the situations are narrow and the chemistry is invisible.

So then, chemistry is present in the output of newspapers and magazines, but only in a heavily disguised form. That form consists of specific situations or themes that are thought to be of immediate appeal to readers. Any narrative provided is brief, perhaps inevitably, and tied to specific issues.

Chemistry on television

Television, as a medium, is capable of carrying a great range and flexibility of programming style. A large amount and diversity of information, in words and many forms of image, can be conveyed in a short time. It is thus capable of portraying chemistry simultaneously both as a human endeavour, as relating to diverse situations, and as having a disciplined structure built on an extensive factual base. But does it do so? The different genres of television seem to be successful to differing degrees.

Documentary programmes on television provide a splendid opportunity for an idea to be explored and developed. Chemistry, no less than any other science, has examples of televised themes that have, on occasion, been showcased in an exemplary manner. Best practice includes *Out of the Fiery Furnace* (CRA 1984), an Australian seven-part series about metals from earliest times through the Industrial Revolution. The narrator (the constructor of the narrative!), Michael Carlton, conveyed his own passion for the subject and had a fascinating set of anecdotes about the major players in the unfolding drama. Its heavy historical and metallurgical theme may not have appealed to everyone, but the story was magnificently and authoritatively told. There was a clear and absorbing narrative, augmented by interesting situations. Audiences for documentary programmes of this type generally are limited to the 'converted' but the programmes themselves can be engrossing, informing and entertaining.

In mainstream television, however, chemistry often features in news items or current affairs programmes about issues of public concern. Unwanted chemicals in foodstuffs, oil spills and so on hit the headlines briefly and disappear just as rapidly. The greenhouse effect, however, together with problems of the ozone layer, are long-running issues which dominate environmental concerns. Many scientists privately take issue with the general public's lack of comprehension of the chemistry involved in these two distinct processes (e.g. Public Understanding of Science 2000; Wilson 2000; Zehr 2000). Extensive efforts have been made to address the evident general confusion, particularly in respect of the actual chemistry and physics involved. Yet here is evidence (Bulkeley 2000) that the public is well able to understand the changes in behaviour which are required to address these issues, even if they have a less-than-perfect grasp of the underlying chemistry. Providing the audience with a narrative, or at least a distinctive situation from which to construct a context and hence a narrative, could lead to better understanding of the chemical ideas.

Perhaps the most confusing presentation of chemistry to the general public occurs through commercial presentations on television. Many of the advertisements included in commercial TV programmes purport to have a 'scientific' base. McSharry and Jones (2002) recently carried out a study into the extent and impact of such advertisements. A total of 1,026 advertisements carried on two of the terrestrial television commercial channels in the UK (ITV 1, Channel 4) were watched over a 10-day period. They found that 65% of the advertisements had some relevance to, or made some claim based upon, science/technology. They grouped these into: general (paints, garden products, cleaning products, hair products); domestic products (e.g. communication/electrical equipment, washing products, pet care, clothing, furniture); food and alcohol; healthcare; cars. Note the high proportion that has a basis in chemistry. McSharry and Jones then surveyed an opportunity sample of 200 people spread across the age range of 7–40 years. Of the list of science-based advertisements described to the sample, only 26% were recognised to have a 'science' content. Whilst it was not clear whether all members of the sample had seen all of the advertisements, it does seem that, despite the science education that people receive, their understanding of scientific aspects of everyday television advertisements is limited. Where the link between the situation and the chemistry is obscure, as in such cases, little connection is made between the two contexts. Narratives are evidently absent.

Educating people through (or in order to deal with) advertisements would be a formidable task, even assuming that the advertisers wish this to happen. Many major products are advertised on television. This sets the scene for the reaction of the shopper on finding the product on the supermarket shelf. What then has to be contended with? Consider this example: 'A scientifically advanced bodyguard defence system of antioxidant vitamins, amino acids and botanical herbs, which assist your body's defences by scavenging for potentially dangerous free radical cells and neutralising them.' This was printed on a widely advertised can of soft drink from New Zealand. There is a list of no less than 45 chemical ingredients on the same can. Purchasers are encouraged to consume up to five cans a day! How would a person in Auckland be educated to make sense of these emotive claims? Here's another kind of confusion:

> *Ingredients:* water, unrefined cane sugar, glucose solids, apple juice (sweetener), raspberries, stabilisers, xantham gum, guar gum, carageenan, locust bean gum, carboxy methylcellulose, emulsifiers, mono and diglycerides of fatty acids, lacto glycerides, propylene esters of fatty acids, citric acid. Free from artificial flavours, colours and animal products...

These are the contents of an English fruit sorbet. How could a person in London be educated to understand the meanings and significance of these (alchemist's?) ingredients? Even processes can be made to sound alien. For example, a bottle of 'ultra pure water' from Australia states: 'purified by a complex process involving reverse osmosis and carbon filtration plus ultraviolet and ozone sterilisation...'

When we come to cosmetics, of course, the chemical language is frightening. In Australia at least, the buying public does seem well aware that the scientific jargon on cosmetics is designed to impress and, broadly, they are not taken in by this strategy.

In general, salespeople and advertisers believe that their jargon dignifies the product and that the public is, on the whole, influenced by it. It is chemistry that is the main tool here, as the examples above reveal. Equally, it is chemistry as a branch of knowledge that is the main casualty, through alienation by the public as they feel unable to access or believe what is being written. Ingredients like 'antioxidants' become buzz-words – indeed one multinational purveyor of vitamin pills advertises simply: 'Free radicals are bad. Antioxidants are good'. The chemistry of advertisements encourages the public to feel as if they know and understand what is being said, by presenting them with situations that seem interesting, while in practice further distancing them from the underlying science. Much work by chemists, communicators, and indeed those concerned with morality in everyday life, will be needed if chemistry is to be rescued from the deep pit of public scepticism to which advertising may consign it.

Live science shows

The tradition of chemists presenting chemistry shows to the public has a long history, reaching back to Humphrey Davy, Michael Faraday and the Christmas Lectures and Friday Evening Discourses at the Royal Institution in the United Kingdom. 'Success in the lecture theatre was until the last quarter of the nineteenth century a necessity for any scientist...Faraday's lectures were a communal experience, not just an individual one' (Forgan 1985: 62–63).

As with all great science communicators, Faraday appears to have had a theatrical gift for presentation: 'Faraday appealed principally to the imagination, both public and scientific, quickening his audiences to a new capacity for experience' (Forgan 1985: 63).

His lectures were aimed at explanation – what we might, today, term the promotion of public understanding – but his capacity to engage an audience meant that he himself was the main attraction. His public was enthralled: we think that this was because he provided both a personal narrative and a narrative of the subject.

The Royal Institution has continued this tradition. Recently, however, many chemistry shows elsewhere have altered in their character. They are no longer aimed mainly at explanation, but at entertainment at the expense of explanation. The trend in shows today is to focus on the marvellous and the mystical. 'The magic of chemistry' is a popular title: rapid colour changes and loud explosions are an outstanding characteristic of such presentations. For example, an article published some years ago seems to have spawned a number of imitators. An analysis of its content illustrates the principle of 'chemistry magic'. The article, called *Stimulating Students with Colourful Chemistry* (McNaught and McNaught 1981), is a series of carefully laid out recipes for school laboratory demonstration practicals that give dramatic colour changes in solutions. The general characteristic of the reactions considered is that they are complicated, requiring a great deal of university-level chemistry to explain them fully. Whether they would 'stimulate students' at school level is a matter of opinion (there is no evaluative support for this statement supplied in the article). They would certainly baffle adults in a general audience. Overall, the emphasis is on 'chemistry as a series of conjuring tricks that defy ready understanding' rather than 'chemistry as providing insights into events of interest'. The only exception to this critique, and the only one where a strong narrative is readily evident, is a 'demonstration of chemiluminescence', a phenomenon of great importance to an understanding of life in the depths of the oceans. The 'magic' theme is echoed in a more recent book (Ford 1993) that goes so far as to exhort the prospective demonstrator to keep the explanations *secret* from the audience so as to preserve the magical illusions.

A recent audit (by the authors) of a number of videos of chemistry presentations given to diverse live audiences in Australia, the United Kingdom and the United States showed that most of them are of hour-long shows. They range over a wide variety of topics which include acids and bases, redox reactions, polymers, chromatography, spectra and so on. The 'theme' in each case appeared to be a focus on astonishing visual or sound effects, rather than on the underlying ideas. Certainly the expressions on the faces of the audiences reflect this astonishment and they are clearly enjoying themselves. When considering improvements in public awareness of chemistry, however, what are these shows actually achieving? We feel inclined to suspect that they are largely for the personal aggrandisement of the presenter. Harsh? Perhaps – but their objectives clearly cannot be linked to greater awareness or understanding of the relevance and importance of chemistry. The personal narrative of the narrator has overwhelmed any narrative of the subject.

In the hands of a broadly educated narrator with a well-developed but controlled ego, live chemistry shows, and their video spin-offs, offer a very promising way to convey narratives. The limiting factor is the availability of such people: who are the Faradays of today?

Chemistry lectures and chemistry in science festivals

Public lectures provide an opportunity for access to situations with chemical explanations which are of topical interest. They are often held in or near university towns, for that is where chemists tend to be found, and probably attract an informed audience to events usually held in the evening. This is not to dismiss them lightly, but it is to conjecture that their impact on the wider public awareness of chemistry may often be slender. Greater success can best be obtained by merging the genre with 'live science shows'. This would involve building the lecture around a coherent narrative, always employing a speaker trained in the arts of verbal communication, and including demonstrations in which some element of public participation is required, for instance, predicting what will happen before a demonstration takes place.

At science festivals, however, the lecturer reaches a broader audience – but festivals are few, on a world scale, and they are often only annual events. Even on such occasions, the main opportunity for the public to get closer to the world of chemistry is likely to be provided by industrial exhibitors who are essentially publicising their particular business. The exhibition also offers few opportunities for interaction with individual exhibits by members of the public. Who has not gazed with some indifference at display panels showing industrial chemistry as a series of flow charts and other complex diagrams, with the only concession to the audience being the employment of graphic designers? They are, in truth, most often monumentally boring. Situations familiar to the audience, or narratives that cut across people, events and ideas, might be more successful.

The 2001 Australian Science Festival in Canberra featured almost 200 events. The only ones with a chemical flavour were part of the Gourmet Science series that included the chemistry of beer, of wine, of cheese and of coffee. They were highly popular events with a general public audience and dealt with chemical reactions and processes with some practical tasting along the way. Other kinds of chemical presentations were absent, unless we include bubble shows and the like which incorporate a little chemistry in their explanations.

Chemistry in science and technology centres

Interactive science and technology centres have wrestled with the presentation of chemistry for some years. Exhibits based on chemical themes are difficult to design. They tend not to be as robust as physics-based exhibits, in that they require the refurbishment of chemicals. The time spans for phenomena that they might display do not usually occur within the attention span of the average visitor. These phenomena often do not have a visual impact. Where interaction is possible, this involves the manipulation of apparatus, even if under remote control, which requires skills that are not learned in a few seconds. For success, solutions of chemicals have to be readily mixed, any observations have to be safe, visibly clear and unambiguous. The consequences of these prerequisites are that interactive exhibits with an overtly chemical content are badly under-represented in the many science and technology centres to be found in most technologically advanced countries.

There has been one major effort in recent years to remediate this situation. The group of science and technology centres known as ECSITE (European Collaborative on Science, Industry and Technology Exhibitions) has a special chemistry feature on their website (http://www.chemforlife.org) that was funded by a European Union grant. The 15 collaborators set out to establish: 'how chemistry could be presented in museums as attractively, clearly and objectively as possible' (ECSITE website).

This ambition was based on the belief that:

> Over the last two centuries, chemistry has changed our daily lives more than any of the other sciences. Chemistry makes our world more colourful, more efficient, more reliable and safer. At the same time, however, no science is connected with more bad emotions, refusal and anxiety across wide sections of society. The object (of the discussions by ECSITE) was to develop suitable measures designed to break down people's fears and promote their trust across a broad spectrum of public opinion. Scientific principles and objective information need to be put across clearly and accessibly. Who better to meet this need than the science centres and technology museums all over Europe?
>
> (ECSITE website)

The project, called 'chemistry for life' incorporates eight 'global themes' which are intended to 'strike up an evocative and emotive direct link to everyday life; they involve, inform and clear up misunderstandings'. These themes have assertive titles such as:

- You are chemistry!
- Chemistry invents new matter à la carte!
- In chemistry there are no copies of molecules, only identical originals!
- Chemistry provides solutions to its own problems!
- Beethoven, Dante, Velasquez, Lavoisier...!

A major outcome of this collaboration was the design of a series of chemistry-based interactive exhibits, summaries of which are displayed on the ECSITE website, the full designs being purchasable through the same source. We downloaded some of these design summaries. Topics include a range of traditional chemistry themes such as *Corrosion*, *Build a Battery*, *Distillation* and *Gas Chromatography*. Our comment on these exhibits is that they all start from the *ideas of chemistry*, seeking in some cases to illustrate them with situations drawn from everyday life. The primary drive is from within chemistry itself, with the notions of context and narrative either absent or included as an afterthought. Moreover, most of these experiments and demonstrations have been in common use in secondary schools in the United Kingdom, Australia and no doubt elsewhere, for the last 40 years, in many cases much longer. Indeed, they may be partially responsible for the alienation of young people from the subject that is often the outcome of school chemistry classes. We find it very difficult to believe that any of these pedestrian exhibits will in any way meet the ambitions of the ECSITE group.

Given the distinct lack of imagination shown in all these exhibits, we find it hard to understand the grounds for the praise heaped on this project (in the website) by eminent scientists. We can only conclude that they had not seen the specifications before doing so. We understand that this website is to be withdrawn.

The alternative to such exhibits might be a thematic exhibition. In 1998, the Powerhouse Museum in Sydney had an exhibition called *Chemical Attractions* which featured sub-themes including 'chocolate', 'fragrance' and 'colour in fireworks'. Its purpose was all embracing, to: 'make people realise that nearly everything was made of chemicals and that commonplace distinctions between "natural" and "chemical" were not meaningful' (Shore 1999).

All these themes had visual displays and some interactive exhibits, which for 'chocolate' involved tasting and for 'fragrance' incorporated smelling various odours. The approach to flames and colour involved a traditional flame test as well as an interactive about the composition of fireworks. Unfortunately this exhibition was not formally evaluated. The comment from the designer was that people appeared to be having 'a good time'. We suggest that these sub-themes have merit but that, once again, they lack an overall narrative that will paint a bigger picture for visitors.

Summary of the problem and the solution

The core problem with much current provision for informal chemical education is that it starts overtly from the ideas of chemistry and then contrives to make them interesting. Traditional classroom ideas are translated into experiments, demonstrations or explanations for real-world phenomena (red cabbage indicator, make a battery and so on). When the chemistry of the laboratory is translated into situations that are thought to be familiar the result is contrived and – dare we say – unengaging. At the end of the exercise, the average person would be quite justified in saying 'So what?'. To some degree, this approach is the reverse of that adopted in other disciplines. There the theme of a problem, a human endeavour, a sequence of effort in a complex social milieu, dominates, with the scientific explanations being included as an underlying and important base for the narrative. Here the 'application' often drives the content and not vice versa. The main attempts to bring traditional chemistry into the public eye all too often seem to be pedestrian and not heroic as in other disciplines. The results are not memorable.

What is needed if informal chemical education is to be effective? First, it must be intrinsically entertaining, in that the public must *want* to participate. The public awareness of science is a personal matter in which the first step must be to engage voluntarily with a scientific experience. Whatever that experience may be, if people are to make links with some target area of chemistry which the populariser wishes to promote, they must be engaged. Further, they are unlikely to participate if the subject is irrelevant and out of context. Certainly an audience might enjoy the 'Magic of Chemistry' but the notion of genuine enhancement of personal awareness through such an encounter is unrealistic. For real promotion of personal awareness of chemistry, personal relevance must be clear. In addition, it is likely that just one experience will not be enough to achieve complete access to the desired target. Iterative processes must be employed.

It is widely believed, if only amongst those especially intersted in science education, that science is important to the public for social, historical, cultural and economic reasons. Chemistry should be a vehicle for explaining aspects of these reasons, not as a foreground focus. Careful examination of where the public commonly encounters chemical technology and chemical ideas should offer some clues as to how this might be achieved. The 'short-term fix' deplored by Dr Inch (House of Lords 2000) is contrary to all our knowledge about how people learn as adults. Our research (Gilbert and Stocklmayer 2001; Gilbert *et al.* 1999) has indicated that scientific awareness grows slowly and haphazardly, but that the critical issues are of interest and relevance.

Irwin and Wynne (1996) have demonstrated that the public will engage seriously with complex scientific issues when their lives are deeply affected. Often, this process is painful and difficult. Effective promotion of chemistry should not wait for such desperate times. Those of us who have pursued this discipline in our professional lives understand its appeal. We should not find it so difficult to share our enthusiasm for chemistry with the rest of the community. We must be ambitious in setting out our aims for informal chemical education.

Informal chemical education: is it really necessary?

We could find no comprehensive treatment of the desirable outcomes that are possible for informal chemical education. The nearest we got was a discussion of possible aims for the 'chemical literacy' of young people before they leave school (Nuffield 2001). These seem a reasonable basis for the informal chemical education of the whole population.

Nuffield (2001) gives three justifications for learning 'chemical explanations'. First, chemical explanations have significance for everyday life, for instance, with respect to the ability to interpret hazard labels on chemicals. Second, chemical explanations can help people take part in democratic debates, for instance, concerning response to new technologies (e.g. the production of a new pharmaceutical drug). Third, chemical explanations constitute a major cultural achievement that is having an impact on how humanity sees itself and the world it inhabits.

Areas of classroom interest suggested in Nuffield (2001) apply to chemistry syllabuses world wide. These areas include chemistry in health and medicine, in food and so on. We have applied the terminology of narrative and situation to these examples. These suggest narratives for everyday and democratic life and, through questions, to illustrative situations where chemical explanations may be needed. Thus:

- health and medicine (e.g. how can chemical disorders cause disease?);
- food (e.g. how do we make food taste nice?);
- materials (e.g. how do new materials help us run, swim or drive faster?);
- cleaning, mending, preserving, decorating (e.g. are some shampoos and conditioners more effective and safer than others?);
- agriculture (e.g. should we use artificial fertilisers when we grow food?);
- environments (e.g. are biodegradable plastics a good idea?);
- alternative fuels (e.g. could renewable resources ever meet our needs for energy?).

Narratives within the cultural justification for chemical literacy are concerned with the history of chemistry. It is here that the best opportunity exists for the heroic narratives suggested earlier. Linking these contextually to the themes suggested above can provide depth and richness. Many excellent themes for such narratives exist: one that occurred to us links the development of synthetic pigments and their use in painting. Such a narrative might address three aspects:

- The progressively greater impact of systematic chemistry on the availability of pigments (see Bomfield *et al.* 1990; Tate Gallery 1982). The first such colour was Prussian Blue, produced by Diesbach in about 1704, with major developments being of artificial ultramarine, by Guimet in 1828, and viridian, produced by Pannetier in 1838. The range of artificial cobalt, cadmium and chromium-based colours, all inorganic in nature, increased steadily from then on. A milestone was the introduction of organic chemistry into the field with the production of mauve as a derivative of coal-tar by Perkin in 1856. This spawned a whole rainbow of new tints that had an impact on dress, furnishings and painting (see Garfield 2000). Indeed Garfield's account, linking the biography of Perkins (the chemist) with mauve (the chemistry and its social impact) could serve as an exemplar for a major section of such a narrative.
- The development of technologies for the widespread distribution of the new colours. The introduction of collapsible metal paint tubes in the 1840s enabled paints to be taken outside the studio, whilst machine-grinding of pigments and the use of fillers enabled finer-hued pigments to be reliably produced (see Bomfield *et al.* 1990).
- The adoption of the new pigments by the Impressionists, for example Monet and Renoir, from the 1870s onwards. These pigments enabled them to portray both the brighter colours met in natural sunlight and the graduations of colour resulting from variations in light intensity. The availability of the new colours enabled them both to work outdoors and to portray subtle optical effects. This was not without criticism in the art world, for traditionalists regretted the trend (Gage 1995).

There are many other potential themes: the story of the double bond; the benzene ring; the fascinating world of forensic chemistry – the reader will be able to suggest many more. The image of chemists themselves is not often considered in popular literature. Consider the heroes (and few heroines) of physics of whom every school child is told. Their counterparts in chemistry are not made visible in school chemistry. After Dalton and Faraday, where did they all go? What about Frankland, of Imperial College, or Ingold, of University College, to take two examples widely separated in time (1850, 1960) if not by place of work.

Chemical frameworks such as Nuffield (2001) point to the *chemical ideas* that could underpin the narratives and situations suggested by all three of the justifications for chemical literacy. One major section of these ideas is concerned with the nature of matter and its relationships. The second major section of these ideas is concerned with the processes of chemistry, with what chemists do. These two sections of chemical literacy can be brought together within and outside the classroom to produce an understanding of a given narrative or situation by using: critical analytical research skills; hands-on practical skills; experimental and investigative skills; making and testing models; communication skills.

Until formal chemical education is based on the development of chemical literacy as defined in these terms, it will be up to the informal sector to make appropriate provision. How far this is possible will be established only through much more effort than is currently being made. The notions of narrative, situation and context should, however, provide a framework for the design of a provision that is educationally defensible. Only when that provision is in place there will be a possibility to conduct research into the impact of informal chemical education on public knowledge of and attitudes towards chemistry and its applications.

References

Barham, P. (2000). *The Science of Cooking*. Berlin: Springer.

Bomfield, D., Kirby, J., Leighton, J. and Roy, A. (1990). *Art in the Making: Impressionism*. London: National Gallery (pp. 51–75).

Bulkeley, H. (2000). Common knowledge? Public understanding of climate change in Newcastle, Australia. *Public Understanding of Science*, 9, 313–334.

Clandinin, D.J. and Connelly, F.M. (2000). *Narrative Enquiry*. San Francisco, CA: Jossey-Bass.

CRA (1984). *Out of the Fiery Furnace* (video). Melbourne: Conzinc Riotinto Associated.

D'Amico, J. and Drummond, K.E. (1995). *The Science Chef*. New York: Wiley.

Davies, P. (1995). *Superforce*. London: Penguin.

Dawkins, R. (1996). *Climbing Mount Improbable*. London: Viking.

Emsley, J. (1994). *The Consumer's Good Chemical Guide*. Oxford: Spektrum.

Emsley, J. (1998). *Molecules at an Exhibition*. Oxford: Oxford University Press.

Ford, L.A. (1993). *The Magic of Chemistry*. New York: Dover.

Forgan, S. (1985). Faraday – from servant to savant: the institutional context. In: D. Gooding and F.A.J.L. James (eds), *Faraday Rediscovered*. Basingstoke: Macmillan Press (pp. 51–68).

Gage, J. (1995). *Colour and Culture: Practice and Meaning from Antiquity to Abstraction*. London: Thames and Hudson (pp. 221–224).

Gardiner, A. and Wilson, S. (1998). *The Inquisitive Cook*. New York: Henry Holt.

Garfield, S. (2000). *Mauve*. London: Faber and Faber.

Gilbert, J.K. and Stocklmayer, S.M. (2001). The design of interactive exhibits to promote the making of meaning. *Museum Management and Curatorship*, 19, 41–50.

Gilbert, J.K., Stocklmayer, S.M. and Garnett, R. (1999). Mental modeling in science and technology centres: what are visitors really doing? In: S. Stocklmayer and T. Hardy (eds), *Proceedings of the International Conference on Learning Science in Informal Contexts*. Canberra: Questacon – the National Science and Technology Centre (pp. 16–32).

Gould, S.J. (1985). *The Flamingo's Smile*. London: Penguin.

Gould, S.J. (1991). *Ever Since Darwin*. London: Penguin.

Hillman, H. (1989). *Kitchen Science*. Boston: Houghton Mifflin.

House of Lords (2000). *Report of the Select Committee on Science and Society*. London: House of Lords.

Irwin, A. and Wynne, B. (1996). *Misunderstanding Science? The Public Reconstruction of Science and Technology*. Cambridge: Cambridge University Press.

McNaught, I. and McNaught, C. (1981). Stimulating students with colourful chemistry. *School Science Review*, 66, 655–666.

McSharry, G. and Jones, S. (2002). Television programming and advertisements: help or hindrance in effective science education. *International Journal of Science Education*, 24(5), 487–498.

Marks, L.V. (2000). *Sexual Chemistry: A History of the Contraceptive Pill*. New Haven, CT: Yale University Press.

Nuffield (2001). *Chemical Literacy: Towards a Working Definition*. London: Nuffield Foundation.

O'Hare, M. (ed.) (1998). *The Last Word: New Scientist*. Oxford: Oxford University Press.

Public Understanding of Science (2000). Special Edition: *Global Climate Change and the Public*, 9(3), 197–346.

Rennison, N. (2000). *Waterstone's Guide to Popular Science Books*. Brentford: Waterstones.

RSC (Royal Society of Chemistry) (2001). *Science and the Public: Learning for the Future*. London: Royal Society of Chemistry.

Schwarcz, J. (1999). *Radar, Hula Hoops and Playful Pigs*. Montreal: McGill.

Selinger, B. (1998). *Chemistry in the Marketplace* (5th edn). St Leonards, NSW: Allen & Unwin.

Selinger, B. (2000). *Why the Watermelon won't Ripen in your Armpit*. St Leonards, NSW: Allen & Unwin.

Shore, J. (1999). Chocolate, fireworks, dollars and scents: chemical informalities. In: S. Stocklmayer and T. Hardy (eds), *Proceedings of the International Conference on Learning Science in Informal Contexts*. Canberra: Questacon – the National Science and Technology Centre (pp. 112–118).

Sobel, D. (1995). *Longitude*. London: Fourth Estate.

Sobel, D. (2000). *Galileo's Daughter*. London: Penguin.

Strathern, P. (2000). *Mendeleyev's Dream*. London: Hamish Hamilton.

Tate Gallery (1982). *Paint and Painting*. London: The Tate Gallery (pp. 10–23).

Wallace, J. and Louden, W. (2000). *Teachers' Learning: Stories in Science Education*. Dordrecht: Kluwer.

Webster, S. (1996). *Public Perceptions of Chemistry: A Public Relations Campaign in Huddersfield: Pre- and Post-survey: Summary of results*. London: Royal Society of Chemistry.

Wilson, K.M. (2000). Drought, debate, and uncertainty: measuring reporters' knowledge and ignorance about climate change. *Public Understanding of Science, 9*, 1–14.

Zehr, S.C. (2000). Public representations of scientific uncertainty about global climate change. *Public Understanding of Science, 9*, 85–104.

CHAPTER 11

THE DESIGN OF INTERACTIVE EXHIBITS TO PROMOTE THE MAKING OF MEANING

Gilbert, J.K. and Stocklmayer, S. 'The design of interactive exhibits to promote the making of meaning'. *Museum Management and Curatorship*, 2001, 19(1): 41–50

It is argued that the place of interactive exhibits in science and technology centres will only be assured when their design and use is based on an empirically justified model which encompasses both entertainment and learning. In the light of research at Questacon, the Australian National Science and Technology Centre, a model for the Personal Awareness of Science and Technology (PAST) is put forward here and an application is made of PAST to existing interactive exhibits. The ability of interactive exhibits, designed using the model, to withstand current criticisms is evaluated.

Introduction

The rapid increase in the number of interactive science and technology centres (STC) in recent years in many countries (ASTC 1999) suggests that this institutional genre is prospering. However, a closer examination reveals that the number of visits made per centre is not rising correspondingly. When aggregated with over-optimistic targets for visitor numbers, there is evidence that many institutions are under such financial pressure that some may close (*The Guardian* 2000). Bradburne (1998: 237) takes an even more pessimistic view, noting that: 'the science center as it is presently constituted is a dinosaur threatened with extinction in the not too distant future'. Bradburne (1998) has six main criticisms of STCs. First, that the genre is no longer socially relevant, having been conceived of as a way of providing information within a framework of advocacy of science and technology. Second, that the current institutional model is that of the collection maintained by a curator, with phenomena being pre-selected for display. This leads to limited use of even the most interactive exhibits, with a consequent emphasis on maintaining the volume of new visitors. Third, that the emphasis is only on the outcomes of science and technology, thereby misrepresenting the processes involved in their generation. Fourth, that ideas, phenomena and artefacts, are presented out of the social context of their invention, discovery or use, thus inhibiting visitors' transfer of understanding into specific contexts. Fifth, that alternative forms of interactivity, particularly those provided through computers e.g. video games, the Internet, are now available. Sixth, that the STC is very expensive in terms of initial capital outlay, operating costs and the need for renewal in the face of changing demands. Any rebuttal of these criticisms must be based on a modern and coherent view of visitors' use of STCs. This view can, we argue, be then used to pay closer attention to the design of the interactive exhibits which are the core provision made.

The significance of science and technology centres

The literature on STCs shows deep divisions in perceptions of their significance. When viewed as providing education (e.g. by Stevenson 1991), STCs have often been conceived of in terms of contributing to the 'public understanding of science' (PUS). This phrase has two implications: that the 'public' is an *undifferentiated whole* in terms of understanding, and that 'technology' is merely the outcome of the *application of science*. From this perspective on education, visitors should, by using the interactive exhibits, acquire a more acceptable understanding of the key concepts which science has produced, the methods of enquiry used in science and the social processes by which science takes place (Millar 1996). This 'deficit model' (Wynne 1993) has recently been sharply criticised as being too restrictive (House of Lords 2000), and is certainly inappropriate, for many studies show that such learning does not *immediately* take place (e.g. Borun *et al.* 1993; Feher and Rice 1985; McClafferty 1995). With this evidence of the failure of STCs to educate, in PUS terms, they have been viewed primarily as a source of entertainment (Shortland 1987). We propose a view of STCs which unifies the two perspectives of education and entertainment. It is built on a notion of the 'personal understanding of science and technology' which we have incorporated into a model able to be used as the basis for the design of interactive exhibits.

A model of the use of an interactive exhibit

The model has four components: the *target*, the phenomenon or idea which is of interest to science or a technological artefact which is of interest to society; the *experience* which is the activity provided by the interactive exhibit; the *Personal Awareness of Science and Technology (PAST)* which a visitor brings to bear on an exhibit and is influenced by it; and the *remindings*, recalling memories of events or circumstances which are similar to those provided by the experience. We will take each of these in turn.

The target

There would appear to have been little significant discussion in the literature by exhibit designers about the *intrinsic* reasons for choosing a particular target, those stemming from the nature of that target. We may infer that the major reasons are that the phenomenon, idea or artefact known to the designer, can be represented in an exhibit which is within the capabilities of construction of the staff of the STC, and will yield an exhibit which will be robust enough when in continual use. For example, Miles (1988) only deals with how the target can be analysed rather than with its origin, whilst Hein (1998) has nothing to say directly about the basis of choice of the subject.

The main reasons currently adopted for target selection seem to be predominately *extrinsic* to their natures. In so far as a major purpose may be to provide support for school learning, those targets selected will be derived from an analysis of the curriculum of science (Ramey-Gassert *et al.* 1994). A major reason for choosing a technological product as a target with the schools' market in mind is that it provides a convenient entry point for specific scientific ideas in the curriculum (Gilbert 1992). In so far as the needs of the family group are the focus of attention, a target will be selected as a basis for exhibit design in terms of the social and psychological characteristics of the resulting exhibit. These will be conceived by reference to the family 'visit agenda' items which it may address (Dierking and Falk 1994; Falk *et al.* 1998), to the aspiration for high attraction and engagements levels in the use of the resulting exhibit (Boisvert and Slez 1995), and to a likelihood of actual family learning taking place (Borun *et al.* 1996).

The experience

The experience provided by an interactive exhibit is made possible because it is a conveniently manipulable '*analogical representation*' of a real-world phenomenon, or of a consensus model (one approved by scientific society) of a phenomenon, or of a technological product. Gilbert *et al.* (1999: 17) have described such a model as: 'a representation of an idea, object, event, system, or process. A model is formed by considering that which is to be represented (the target) analogically in the light of the entities and structures of something, which it is thought to be, like (a source)'. 'Analogical' representations used in this way have been called '*teaching models*' (Gilbert *et al.* 1998) in that their specific purpose is to facilitate learning. They have been shown, albeit within school settings, to support conceptual change – the learning of the meanings of words which are approved by the scientific community (Dagher 1994, 1995). We deal with the design of exhibits, teaching models, in greater detail later in this contribution.

The personal awareness of science and technology (PAST)

Rather than using the phrase 'the public understanding of science' we think it more realistic to use 'the personal awareness of science and technology' (PAST). This recognizes both the individual nature of the understanding of scientific matters and the status of technology. It may be defined as a set of attitudes, a predisposition towards science and technology, which is based on beliefs and feelings and which is manifest in a series of skills and intentions. The confidence with which the ramifications of PAST are explored by an individual will be governed by two factors. First, the level of skills available to the person with which to access scientific and technological knowledge. Second, the degree to which a sense of ownership of that knowledge can be personally held, as opposed to learning it as an imposition of the values of others. The nature and strength of this confidence will lead, *at some time*, to an understanding of some key ideas/products and how they came about, to an evaluation of the status of some segment of scientific and technological knowledge and to its significance for personal, social and economic life. One's PAST depends upon all of one's previous experiences, and changing one's PAST involves its active use.

Remindings

Underpinning the idea of 'analogical' representations of phenomena, consensus models and technological products, is a large body of research which deals with how people process information and use it to understand new phenomena. This research is based on the tripartite classification of memory into long-term, short-term and working memory. According to many researchers (e.g. Barsolou 1989: 93) mental processing occurs only in working memory, using information from both the other areas to make sense of new information: 'a concept is simply a particular individual's *conception* of a category on a particular occasion. And rather than

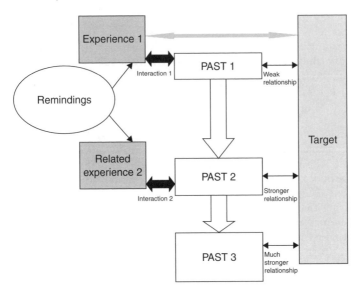

Figure 11.1 A model for PAST.

being definitional – as they are often assumed to be – concepts simply provide an individual with useful expectations about a category based on long term past experience, recent experience and current context.' Memory *retrieval*, therefore, is fundamental to this process. When a visitor uses an interactive exhibit, the *meaning* of the experience is constructed by considering what is being done in the light of both PAST and of any '*remindings*' (Ross 1989; Ross and Bradshaw 1994; Wharton *et al.* 1996) that can be retrieved from memory. These 'remindings' are of specific ideas, objects, events, processes, which seem to bear some resemblance to those being encountered in, or invoked by, the experience. A 'reminding' serves as the 'analogical' basis for an understanding of present experience.

Our model for the use of an interactive exhibit is to be found in Figure 11.1. In this model, a visitor having PAST 1 encounters Experience 1 (a teaching model) which engages PAST 1 in some way. Links between PAST 1 and Experience 1 made during the engagement are therefore quite strong and are informed by 'remindings'. The result is a new level of awareness, PAST 2. The connection made with the desired target (the consensus model) may, however, be quite tenuous at the time of the interaction with Experience 1. A further, probably later, Experience 2 related to the same target will be informed by a new, stronger set of 'remindings' (drawn from this and other, prior, experiences). These will, in their turn, increase the connection made with the target. New or refined scientific or technological knowledge will result. This iterative process can occur indefinitely, resulting in the gradual development of PAST, driven by appeal, need or interest, and this development will be idiosyncratic *for all individuals*. This idiosyncrasy will derive from the wide divergence in personal tastes, interests and needs. For adults, the core of the family visitor category, this divergence will lead to what Layton *et al.* (1986) called the learning of science 'for specific social purposes' and to the use of diverse sources of information in that learning (Hardy 1992).

The model places all the emphasis upon the 'Experiences' since they are the way in which the visitor encounters the underlying science. Good design of such experiences is critical – and good design implies facilitating interaction and understanding.

The design of 'experiences'

Interactive exhibits are constructions produced by designers through the use of analogy. An object which is capable of manipulation by visitors is produced by thinking about the target in terms of something more readily physically and mentally accessible to the designer and (it is hoped) to the visitor – the source. Not all aspects of a source can be used in the representation. Hesse (1966) divided the source of any analogy into three parts: the 'positive analogue'; the 'transferable aspects', for which some similarities are identified between the source and the target; the 'negative analog' or those aspects for which no similarity can be seen; and the 'neutral analogy', those aspects about the similarity of which no decision can be taken. Designers can only safely use the 'positive' aspects if they are to avoid confusing the visitor.

A major model for the operation of analogy, due to Gentner (1989), is based on a perceived similarity between the objects and the relationships between them in the source and in the target. In 'pure matching' – here the construction of the exhibit – all the characteristics of both the source and the target must be known by the designer. In what Gentner (1989) calls a 'mere-appearance match', only the number and attributes of the objects in the source and in the target are related to each other. An exhibit designed on this basis is restricted in what can be learnt about the target through its use. For example, an exhibit based on the phrase 'a wet road is like a mirror' tells you little beyond a similarity in the surface structure of the two. More can be learnt from exhibits based on 'literal similarity', where the similarities between both the objects and the relationships between those objects are drawn between the source and the target. For example, an exhibit based on the statement that 'a fluidized bed is like a current of air blowing through small polystyrene balls' enables both the objects and the relationships between them in the fluidized bed to be discussed in terms of both the visible nature and behaviour of the source. In the most abstract form of similarity, where the range of possible transfers of meaning is the least constrained, what Gentner (1989) calls the 'analogy similarity', only the relationships operating within the source and the target are compared to each other, no mention being made of the objects involved. An exhibit based on the phrase 'heat is like water' enables the behaviour of heat to be compared with that of water without the need to speculate on the objects of which both are composed.

Designers may legitimately use a source which only yields a 'mere appearance' similarity if their purpose is to introduce a phenomenon to visitors which it is believed will be new to them. An example is the use of 'spread of infection' analogy to introduce the idea of radioactive decomposition. A source yielding a 'literal similarity' may also be used if the target is a familiar phenomenon which is intellectually complex and it is desired to begin the facilitation of a full understanding of it. For example, as we will discuss later, the use of mechanical model of forces acting on water droplets to represent a tornado. An 'analogy similarity' will be used if the intention is to focus on the structural relationships within the target, seeking an understanding with a high level of transfer to new contexts. For example, the notion of 'energy' and its 'transfer' into different 'forms' can be represented by an analogy based on the behaviour of water. The point is that the designer must have a clear understanding of what is intended before a source is chosen.

In addition to the choice of source, the care with which the similarity to the target is represented in the exhibit will govern the success of the experience provided. In many cases the target will not be familiar to visitors. They will have to engage in what Gentner (1989) calls 'pure carry-over', attempting to explain the target only in terms of the source through the mediation of teaching model. It is therefore important that the source is familiar, indeed very familiar, to the exhibit user so that the analogy that has been drawn (the teaching model) can be understood. Even if the source is familiar to them, understanding will be helped if the number of points of comparison between the source and the target are roughly equivalent (Duit 1991; Glynn 1991; Treagust *et al.* 1992). It will of course, have to be assumed that visitors understand the principles by which analogies are drawn. There is little evidence on this key issue in respect of adults, research having apparently been concentrated on children (Goswami 1992).

How do these issues work out in the context of actual exhibits and visitors use of them? We will take some short case studies.

Experiences with interactive exhibits

Interactive exhibits can be placed in one of three groups.

Exhibits which provide a simple demonstration of a phenomenon

An example is 'Polarised Light', in which visitors are invited to place a polariser in front of a beam of light and to view various objects, e.g. plastic spoons and forks, with diffraction patterns being seen. Beyond the instructions about what to do, the graphics associated with this type of exhibit tend to be couched in terms of the consensus model, here that of the action of polariser in terms of a wave model for light. They tend to be little used. Some visitors construct meaning for the 'experience' with the help of 'remindings' derived from everyday life, e.g. their memory of polarised sunglasses. In this latter case, links between the 'experience' and the 'target' may be forged and PAST enhanced. However, it is very likely that misconceptions which have their origins in scientific language will be imported and reinforced, e.g. that 'polarisation' has, in a simple sense, to do with the 'poles' of electrostatics. A graphic couched in terms of the composition of white light and the effect of 'stripping out' some of the colours – perhaps with an analogical reference to water waves – might help.

A valuable outcome of the use of such an exhibit would be an appreciation that a phenomenon is found in more than one context and, perhaps, that an explanatory consensus model exists.

Exhibits which closely match both a real-world phenomenon and the consensus model of it

Such exhibits include, e.g. *Tornado*, in which a spiral of air moves circularly and transversely, propelled by asymmetrically placed horizontal air jets, carrying a column of water vapour produced by a generator in the base of the exhibit. This column is sucked upwards by an extractor fan at the top. This is a 'near' analogy, in that the exhibit shows a direct physical similarity to both the phenomenon and to the consensus model of it. Visitors will have strong 'remindings' of the former derived, albeit vicariously, from videos (notably of *The Wizard of Oz* in this case!), and access to the usually clear graphics provided about the consensus model. Strong links will be formed between the 'experience' and the 'target', such that PAST is developed and a better understanding of the 'target' acquired.

A valuable outcome here would be some understanding of the consensus model of how tornados occur and operate.

Exhibits based on the use of a 'far' analogy to represent the consensus model

An example of this group of exhibits is *The Light Harp* which consists of a series of small holes in which photo-sensors are embedded. About 1 m above the holes are a series of light emitters whose beams may be interrupted by the user's hands and arms. As soon as this occurs, musical sounds are heard and the user may 'play' *The Light Harp* by moving the hands across the space above the holes. The sounds are electronically generated and the user has the option to change the nature of the sounds to correspond to various musical instruments. The stated purpose of this exhibit is to demonstrate the rectilinear propagation of light. When using exhibits in this group, the visitor is expected both to understand the analogical representation and make inferences about a target that is 'far' from the source of the analogy.

Meeting a new exhibit and deciding to engage with it involves several factors. That which the visitor already knows about, and is interested in, is the most likely to promote engagement. Other factors initiating engagement may be the aesthetic appeal of the exhibit, curiosity about it and social factors (what other people are doing) (e.g. see Gilbert *et al.* 1999). When the underlying model is not obvious, and the user is not an expert in the field from which it is drawn, it is more likely that the exhibit encourages engagement through aesthetic or social factors. For example, liking the appearance of the exhibit or seeing others using it may result in visitors trying the exhibit for themselves.

Implications for exhibit designers

Our model does suggest that more attention should be paid to intrinsic reasons why particular targets are selected as the basis for exhibit design. If the purpose is to introduce those areas where science itself is currently focused, then the biological sciences (particularly ideas from the biological basis of biotechnology) should have a higher profile, as is the case with ideas from astronomy. Technologies might be more explicitly included because they have an overt significance for personal, social or economic, decision-taking, e.g. diet, genetically modified foodstuffs and the Internet-mediated global economy, respectively. The notions of 'exhibit group' and 'level of analogy', introduced earlier, can serve as the basis for the design of exhibits which have a distinctive purpose and rationale.

The model does permit the 'peaceful co-existence' of educational and entertainment purposes in the construction of interactive exhibits. If overtly 'entertainment' objectives for an exhibit succeed in provoking engagement with it by visitors who would otherwise bypass the subject, all well and good. By using an exhibit, visitors are, at the very least, laying in a store of experience with a phenomenon that may be referred to later. Taste has been altered, interest aroused and a need (albeit inchoate) created. If an entertaining exhibit engages the 'remindings' of visitors, then PAST may be changed, which is education. What is very clear from the model is that PAST, as it is defined, is entirely personal, so that the 'mass indoctrination', so beloved of the PUS movement, makes no sense. Also the educational significance of the use of exhibits arises in the long term, as opposed to the immediate or short term. These conclusions suggest that exhibit designers must be reconciled to an apparent redundancy of many of the exhibits in respect of any single visitor, and that, whilst entertainment is immediate, education may be long term.

Towards a designed future for interactive exhibits

What rebuttal of Bradburne's (1998) criticisms does our research suggest? Taking his points in order.

1   It is possible to reconstruct the advocacy tradition of STCs. They should not be viewed as places to be judged in terms of their ability to change all visitors' knowledge of a fixed agenda of scientific ideas. They can instead be viewed as places in which the tastes, interests and needs of visitors can be idiosyncratically expanded, engaged and met, through entertainment which has or can provoke individual significance.

2   Interactive exhibits have to be designed and built, so inevitably the designer has to take decisions about which phenomena, models and objects to present. We suggest that greater overt emphasis can be given to the intrinsic reasons for that choice. That said, it does seem that exhibits based on more recent scientific ideas are called for.

3   The representation of the processes of scientific discovery in STCs is problematic. There is no definitive 'scientific method' so that will never be possible. If the intention is to show 'laboratory enquiry skills' then this could be done in laboratories attached to an STC. If the intention is to show 'the processes by which scientific ideas gain acceptance' then the notion of 'cognitive reconstruction' (Nersessian 1992) does make historical case study material available: using this would lie beyond the normal remit of an STC.

4   Decontextualisation does seem inherently endemic to interactive exhibits. Embedding an exhibit in a more conventional museum display on the social background to the ideas being portrayed is one possibility. Appropriate graphics could be provided, for those interested.

5   As far as we are aware, nobody has ever claimed that the interactive exhibit is the 'only way' to portray scientific ideas. Associating interactive exhibits with other forms of provision, e.g. video, internet access, seems entirely appropriate. The issue must then be to identify what specific contribution interactive exhibits make to the development of PAST in any given field of interest.

6   Developing interactive exhibits is expensive. What we have suggested would probably increase that expense. It would then be possible, however, to demonstrate the long-term benefits of each exhibit to some, but not all, individual visitors.

References

ASTC (Association of Science and Technology Centers) (1999). *Yearbook of Science Center Statistics*. ASTC, Washington, DC.

Barsolou, L. (1989). Intraconcept similarity and its implications for interconcept similarity. In *Similarity and Analogical Reasoning*, S. Vosniadou and A. Ortonym (eds), pp. 76–121. Cambridge University Press, Cambridge.

Boisvert, D. and Slez, B. (1995). The relationship between exhibit characteristics and learning-associated behaviors in a science museum discovery space. *Science Education*, 79(5), 503–518.

Borun, M., Massey, C. and Lutter, T. (1993). Naïve knowledge and the design of science museum exhibits. *Curator*, 36, 201–219.

Borun, M., Chambers, M. and Cleghorn, A. (1996). Families are learning in science museums. *Curator*, 39(2), 122–138.

Bradburne, J. (1998). Dinosaurs and white elephants: the science center in the twenty-first century. *Public Understanding of Science*, 7, 237–253.

Dagher, Z. (1994). Does the use of analogies contribute to conceptual change? *Science Education*, 78(6), 601–616.

Dagher, Z. (1995). Review of studies on the effectiveness of instructional analogies in science education. *Science Education*, 79(3), 295–312.

Dierking, L. and Falk, J. (1994). Family behavior and learning in informal science settings: a review of the research. *Science Education*, 78(1), 57–72.

Duit, R. (1991). On the role of analogies and metaphors in learning science. *Science Education*, 75(6), 649–672.

Falk, J., Moussouri, T. and Coulson, D. (1998). The effect of museum visitors' agendas on museum learning. *Curator*, 41(2), 107–120.

Feher, E. and Rice, K. (1985). Development of scientific concepts through the use of interactive exhibits in a museum. *Curator*, 28, 35–46.

Gentner, D. (1989). The mechanism of analogical learning. In *Similarity and Analogical Reasoning*, S. Vosniadou and A. Ortony (eds), pp. 199–241. Cambridge University Press, Cambridge.

Gilbert, J.K. (1992). The interface between science education and technology education. *International Journal of Science Education*, 14(5), 563–578.

Gilbert, J.K., Boulter, C. and Rutherford, M. (1998). Models in explanations, Part 1: horses for courses. *International Journal of Science Education*, 20(1), 83–97.

Gilbert, J.K., Stocklmayer, S. and Garnett, R. (1999). Mental modeling in science and technology centres: what are visitors really doing? In *Proceedings of the International Conference on Learning Science in Informal Contexts*, S. Stocklmayer and T. Hardy (eds), pp. 16–32. Questacon, The National Science and Technology Centre, Canberra, Australia.

Glynn, S. (1991). Explaining science concepts: a teaching-with-analogy model. In *The Psychology of Learning Science*, S. Glynn, R. Yeany and B. Britton (eds), pp. 219–240. Erlbaum, Hillsdale, NJ.

Goswami, U. (1992). *Analogical Reasoning in Children*. Erlbaum, Hillsdale, NJ.

Hardy, T. (1992). Adult experiences of science and technology in everday life: some educational implications. *Research in Science Education*, 22, 178–187.

Hein, G. (1998). *Learning in the Museum*. Routledge, London.

Hesse, M. (1966). *Models and Analogies in Science*. Sheen and Ward, London.

House of Lords (2000). *Third Report of the Select Committee on Science and Society*. House of Lords, London.

Layton, D., Davey, A. and Jenkins, E. (1986). Science for specific social purposes (SSSP): perspectives on adult science literacy. *Studies in Science Education*, 13, 27–52.

McClafferty, T. (1995). Did you hear, Grandad? Children's and adults' use and understanding of a sound exhibit in an interactive science centre. *Journal of Education in Museums*, 16, 12–16.

Miles, R. (ed.) (1988). *The Design of Educational Exhibits*, 2nd edn. London, Unwin Hyman.

Millar, R. (1996). Towards a science curriculum for public understanding. *School Science Review*, 77(280), 7–18.

Nersessian, N. (1992) How do scientists think? Capturing the dynamics of conceptual change in science. In *Cognitive Models of Science – Volume XV of Minnesota Studies in the Philosophy of Science*, R. Giere (ed.), pp. 3–44. University of Minnesota Press, Minneapolis, MN.

Ramey-Gassert, L., Walberg III, H. and Walberg I, H. (1994). Reexamining connections: museums as science learning environments. *Science Education*, 78(4), 345–364.

Ross, B. (1989). Remindings in learning and instruction. In *Similarity and Analogical Reasoning*, S. Vosniadou and A. Ortony (eds), pp. 438–469. Cambridge University Press, Cambridge.

Ross, B. and Bradshaw, G. (1994). Encoding effects of remindings. *Memory and Cognition*, 22(5), 591–605.

Shortland, M. (1987). No business like show business. *Nature*, 328, 213–214.

Stevenson, J. (1991). The long-term impact of interactive exhibits. *International Journal of Science Education*, 13, 521–531.

The Guardian (2000). 'Millennium projects fall foul of optimism', *The Guardian*, August 8, 5.

Treagust, D., Duit, R., Joslin, P. and Lindaur, I. (1992). Science teachers' use of analogies: observations from classroom practice. *International Journal of Science Education*, 14(4), 413–422.

Wharton, C., Holyoak, K. and Langer, T. (1996). Remote analogical reasoning. *Memory and Cognition*, 24(5), 629–643.

Wynne, B. (1993). Public uptake of science: a case for institutional reflexivity. *Public Understanding of Science*, 2(4), 321–337.

CHAPTER 12

MODELS AND DISCOURSE*
A primary school science class visit to a museum

Gilbert, J.K. and Priest, M. 'Models and discourse: a primary school science class visit to a museum'. *Science Education*, 1997, 81(6): 749–762

A critical incident is viewed as an event which is sufficiently coherent and apparently significant, as reflected in the discourse which takes place between individuals in a group, to permit inferences to be made about the formation, use, and development of mental models. This approach was used to analyze the discourse which took place during a visit by a class of 8–9-year olds to a gallery concerned with food at the Science Museum, London. Data were collected by observation and interview before, during, and after the visit. Events that led to the initiation, continuation, and closure of discourse were identified, broad classifications produced, and inferences made about their impact on pupils' mental models. The conditions which produced evidence of mental engagement with exhibits included: The linking of the visit to the curriculum currently being followed at school; the pupils being able to follow their own itinerary in the museum; the pupils being able to work in small friendship groups; the accompanying of each group by an adult with some knowledge about the topic of the exhibits. It is suggested that a critical incident approach may be helpful more generally in analyzing the learning which takes place in groups.

Introduction

The number of organized visits by classes, particularly from primary schools (grades 1–7), to museums of all kinds has increased steadily in many countries over the past decade. Such visits enable the social value of museums to be demonstrated to the influential constituency of parents. Museums which emphasize science and technology have been exceptionally busy, perhaps because of the increasingly central place allocated to the subject in the curriculum in many countries. McManus (1992) has distinguished between those science museums organized on a thematic basis, where education is seen primarily as the provision of information (Chambers 1990), and those organized on a conceptual basis, where the intention is to support the development of specific ideas, in "science centers." While the latter type always involves some physical action by the visitor, there has been a subdivision (Rennie and McClafferty 1996) into those composed of "hands-on" exhibits, which offer no response to the visitor, and "interactive" exhibits, which offer a response and then suggest further activity. It does seems that, in the UK at least, science museum educators are designing galleries that are comprehensive in that they mix the thematic and conceptual approaches and include both hands-on and interactive exhibits.

The wide range of resources held by science museums, relevant to many topics in the curriculum, makes organized visits very attractive to schools. Unique, unusual, and even ordinary objects, replicas, and models are of great potential educational value (Durbin *et al.* 1990; Gilbert 1995). The challenge to teachers and museum educators alike is how to realize that potential, and how to improve the quality of learning achieved by pupils. Guidelines for the design, conduct, and follow-up of effective visits to museums, the prerequisites for effective learning, are well established (Bitgood 1989; Rennie and McClafferty 1995). While there are many reasons for organizing a school visit to a science museum (e.g. improving attitudes toward science, Boyd 1990), the major issue is how to conceptualize the direct contribution to learning made by such visits. Resnick (1987) differentiates sharply between the nature of "school learning" and "other learning," of which learning in a museum would be an example. Similarly, Wellington (1990) distinguishes "formal" learning, which takes place in school, from "informal learning," which takes place in museums. Dierking (1991), on the other hand, argues that such sharp distinctions are inappropriate, seeing the nature of the learning achieved to be governed by a number of factors, of which the physical setting is only one. If teachers in schools and adult companions during museum visits both see themselves as promoting meaningful activity by means of focused conversation, then it does seem very likely that the learning taking place would be similar in type and quality.

Reviews of the significance of museum (and other) visits for learning tend to combine studies conducted on recreational (family) visitors and on school parties. Studies in thematic galleries, science centers, and in comprehensive galleries are similarly combined (see, e.g. Ramey-Gassert *et al.* 1994; Rennie and McClafferty 1995, 1996). This aggregation may arise because the overall number of studies conducted internationally is relatively small. It may also suggest an implicit acceptance of the premise that learning is a general phenomenon. Such studies have focused on learning-associated behavior (e.g. Falk 1983), on learning outcomes (e.g. Tulley and Lucas 1991), and on the processes by which learning takes place (e.g. McManus 1987).

The Falk and Dierking (1992) model of learning in museums suggests that the study of these processes could concentrate on the physical context (e.g. the layout of the gallery), on the personal context (e.g. the learning style of individual visitors, McCarthy 1980), or on the social context in which learning takes place. While the study of the social context has so far tended to concentrate on the modeling of social behavior that takes place (see Falk and Dierking 1992), another fruitful line of inquiry is the social construction of knowledge (Osborne 1996).

In recent years there has been an increased concern to provide "authentic" science education based on activities that are as similar to those experienced by scientists as possible. The characteristics of scientists' activities are that they involve inquiry into ill-defined problems by a group composed of individuals of differing experience working together as a community by means of a discourse based on the use of shared knowledge and practices (Roth 1995). Museums can provide contexts that more nearly approximate those of scientific laboratories, than do most classrooms. Exhibits typically represent an address to broadly conceived, often implicit, problems from which visitors must construct personal or group concerns. A group of visitors, often composed of individuals of varying experience of the phenomena involved, are able to share prior and present understanding through focused conversation, thus engaging in the social construction of knowledge. Hands-on, and particularly interactive, exhibits provide opportunities for physical activity. It is the element of focused conversation by a group at an exhibit, the conduct of one form of discourse (Lemke 1995), which can make museums so valuable in this regard.

If learning in museums takes place to an appreciable extent through the social construction of knowledge, what is actually involved for those concerned? The study reported in this article views learning as the development and use of mental models by individuals. Processes involving mental models are shaped by the social context in which they occur. Their production and use is seen to be triggered by events, which, because they have this consequence, are called critical incidents.

Models and critical incidents

Models

A model is, in general, a representation of an idea, object, event, process, or system. A model is formed by considering that which is to be represented (the target) analogically in the light of the entities and structures of something which it is thought to be like (the source, Duit and Glynn 1996). It is used to make predictions, which are then empirically tested, such that the model is subsequently either confirmed, modified, or abandoned. A terminology for the various meanings of "model" has been developed and is used here (MISTRE 1996).

A *mental model* is an internal representation of an object, states of affairs, or a sequence of events or processes, of how the world is, and of physiological and everyday social actions. It enables individuals to make predictions and inferences, to make decisions, and to control their execution. By their very nature, mental models can only be directly appreciated by the individuals having them. This means that their nature cannot be directly probed. Inevitably, there is much debate about the nature of mental models. While Gentner and Stevens (1983) argue that there are ways in which people reason in specific domains of knowledge, Johnson-Laird (1983) thinks of them as structural analogs of the world-as-experienced, with de Kleer and Brown (1981, 1983) going further and talking of envisionment in the mind's eye as a way of describing their operation. This latter view proves to be useful to science educators, especially with respect to research in museums (Gilbert 1995). An *expressed model* is that version of a mental model expressed by an individual through action, speech, or writing, and which is therefore available for interrogation. The main characteristic of an expressed model is that it is in the public arena and therefore available for anyone to form a mental model of it.

A *consensus model* is an expressed model that has been subjected to testing by the academic community associated with a given subject and which has been socially agreed upon by at least some members of the community as having merit for the time being. Thus, consensus models are one of the main products of socially organized scientific activity. One of the main purposes of science education is to have students develop mental models of the major consensus models of science. A *teaching model* is a specially constructed expressed model used by both teachers and students to aid the understanding of a given consensus model (see, e.g. Treagust *et al.* 1995). Expressed, consensus, and teaching models, all analogies of varying status, may be put forward in any one, or combination of, the modes of representation: visual, verbal, symbolic, or material. While consensus models are the outcomes of social construction within the academic community, expressed and teaching models are subject to less stringent social verification.

Critical incidents

An individual is continuously forming, using, and revising mental models. While those models of everyday relevance are activated frequently, those that concern less significant issues will only be used infrequently. Circumstances that provoke or entail such an activation continuity of experience and will be critical in bringing about such an activation.

The use of critical incidents in educational research stems from the work of Flanagan (1954), who defined a critical incident in the following way:

> By an incident is meant an observable human activity that is sufficiently complete in itself to permit inferences and predictions to be made about the person performing the act. To be critical, an incident must occur in a situation where the purpose or indent of the act seems fairly clear to the observer and where its consequences are sufficiently definite to leave little doubt concerning its effects.
>
> (p. 327)

Flanagan's (1954) work was based on a clinical methodology set within the paradigm of behavioral psychology and led to numerous inquiries, as he recounts. More recently, in the light of the growing use of the cognitive science paradigm in educational research, new definitions have emerged. For example, Nott and Wellington (1995), within the context of a study of science teachers' thinking during the conduct of class practical work, decided to: "define a critical incident as an event which confronts teachers and makes them decide on a course of action which involves some kind of explanation of the scientific enterprise" (p. 41). Building on these perspectives we have taken the view that a critical incident is an event that is sufficiently coherent and apparently significant, as reflected in the discourse which takes place, to permit inferences to be made about the formation, use, or development of mental models, as presented in the form of expressed models, by individuals in a social group.

This definition represents a significant shift from that adopted by Flanagan (1954). In Flanagan's work, what constituted a critical incident was identified by the individual concerned, whereas here this is done by a researcher. Moreover, the present use is concerned with events that seemed coherent and perhaps significant rather than only those that constituted a major turning point or watershed.

Within the definition adopted here, a critical incident is part of a text, a bounded piece of language or discourse produced by an individual or group, which may contain spoken, gestural, or perhaps written elements (Boulter and Gilbert 1996). The purpose of the present study is to explore the nature of critical incidents which play an important part in the formation, recall, development, and use of mental models during a well-organized visit to a science and technology museum.

It was decided to conduct the inquiry in a comprehensive gallery of a science museum; that is, one which contained "thematic" elements together with "conceptual" elements (McManus 1992). This would allow the maximum scope for the deployment of different learning styles by the visiting pupils (McCarthy 1980). The visit was to be closely integrated into the science curriculum currently being experienced by the pupils in school (Bitgood 1989). It was considered important for this inquiry that the class teacher involved held the view that good teaching and learning were common in style and form to the school and the museum (Dierking 1991). By choosing the same personal and physical contexts, in that a class would experience the same curriculum and exhibits, it then would, within the Falk and Dierking (1992) model, be possible to inquire into the significance for learning of the social interactions taking place. The research would, inevitably, focus on expressed models, while seeking to make inferences about mental models (MISTRE 1996).

Context and study

The study centered on a visit, by a class of 30 pupils aged 8–9 years (grade 4) and their teacher from a state primary school located in a private housing suburb of a small city in the east of England, to the Science Museum in London. The class was just finishing the study of "healthy eating," a topic in the compulsory National Curriculum for Science in England and Wales (Department for Education 1995) for that age group. The inquiry was a case study produced by a researcher (MP) as participant observer, collecting data by interview and observation.

Preparation

The teacher followed her normal sequence of planning for any class visit to a museum. She went to the Science Museum in London and met with the education officers. It was decided to have the pupils visit that section of the Sainsbury "Food for Thought" gallery which deals with all aspects of the nature, production and marketing of bread. The decision was taken because the school is located in a wheat-growing area, so that the pupils would bring a range of extracurricular experience to the visit, as well as because the gallery contained a good collection of relevant objects and consensus models in material, visual, and symbolic forms. The class was not to be required, as is often the case, to complete a worksheet assignment during the visit, but rather allowed to experience the gallery in friendship groups of four each accompanied by an adult. The teacher, a curriculum development officer, museum education officers, an explainer, and the researcher would be available for this purpose. This arrangement would, it was felt, give pupils the maximum opportunity to express and develop their ideas within purposeful groups.

The teacher prepared the pupils for the visit by means of a special science class activity. Each pupil was given a large piece of paper, told to divide it into four numbered quadrants, and to both write and draw responses to the four questions "What would you expect to see or find out: in a museum/in a science museum/about food in a science museum/about bread in a science museum?" This activity provided a general orientation for the visit: the data arising from it are not considered here. However, it is mentioned because it did frame the expectations that the pupils had of their visit.

The visit

On arriving at the museum, the pupils were taken to a whole-class activity, staffed by an explainer, which had them examine a grain of wheat closely and explore the properties of flour, particularly the separation of gluten in water. The small groups of pupils and their accompanying adults were then free to visit the six chosen stations (exhibits) in the gallery for about 1 hour in any order that was collectively decided upon within a group at the time: station 1 (a hand-operated milling machine, or quern); station 2 (a glass case containing the ingredients of bread and a longitudinal section of a wheat grain); station 3 (an old fashioned roller mill); station 4 (a set of millstones of varying ages); station 5 (a replica of a modern in-store bakery of the type operated by the Sainsbury Company); and station 6 (various objects from the history of baking-handling implements, laws about bread, weights, and an ancient clay oven).

It had been previously agreed upon that the accompanying adults, while answering any questions asked, would not attempt to instruct the pupils to any significant extent. One of the adults (the researcher) carried a pocket tape recorder to collect as much of the resulting conversation among and with the pupils was possible in a crowded, and inevitably noisy, gallery. The transcriptions of these recordings are used later in this study.

Follow-up

Immediately after returning to school, the teacher made notes of events that could be used as the basis of follow-up activities with the pupils. The teacher's notes were made available for the inquiry. The pupils were interviewed (by MP) in an informal way during their science class about their experience of the visit as well as about the follow-up activities. These data were transcribed.

Finally, the teacher organized an evening event at the school to which the other pupils at the school, parents, and the researchers were invited. Most members of the class had produced a poster about what they had learned during the visit and subsequently. Additionally, a few of the pupils gave verbal accounts of their experiences to the assembled audience, based on their posters. Transcripts of the presentations and the posters were included in the dataset.

A full report of the inquiry is available (Priest and Gilbert 1996). Threads were drawn from the dataset so that the particular themes of this study could be addressed.

Critical incidents, discourse, and mental models

The notion of critical incident was used to analyze the discourse about food, and especially about bread, that took place both during and after the visit. Critical incidents were identified by seeking discontinuities, changes of pace and focus of attention, in that part of the overall discourse concerned with bread. This analysis is presented in terms of the discourse initiation, discourse continuation, and discourse closure, instigated by the pupils. Their possible relation to mental modeling was then explored. The examples of discourse quoted use the following convention: T (the teacher), E (the explainer), R (the researcher), and P (a pupil). Pupils are distinguished by number within a quote to show the origin of contributions: there is no continuity of the numbering system between quotes. For convenience of reference, each example of a critical incident is allocated a number. It is not claimed that the categorization is definitive, but rather that it seems to represent the range of critical incidents occurring with these pupils on this occasion.

Discourse initiation

Three types of incidents were found to be critical in allowing a section of discourse about bread involving the pupils to begin. One example, drawn from a range of examples of each type, is given.

Recognizing an object as being familiar. Recognizing an object in the gallery as something that had been encountered elsewhere led to the establishment of discourse. For example, two pupils were talking to the researcher about some objects in a display case adjacent to station 5:

CI.1

P_1: My gran's [grandmother's] got some of those...weights [pointing at a display case].

R: Does she use them?

P_1: No, she uses them as decoration. My gran's mother used to use them...

P_2: You get some scales...you put those weights on...and the bread...and see how much it weighs...so that you can then sell it.

A discourse about the relation between weight and the sale of bread has been established. This resulted from the recall by one of the pupils of a family situation with which the other pupils readily identified. Mental models seem to have been established about the basis for the division of bread for sale.

Introducing an element of surprise and providing an associated task. The whole-class activity provided by the explainer produced one major discourse, that concerning gluten. The explainer started by throwing handfuls of wheat at the class, which was most unexpected by the pupils. She then told them to examine a wheat grain closely, including its inside structure. This done, she asked the class, as small groups, to mix some flour (which she provided) with water in a bowl. This invitation produced a buzz of interest, which soon settled down in one group into a dialogue involving several people:

CI.2

P_1: There's white powder at the bottom...it looks like flour, but there's this gray horrible sticky stuff on my fingers and it won't come off.

E: What about this [holding and stretching the sticky stuff]?

P_2: Gluey.

E: ...It's called gluten and the white powder is called starch [the class continues with the task for some time].

E: Right, now we have divided the flour into starch and gluten. Did you know that gluten makes some people quite ill?

T: Oh, my mum has that and she never eats anything made from ordinary flour. She has to buy special flour...with no gluten in it.

P_3: What does it taste like, Miss?

T: We'll ask her when we get back. She only lives up the road.

E: The disease has a name. It is called coeliac disease.

This incident was critical both because it allowed new, dramatically different, information to be inserted into the dialogue and because it produced a suggestion for a postvisit activity. The pupils shared a new experience, from which mental models would seem to have been constructed, and were provided with additional information which evidently consolidated the learning that had taken place. Mental models about the divisibility of flour into starch and gluten and about limits to the nutritional value of ordinary bread were perhaps involved.

Inserting a question to focus pupils' attention. At station 1, small groups of pupils were allowed to grind wheat in a handmill (or quern). Initially, they placed all the emphasis on the speed with which they could turn the handle, with scant attention being paid to the production of flour. The explainer caused the emphasis to change:

CI.3

P_1: ...It's really hard work, you've got to do it fast.

P_2: It's like an exercise machine.

P_3: You have to use the wooden bit sticking up, a handle, really quickly...
it's just like an exercise machine for your hands.

E: Why do you use exercise machines?

P_4: To build up muscles and make you fit.

E: What do you need to do that?

P_4: Energy.

E: So you need energy to turn it and make flour.

P_4: Oh, I see....

The explainer moved away, leaving the pupils looking at the amount of flour produced at different speeds of mill rotation. A discourse about energy investment in food production was established. Experience had been provided, but, with the help of the explainer's question, mental models seemed to have been constructed that linked these key ideas.

Discourse continuation

Incidents were critical for the continuation of a discourse either because they enabled greater meaning to be attributed to past experience or because they acted as a simple bridge to later activity. Some incidents served both purposes. Five types were identified. Again, one example, drawn from a range, is given for each:

Ideas for postvisit activity are suggested.   At the end of CI.1 one boy said:

CI.4

P_4: When I get back to school I could try a fair test to see how stretchy gluten really is.

This "fair test" took place, for the word-processed text included in P_4's poster at parents' evening said (with some tidying-up of spelling to aid intelligibility):

We wanted to see how we could stretch a piece of gluten. We made some dough using flour and water. Then each person took a small piece of dough and tried to wash the starch out of it by squeezing it into a bowl of water. Starch is a carbohydrate and the gluten that is left is a protein...[together] we were left with 700 grams of gluten. Next we tried to stretch the gluten. It stretched to 91 cm in 1 minute 3 seconds before breaking.

P_4 had been allowed to retain ownership of the idea in that it was he who presented the results to the parents. He had also managed to acquire some additional information, which was included in the text. Mental models, perhaps initiated at the museum, were extended by providing additional, more focused experience, together with information derived from the relevant consensus model.

Other follow-up activities included: a visit by the teacher's mother to the class; a class visit to a working bakery at the local Sainsbury store; and class practical work arising from the visit.

The generalized and the particular are linked. Theory encountered in the museum was linked to practice as encountered in the everyday world. For example, the following dialogue took place during the postvisit interviews:

CI.5
P_1: [at the Sainsbury's in-store bakery] They said that after you've cooked [the bread] you have to check the weights...and put on the label.
R: What started this off at the museum?
P_2: I think it was plastic bread, some weights and some bags to put the bread in, and there were weights [printed] on the bags.
P_1: ...They have to put the right things on the label [at Sainsbury's] because they can't lie about what they've put on it...One of the bakers told me about the rules from Europe.

Mental models based on a rather generalized sequence of events have evidently been linked firmly by an experience to a particular instance of that same sequence.

Objects linked together. Pupils were able to link objects through perception of their common purpose. For example, having ground wheat in the quern at station 1, they later passed on to station 4, a roller mill:

CI.6
P_1: [lifting the lid of the roller mill] Oh, look, there's a roller in there. How does it work? Where's the handle? Sarah's [the explainer] machine had a handle to turn.
P_2: I think it's for grinding wheat and that's an engine [running her hand over the roller].
P_1: It could be clockwork or something mechanical [pointing to the rollerbox]. That's where the wheat goes.
P_2: That's the grinder isn't it?

Mental models have evidently been extended by using them to explain multiple, significantly divergent, realities.

Sustained attention provoked. At station 2, while most of the group members were initially discussing the ingredients of bread as laid out in a display case, one boy was observed looking at a split grain of wheat, which he had retained from the demonstration. He compared it with a diagram of the structure of an ear of wheat on the wall and gradually attracted the attention of the whole group:

CI.7
R: What are you looking at?
P_1: My grain of wheat...look [holding it up].
P_2: Look, here [moving away from the display case on ingredients]...there's a picture of one here.
P_1: It's bigger. It's an enlarged one, that.
R: What do you mean by enlarged?
P_3: They looked at it under a microscope and then sketched it big.

The example of one pupil leads to sustained attention, and hence to more learning than might otherwise have been the case.

Text is successfully consulted. Following on from CI.7, the group in question (by now unified in looking at the structure of a wheat grain) consulted the adjacent wall-mounted diagram and text more closely:

CI.8

P_1:   [pointing at the diagram] This is the bran and this is the wheat germ...
and this is the exsperm [endosperm]

R: What's the endosperm?

P_1:   It's the white flour and the bran makes it into brown flour.

P_2:   There's no gluten in the picture.

P_3:   There can't be... you have to make flour and wash it to get the gluten out.

The diagram enabled students to systematize their learning. Mental models seem to have again been developed further by comparing an object, their expressed model, and a consensus model of it.

Discourse closure

Inevitably, most of the bread-related discourse which started in school before the visit, during the visit, or even in the postvisit activities, came to an unremarked and unobserved close. This would, we suggest, have been because the natural curiosity of the pupils was satisfied or because the teacher decided that the particular section of curriculum had been covered to an appropriate extent. What were much more evident were examples in which discourse came to a premature close, where some event very evidently stopped their continuation. One major type of critical incident seemed to have the following effect.

Unsatisfactory nature of accompanying text.   When students approached an exhibit during the visit, their general strategy was to look, perhaps to touch, and only then to look around for information from an accompanying text. Even then they often asked their accompanying adult to read it to them while they continued their exploration. As they handled objects while reading took place, the pace of their activity dropped and they seemed to be formulating questions. They expected the text, however read, to answer these questions. In many cases, this strategy evidently worked well: useful answers were acquired and the discourse continued.

However, sometimes the text was not readily accessible to them for technical reasons. It may have been that the gallery was designed with an adult-size audience in mind, for during the postvisit interviews it was said that:

CI.9a

P_1:   Miss X [the teacher] read some of the labels out... she is a bit taller and could see it.

R: Was it hard to see?

P_2:   Yes, especially for Peter... he's about that big [indicating about half her height].

Some texts were inconveniently positioned:

CI.9b

P_1:   The writing wasn't near the display enough... Peter was looking at the display and at the nearest thing... it wasn't right... it was too far away from the things.

Some anticipated texts were absent. Thus, at station 4, which was a collection of millstone, the pupils turned to the accompanying diagram to find out more about each of them. In fact, it was a diagram of a waterwheel and, because it had a collection of large stones in the foreground, the pupils mistakenly thought that the display and the diagram were related, to their evident confusion. The situation was made even worse by the fact that, while the diagram had a letter-code labeling system, the key to it was absent. In both cases, additional data, needed if mental models were to develop further, was not readily available.

While these inconveniences might have slowed the pace of the discourse, there were a few instances in which the text actually interfered with the substance of the discourse. For example, at station 6, a discourse about bread-related laws was proceeding well when the pupils felt the need to consult a panel. This contained both text and a cartoon picture. What happened was as follows:

CI.10
P_1: Look, there's something about weights here and … laws. [P_1 is interrupted by P_2 looking at old prints and associated text on the wall.]
P_2: Oh, look at this funny picture. He's in the stocks and this one is being taken away in a cart [P_1 and P_2 laugh together and move on.]

The data displayed will certainly have met the needs of other visitors. Here, rather than providing information with which the pupils could develop their mental models further, it caused attention to be deflected from those models.

Discussion

The benefits of organizing the visit around small groups of pupils, each accompanied by an adult who knew something about bread, were evident in the patterns of discourse that resulted. The adults provided information where this seemed needed (e.g. CI.2); inserted questions that shaped the dialogue (e.g. CI.3); and, most importantly for this group of young pupils, read aloud from text panels (e.g. CI.9). These experts contributed to the putative "community of practice" (Roth 1985: 29). The interactions between the pupils themselves played a major role in group learning. Thus, individuals suggested ideas for postvisit activities (e.g. CI.4), acquired information needed in the discourse (e.g. CI.7), and answered questions from their peers about personal knowledge or experience (e.g. CI.6). The social construction of knowledge, mainly mediated by verbal discourse, evidently took place.

Critical incidents seemed to play roles in relation to expressed models and, therefore, by the inferences made earlier, to mental models. These critical incidents, which initiated discourse, involved the provision of a new activity, focusing an activity onto a meaning which supported the experience being had, or recalling and sharing an established mental model. The raw materials for the mental modeling of experience and the questioning of such models were provided. The critical incidents that facilitated the continuation of discourse were about links. Links were forged between activities in the school and in the museum, between the experiences being had at the museum, between objects on display and consensus models of them, and between present experiences and possible future activities. The development of mental models seemed to involve the formation of links between experiences and their interpretation: ideas were both aggregated and differentiated. The critical incidents that led to the closure of narratives involved an inability to acquire needed information or a deflection from an evident purpose. Perhaps the development of mental models was thereby inhibited.

The conditions provided for the pupils before, during, and after the visit seemed conducive to the production and exploration of expressed models. They had the benefit of the sustained, close, yet nondirectional, participation of an adult with some knowledge of the subject being studied (i.e. bread). In Granott's (1993) terms, because there was a considerable degree of symmetry of level of expertise about bread among the members of groups, there was a high level of cooperative learning. Interviews with pupils conducted after the visit, referring to aspects of the visit, coupled with the requirement that pupils produce a poster about the visit for their parents and peers, effectively constituted a stimulated recall of the experiences had. Most important, pupils had multiple and different opportunities to speak, write, and draw about their experience and its significance. The response by pupils to this continuity of experience between school and museum lends weight to the assertion that the types of learning occurring in each are similar (Dierking 1991).

In recent science education research literature, "teaching models" (or "teaching analogies" as they have otherwise been called) have been portrayed as models specially produced by teachers (or, indeed, pupils), drawing by analogy on objects and events known to the pupils, as an aid to understanding consensus models (see, e.g. Treagust *et al.* 1995). The present study enables this category to be expanded in two ways. First, many of the wall-mounted texts which accompanied exhibits provided a different type of explanation to that evident from the immediate labels attached to objects. Although these young pupils were not able to display "text echo" (McManus 1989), the reading of these wall panels by the accompanying adults did seem to give pupils access to a different interpretation of events. Second, the authentic objects or replicas on display did seem to act as "exemplary models" in that the more abstract representations of objects, events, or systems (e.g. form of diagrams) were exemplified in material form.

The use of a comprehensive gallery, in terms defined earlier, does seem to have been an important ingredient in eliciting on-task discourse during the visit, and hence in promoting the formation, use, and development of mental models. Pupils paid various levels of attention to different exhibits, indeed to different parts of any one exhibit. This may have been because different pupils had different learning styles (McCarthy 1980). To varying degrees, within and between individuals, there may have been preferences to learn by emotional response through concrete experience, by watching through the reflective observation of others, by thinking through the conceptualization of events, or through doing by experimentation.

A fuller exploration of the social construction of knowledge in museums will entail a closer examination of some of the factors that emerged in this study. It would undoubtedly be insightful to follow one small group of pupils in close detail through their preparation for, experience of, and follow-up to such a visit. Within such a study it would be possible to explore the significance of any discernible mixtures of preferred "learning styles," in McCarthy's (1980) terms, for the choice of exhibit made and for the ensuing discourse. Most interesting of all, it would be instructive to see if the above way of viewing learning in school groups had value for explaining the learning of recreational visitors. Last, a stimulated recall approach, built around the use of video, would enable visitors to directly identify which were the critical incidents for them and, it is hoped, how their thinking changed as a result.

Acknowledgments

We are most grateful to the governors, staff, parents, and pupils, of Dale Hall Primary School, Ipswich, and to the staff of the Science Museum, London, for their permission to conduct the study and for their willing participation.

Note

* The use of the term "discourse" in this article denotes conversation rather than a sociolinguistic use of the term.

References

Bitgood, S. (1989). School field trips: an overview. *Visitor Behaviour*, 4, 3–6.
Boulter, C. and Gilbert, J. (1996). Texts and contexts: framing modeling in the primary science classroom. In G. Welford, J. Osborne, and P. Scott (eds), *Research in Science Education in Europe* (pp. 177–188). London: Falmer Press.
Boyd, W.L. (1990). Museums as centers of learning. *Teachers College Record*, 94, 761–770.
Chambers, M. (1990). Beyond 'aha': motivating museum visitors. In B. Serrell (ed.), *What Research Says about Learning in Science Museums* (pp. 10–11). Washington, DC: Association of Science–Technology Centres.
de Kleer, J. and Brown, J. (1981). Mental model of physical mechanisms and their acquisition. In J.R. Anderson (ed.), *Cognitive Skills and their Acquisition* (pp. 258–310). Hillsdale, NJ: Erlbaum.
de Kleer, J. and Brown, J. (1983). Assumptions and ambiguities in mechanistic mental models. In D. Gentner and A. Stevens (eds), *Mental Models* (pp. 155–190). Hillsdale, NJ: Erlbaum.
Department for Education (1995). *Science in the National Curriculum*. London: Department for Education.
Dierking, L. (1991). Learning theories and learning style: an overview. *Journal of Museum Education*, 16, 4–6.
Duit, R. and Glynn, S. (1996). Mental modeling. In G. Welford, and J. Osborne, and P. Scott (eds), *Research in Science Education in Europe* (pp. 166–176). London: Falmer Press.
Durbin, G., Morris, S., and Wilkinson, S. (1990). *A Teacher's Guide to Learning from Objects*. London: English Heritage.
Falk, J. (1983). Time and behaviour as predictors of learning. *Science Education*, 67, 267–276.
Falk, J. and Dierking, L. (1992). *The Museum Experience*. Washington, DC: Whalesback.
Flanagan, J.C. (1954). The critical incident technique. *Psychological Bulletin*, 51, 327–358.
Gentner, D. and Stevens, A. (1983). *Mental Models*. Hillsdale, NJ: Erlbaum.
Gilbert, J.K. (1995). Learning in museums: objects, models and text. *Journal of Education in Museums*, 16, 16–18.
Granott, N. (1993). Separate minds, joint effort, and weird creatures: patterns of interaction in the co-construction of knowledge. In R. Wozniak and K. Fischer (eds), *Development in Context: Acting and Thinking in Specific Environments*. Hillsdale, NJ: Erlbaum.
Johnston-Laird, P. (1983). *Mental Models*. Cambridge: Cambridge University Press.
Lemke, J. (1995). *Textual Politics: Discourse and Social Dynamics*. London: Falmer Press.
McCarthy, B. (1980). *The 4MAT System: Teaching and Learning Styles with Right/Left Mode Techniques*. Chicago, IL: Excel.
McManus, P.M. (1987). *Communications with and between Visitors to a Science Museum*. Unpublished PhD thesis, King's College, University of London, UK.
McManus, P.M. (1989). Oh, yes, they do: how museum visitors read labels and interact with exhibit texts. *Curator*, 32, 174–189.
McManus, P.M. (1992). Topics in museums and science education. *Studies in Science Education*, 20, 157–182.
MISTRE (1996). *The Models in Science and Technology: Research in Education Group*. Reading, PA: Department of Science and Technology Education, University of Reading.
Nott, M. and Wellington, J. (1995). Critical incidents in the science classroom and the nature of science. *School Science Review*, 76, 41–46.
Osborne, J.F. (1996). Beyond constructivism. *Science Education*, 80, 53–82.
Priest, M. and Gilbert, J.K. (1996). *Some Contributions to the Formal Learning of a Group of Key Stage 2 Pupils made by a Visit to the Sainsbury "Food for Thought" Gallery at the Science Museum. London*. Reading: Department of Science and Technology Education, University of Reading.
Ramey-Gassert, L., Walberg, H., III, and Walberg, H. (1994). Reexamining connections: museums as science learning environments. *Science Education*, 78, 345–365.

Rennie, L. and McClafferty, T. (1995). Using visits to interactive science and technology centres, museums, aquaria, and zoos to promote learning in science. *Journal of Science Teacher Education, 6,* 175–185.

Rennie, L. and McClafferty, T. (1996). Science centres and science learning. *Studies in Science Education, 27,* 53–98.

Resnick, L.B. (1987). The 1987 presidential address: learning in and out of school. *Educational Researcher,* 13–19.

Roth, W.-M. (1995). *Authentic School Science.* Dordrecht, The Netherlands: Kluwer.

Treagust, D., Venville, G., Harrison, A., Stocklmayer, S., and Thiele, R. (1995). *Teaching Analogies in a Systematic Way.* Perth, Australia: Science and Mathematics Education Center, Curtin University.

Tulley, A. and Lucas, A. (1991). Interacting with a science museum exhibit: vicarious and direct experience and subsequent understanding. *International Journal of Science Education, 13,* 533–542.

Wellington, J. (1990). Formal and informal learning in science: the role of interactive science centres. *Physics Education, 25,* 247–252.

PART 4

ALTERNATIVE CONCEPTIONS AND SCIENCE EDUCATION

CHAPTER 13

ELICITING STUDENT VIEWS USING AN INTERVIEW-ABOUT-INSTANCES TECHNIQUE

Gilbert, J.K., Watts, D.M. and Osborne, R.J. 'Eliciting student views using an interview-about-instances technique'. In C. West and A. Pines (eds), *Cognitive Structure and Conceptual Change*. Orlando, FL: Academic Press, 1985, pp. 11–27

Introduction

The realization that students bring idiosyncratic meanings for words which are commonly used in science to the science classroom is of long standing. The possible consequences of this for teaching have also been established for some time – Ausubel (1968), for example, refers to them as preconceptions (that are) amazingly tenacious and resistant to extinction, implying that such views are, in some essential way, not only wrong but bad. More recently, however, another interpretation has emerged: Driver and Easley (1978), referring to them as alternative interpretations, observe that "In learning about the physical world, alternative interpretations seem to be the product of pupils' imaginative efforts to explain events and abstract commonalities they see between them" (p. 62). Whatever response teachers decide to make to them, private meanings are a genuine part of children's culture.

If a word's meaning is held before that word is met in formal science classes, we would refer to it as being part of children's science. Across a population, a range of meanings would exist for a particular word. Every child would employ a range of words, the sum of which would constitute a large part of the individual's children's science (S_{ch}). That meaning for a word which was held by a consensus of the scientific community we would describe as being part of scientists' science (S_{sc}). That version of scientist's science which is selected by curriculum planners for inclusion in a syllabus or is enshrined in a textbook could be called curricular science (S_{cr}). The interaction between curricular science and a teacher produces teacher's science (S_t). In classrooms, children's science and teacher's science interact to produce student's science (S_{st}). These representations of knowledge can be articulated (Zylbersztajn 1983) into the sequence shown in Figure 13.1.

If children's science is a major influence on the curriculum, then attempts to elucidate its characteristics are called for. Investigators seeing their task as the measurement of the extent to which students have learned the correct – or scientist's science – version of a word meaning will use comparative methods, for example, multiple-choice and open response questions. At the opposite end of the scale, investigators taking an ethnographic, or noncomparative approach, will wish to examine the use of words in real-world settings. The most usual and complex of such settings is the normal school classroom: both Tasker (1980) and Zylbersztajn (1983) are conducting such studies. The simplified classroom, that is, one where two or three students are given lesson tasks to perform in a separate room,

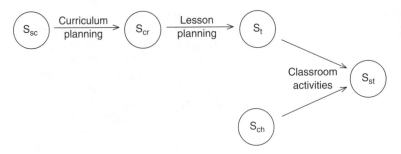

Figure 13.1 Transformations of scientific knowledge.

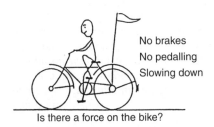

No brakes
No pedalling
Slowing down

Is there a force on the bike?

Figure 13.2 An example of an IAI card.

has also been used (e.g. by Tiberghein 1980). The middle ground between these extremes, which is represented by methodologies capable of adaptation to either a comparative or noncomparative mode, seems potentially fruitful. Certainly open response questions can be used in this way (e.g. Viennot 1979), as can a modified multiple-choice approach (e.g. Helm 1980). It is into this middle ground that we have introduced the Interview-about-Instances (IAI) technique (Gilbert and Osborne 1980b; Osborne and Gilbert 1980).

In outline, the IAI technique consists of tape-recorded dyadic discussions between the researcher and a student, using a deck of cards and focusing on the applications of a single word. A card consists of a linedrawing of a situation which may, or may not, represent an example of the application of the word. Whatever the student's response, reasons are sought. This method elicits a range of responses, as an example will show. Take, as an illustration, the card about cycling from a deck of cards dealing with force (Figure 13.2).

Typical answers are

> Yes... the wheels are still going so that there would be a force from that.
>
> (aged 9)

> It is just putting force on by itself... from the force you gave it before.
>
> (aged 11)

> Yes... the speed that he [sic] has already got up.
>
> (aged 19, Teachers College student who passed
> UK "O" Level in Physics at 16)

> There is a force because of the bike's own mass... the mass of the bike has come (to) such a speed that it won't just stop straight away... the force is still

in there...in the bike...the force was transferred from the person pedalling and it is now still adherent in the bike...the bike still moves forward.

(aged 20, Teachers College student who passed UK "A" Level in Physics at 19)

A deck of such cards, perhaps 15 in number, will reveal the breadth and manner of use of the chosen word over a range of situations.

Designing a deck of IAI cards

The initial design of a deck of IAI cards may be undertaken by following the algorithm presented below (Watts 1980a,b,c):

Identify the scientists' science meaning of the word

For force, three component viewpoints can be isolated: "force as pushes and pulls," "resultant force equals mass × acceleration," "forces are interactions of nature." Combined and extended, these produce the conclusion that "a force results from interactions between bodies capable of changing the velocity, size, of shape of the bodies."

Analyze the scientists' science view

This implies an exploration of the accepted meanings of body, interaction, and changing velocity. It also involves analyzing the view for its vagueness and ambiguity after the manner of Quine (1960) and Lachenmeyer (1971).

The vagueness of a view is a function of the extent to which the range of object predicates, which form its referential meaning, have been specified; that is, the distance between its connotative meaning and its denotative meaning. For force, the level of vagueness is low: its meaning rests partly on that of interaction which can be specified in one of four distinctive modes: strong, weak, electro-magnetic, and gravitational. The ambiguity of a view depends upon the extent to which it has multiple, equally legitimate meanings. In a physics context, the ambiguity of force is low, but for its societal meaning, the value can be said to be higher. For example, "a person was forced to agree with another," and "a person was forced to do what another person wished."

The analysis also involves clarifying the distinction between basic quantities and derived quantities, the latter being derived from, and defined by, the former. However, the base which is chosen is usually a matter of convenience. This analysis recognizes the convenience of defining force in terms of mass and changing velocity.

Identify the criterial attributes of the word in S_{sc}

Criterial attributes are the essential qualities, all of which must be recognized if the word is to be used in a way acceptable to scientists. For force, these involve specifying

1   its magnitude and direction
2   the body on which it acts
3   the body that exerts the force
4   the nature of the force, for example, the type of interaction occurring, and
5   the effects of the force, for example, changes of direction, size, shape.

Identify the non-criterial attributes of the word in S_{sc}

These are statements which are sometimes made about situations or circumstances and which involve the word in some way, for example, for force:

1   the weight of an object can be said to act through its center of gravity;
2   tension is the condition of a body subjected to equal but opposite forces which may lead to an increase in linear dimensions;
3   forces can be measured in Newtons;
4   surface tensions can be accounted for in terms of inter-molecular forces.

The number of these statements is usually great (around 30 for force, at school level), the range being decided essentially by the researcher after examining a sample of appropriate curriculum materials.

Identify sources of obvious linguistic confusion

These may be grouped into one of three types: antonyms, synonyms, homonyms. Antonyms for force include weakness, incapacity, enfeeblement, tameness. Synonyms for force include strength, power, impetus, violence, intensity, effort, military might, body of people, drive, population, validity. Homonyms for force include fours (quadruples), fauces (a cavity of the mouth), and faucet (a tap). Any of these might conceivably enter into a potential student's answer: cards should be designed with these possibilities in mind.

Identify sources of invalid use of the scientist's science meaning

The first, and probably major, source is that pool which can be called common usage meanings. These are inevitably numerous; examples for force include force of habit, force of law, forced labor, forced march.

A second group is that which may be termed misuse of the scientist's science meaning. Warren (1979) has identified a collection of misuses of force:

1   the supposition of force where none exists, to account for effects noted in a situation;
2   the omission of vector components;
3   an inadequate description of the bodies exerting the forces in the situations in which the interactions are taking place.

If "examples" of a word may be defined as instances where it could be accurately applied in the S_{sc} meaning, then "nonexamples" may be defined as instances where it cannot be so applied. For force, instances which could be considered nonexamples might be those where it is not possible to specify the following:

1   the forces acting
2   the nature of the forces
3   the magnitude of the forces and
4   the direction of the forces.

That is, where one or more of the criterial attributes is not able to be identified.

Producing cards

The parts in this sequence, which have already been outlined earlier, might be designated as an exploration of potential sources of alternative conceptions. The actual production of cards is a mixture of art and science. What is required is a mixture of examples and nonexamples. Each example would represent all the criterial attributes and some of the noncriterial attributes; for a deck of cards, the intention would be to include most or all of the noncriterial attributes. Each nonexample would have one or more of the criterial attributes absent, yet include some of the noncriterial attributes: in a deck of cards the whole spectrum of combinations of criterial attributes would be omitted, and all the non-criterial attributes included.

Therefore, the design of individual cards will depend on the insight of the researcher. *The overriding concern must be to have cards which simultaneously allow* S_{sc}, S_{ch}, *and* S_{st} *to be demonstrated.* Thus, they must be interesting to the interviewees. Ideas can be sought from the illustrations in textbooks, from television programs, and by the simple expedience of asking students and teachers for challenging situations.

Ordering cards into a deck

For interviews, particularly with students who are shown to have adopted the S_{sc} view, it has been found that the optimum deck contains fewer than 20 cards; with younger and less knowledgeable interviewees, the activity is appropriately truncated in light of the following sequence:

1 Cards 1–8: examples.
2 Cards 9–12: nonexamples. At this point the interviewee is asked for an explanation or definition of the word in his or her own words.
3 Cards 13 and onwards include more difficult instances, likely to be novel to the interviewee but looking rather like textbook physics and including borderline examples.

However, the final identification of cards and their ordering into a sequence is an interactive process forming part of the investigation design.

The process of elicitation

After designing the deck of cards, a series of pilot interviews is conducted with about 5–10 students covering the particular age range to be investigated. The purpose of these interviews is to remove simply anomalies in design, wording, and sequencing of the cards. Then a larger trial of about 15–20 interviews is conducted in which fully detailed transcriptions are made. An inspection of these will show whether individual cards, and the sequence adopted, best facilitates the presentation of the *student's own interpretation*. This leads to a final review of the deck, after which the investigation proper is conducted.

The choice of a sample of students with whom to conduct interviews is an initial challenge. The IAI approach may be seen to be a kind of case study. The problem is, what is the case? If the case is an individual student, who will perhaps be interviewed about a number of words, then any articulate individual is satisfactory. If the case is the whole of a naturally occurring group (e.g. a class of students) then no problem is encountered, for all are interviewed. However, if a selection is to be made from a class, then its basis needs careful consideration. As Pines (1980) has pointed out, interview techniques work best with articulate students and these may not be cognitively representative. If the teacher is asked to select a number of "about average" students to be interviewed, then the articulation factor is compounded by the natural desire to please the researcher by providing a satisfactory student: this often means a good one. The use of auxiliary tests to select students (e.g. Piagetian tests, verbal reasoning tests, and so forth) seems equally problematic because the relation between scores on these and children's science is unclear.

Interviews are best conducted in complete privacy, for only then are participants relaxed (and then only if they feel unthreatened). The amount of extraneous noise on a tape-recording will also be reduced to a minimum. The participants sit side-by-side while the interviewer sets the scene for the interview. We have found it useful to say something like: "You will know from your own experience how teachers sometimes use words in classes which do not agree with your understanding of those words. I want to find out how you use a particular word.... There are no right answers or wrong answers...so that we can get teachers to see your point of view." The interviewer then asks permission to record the discussion, "because I can't concentrate on what you are saying and take notes"; this is almost always acceptable.

The interview takes the form of a discussion. From an initial fairly openended question based on the first card, such as "is there a force here?" the interviewer attempts to identify, as accurately as possible, the student's perceptions. The response of the interviewer is nonevaluative, with supplemental questions being asked until the interviewer has fully grasped the student's ideas. Many of the skills needed, such as the reflective quotation, are standard in the area of counselling. As progress is made through the deck of cards, reference back to earlier cards is made either at the student's behest, to change an earlier response, or by the interviewer, if there is a glaring discrepancy between the answers given to two cards. At the end of the sequence, or whenever the interviewer has decided that further progress is unlikely, the student is offered an opportunity to revisit any of the cards. By this time the student is usually still relaxed but a little tired. The interviewer's final questions (although they can also be asked at the beginning) concern the student's prior academic achievements, age, class, previous experience of science teaching, and interest in physics. At no point is the student given any evaluation of success or failure.

The transcription of interviews

The transcriber faces a number of challenges in this vital task. First, the discussion is discontinuous, being punctuated by gestures, and faltering and irregular pausing; it shows various qualities of voice, and is delivered with a variety of facial expressions and in many body postures. Second, the sequence of conversation is difficult to follow, being full of inconsequence, confusion, pauses, and contradictions. And third, the tape recorder, whilst ignored for the majority of an interview, can occasionally become the focus of the student's attention. In short, it is inevitable that a transcript becomes a translation prepared by the transcriber.

Whilst the interviewer will have made notes on some of these features, which should be immediately related to a transcription prepared very shortly after an interview, the transcriber will need to use some precise conventions. Of these, the two most important concern the *style* and *notation* of the transcript. In style, we have found it convenient not to punctuate the transcript nor to divide the speech into sentences, and to omit fullstops, commas, and the like. The question mark is also omitted, for the verbal inflective associated with questioning is sometimes used to make statements, the truth of which is uncertain to the speaker. Pauses are typed and marked: noises (e.g. laughter) are included. The transcription is prepared with wide margins (for subsequent annotation) and marked with revolution-counter numbers from the tape recorder.

The transcription is best done in two distinct phases. First, the words and noises are presented verbatim. Then the tape is replayed, entering intonations, pauses, and comments. Irrespective of who transcribes the first part, it must be checked by the interviewer, who must undertake the second part.

The notation that we use is as follows:

1 Square brackets [] enclose information and comments added on the right hand side of the page.
2 Parentheses () enclose interpretations, where unavoidable, after the utterance to which they apply.
3 Transcriber's doubt (*) is shown by asterisks in parentheses with as much of the sound included as possible. The number of asterisks indicate the number of words.

Example 1

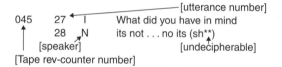

4 Periods indicate pauses. One period is used for a very short pause, thereafter the number of dots reflect the length of pause.
5 The number of seconds are indicated for long pauses by a numeral preceded and followed by two periods: . . 8 . .

Example 2

053 32 N   its a . er . . . force of some sort . . 5 . . I think.

6 A number of colons included in a word show a prolonged, or drawn out, sound.

Example 3

064 46 I   yes if someone tries. . . . well . . fo::rce you to do it.

7 Stressed syllables – this is done by underlining.

Example 4

066 48 N   oh but no that's different. . . . <u>pow</u>er is not the same thing at all.

8 A single square bracket [indicated overlapping utterances or where a second speaker cuts into another.

Example 5

102 63 I   did you. . . . . . ⌈ I mean
     N   ⌊ yes it was easy to keep doing it

Example 6

121 76 N   and the force is this   ⌈ way and
    77 I   ⌊ I see
    78 N   not the way it seems here

9 Rising tone / marks a rising inflection, not necessarily a question.
10 Falling tone \ marks a falling inflection.
11 Exclamation mark ! is used at the end of an utterance considered to have exclamatory intention.
12 Umms and ers are included as appropriate. A whistle or a sigh – the sort you might let out at the end of a hard day's digging in the garden (whew!) – these are included as whhhh.
13 Laughter, a snort, a cough, and so forth are included as (laugh), (snort), (cough), and so forth.

The analysis of a transcript

The first task of the editor (usually the interviewer) is to impose a structure on the transcript by identifying and separating the discreet utterances made. We have identified five categories in these units which are concerned with different types of talk and function in the interview. These are A, Personal; B, Task; C, Card; D, Concept; E, Framework. They are elaborated here.

A. Personal

This is all the dialogue that takes place to relate the individuals in the interview. It includes the greetings, introductions, "What courses are you doing?," "What do you want to do when you leave school?" It includes the institutional, social talk that starts and ends the meeting. The function of the talk is the necessary softening of approaches and opening of communication channels. The function of the category is as a collection of dialogue that does not fall into the other boxes. It is useful, too, for providing the overview to the whole interview. Both the personal and task categories are necessary if all the data are to be accounted for.

In network terms it can be abbreviated to

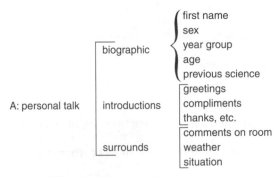

N.B { indicates "and"; [indicates "or".

B. Task

In this category, the talk concerns the task: what to do, the confidentiality of the tape, the nature of the survey, the topic of the cards, an so on. This occurs throughout the interview. The function of the talk is to establish the role-play of the task and to cement the "contract" of the interview. The success of the mission, in particular as it relates to the strategies the students adopt, will depend on this part of the session. The function of the category is, again, as a fairly loose framework for small talk.

C. Card

It is easier to describe the function of this category before describing the type of talk it involves. When a youngster is faced with one of the cards, the concentration fluctuates between what is happening in the card, to what happens generally, and then to the wider scheme of things. We wish to be able to separate local effects and observations from underlying beliefs or "articles of faith." This category we call "card," and it relates together all the talk that we think is a local effect, that is, an immediate response to details or aspects of a particular card. These context-specific responses also have another function. They carry all the evidence for the next category (D: the concepts). They bear all the hallmarks of articles of faith. For example, we quote Jonathan, 14–15 years old. The card concerns a golfer hitting a ball, and the "card" category responses have been italicized.

129	115 J	there'll be quite a few forces acting on that ball
	116 I	yeah what are ⌈*they*
	116.5 J	⌊*they're the thrust of the heating drag of the resistance*

.... yes.... there'll be gravity pushing it down and.... when it hits the ground there will be momentum keeping it going and it will be leaving far more fric by far more kinetic energy by friction because it's going through the air because the air is much much more er.... less dense than the ground its.... so what few forces acting on ⌈there ⌊tell me about

	117 I	the thrust one, you haven't you didn't mention that one before ⌈ *what's the*
	118 J	⌊*thrust is because the push is the actual thing powering it which in fact could be the club making it go so the power is transferred from club to ball and then lost in the ground.*

The student, Jonathan (J) starts with 115 "there'll be quite a few forces acting on that ball." This is a discrete utterance in that the student seems to have communicated a particular piece of understanding. It is also card specific. It means that he has signaled that the card represents his idea of forces. Second, he has singled out the ball, not the golfer. After I 116 "what are they?" he gives "...they're the thrust of the heating drag of the resistance..." That in itself is another meaning unit. It is also card specific, in that he doesn't refer to this topic again in the interview.

D. The concept

This category consists of two kinds of talk. It is explicit talk that generalizes *between* cards, for example, in defining types of forces, giving examples of forces in different situations, making statements about a general concept of *force*, and so on. It is also a combination of the implicit hallmarks that indicate a particular conception, in the above example, an implicit statement lies behind Jonathan's utterance 118. According to the Aristotelian view of forces, the golf ball needs a "driving force" to get it through the air. Jonathan's view is basically Aristotelian-like. When asked about thrust (117) he moves quickly from thrust to power, via the word powering. His entire statement is a meaning unit that says that the ball needs a driving force to make it go. This is a part of specific card talk but also heralds the underlying force/motion conception. It reappears in two more cards and so it is something general to the cards.

The meaning unit has at least two functions. It *labels* some aspect of a card or situation; and it also holds a *meaning potential* – it might be a harbinger to a particular alternative conception. This can be best decided if more than one instance occurs to reinforce a particular view.

E. The framework

Children may have specific answers to specific cards (card talk) and they may have starting points for arguments (concept talk). But they also include a wide range of ideas from outside the cards to substantiate their conceptions. This is framework talk. For example, Jonathan has an extensive framework in that he incorporates the terms "less dense air," "power," "resistance," "thrust," "drag," and so forth, with some semblance of meaning into his explanations. It is this web of ideas – the causes and effects of forces – that form the wider understanding of the pupil.

This category includes two kinds of talk. First, it includes talk that expands the conceptual framework of ideas; and second, it includes all the talk that alludes to projections, speculations, or consequences of the situation depicted in the cards.

In this chapter, we can do no more than provide a taste of the complexities of data analysis. Indeed, one of the major problems in this type of work is to provide a suitable platform for its detailed discussion. The format outlined earlier for classifying the interview discussion is an attempt to do no more than that.

Results obtained by the IAI method

We have identified five distinctive types of understanding of words used in science which are found across a large number of words. A brief sketch of each now follows (Figure 13.3) and includes an explanation of the type, a reduced form of the card concerning the word *force* by which the explanation was elicited, typical quotes from students (their ages are given), and an outline of the scientist's view.

The representativeness of data obtained

In case study work, particularly that concerned with the representation of cognition, it is difficult to draw parallels to the notions of reliability and validity taken from the psychometric tradition of research. Test–retest reliability seems unlikely to be applicable, for the mere process of being interviewed about a word seems, after some reflection, to cause a revision of the understanding held. Parallel-form reliability seems more operable, in that it should be possible to design parallel versions of the same IAI card deck. However, the problem of separating two identical groups of students seems intractable. Face validity seems assured by the group nature of the preparation of card decks. Concurrent validity has not yet been explored; however, work is underway to

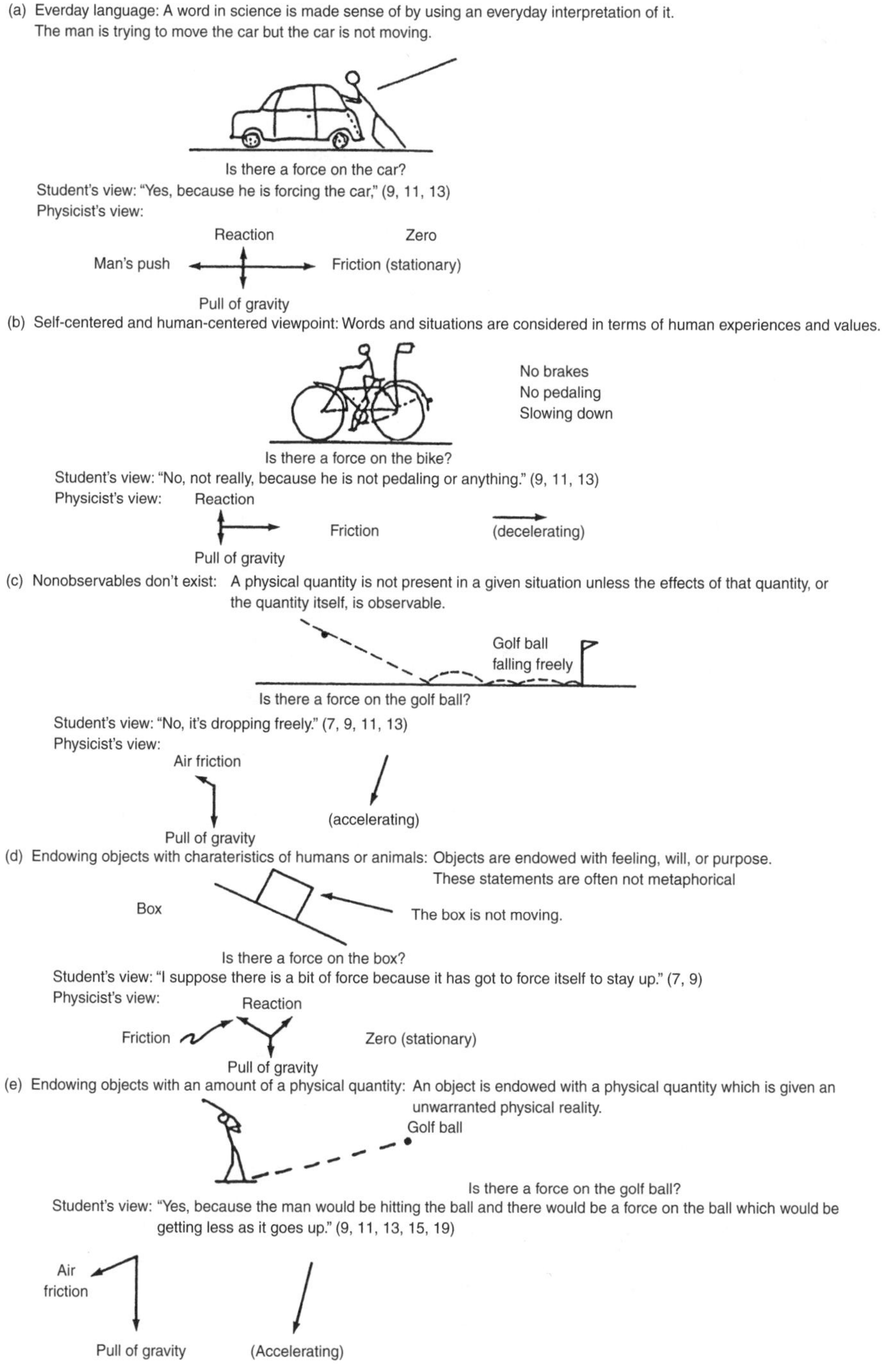

(a) Everday language: A word in science is made sense of by using an everyday interpretation of it. The man is trying to move the car but the car is not moving.

Is there a force on the car?

Student's view: "Yes, because he is forcing the car," (9, 11, 13)

Physicist's view:

Reaction   Zero

Man's push ◄————┼————► Friction (stationary)

Pull of gravity

(b) Self-centered and human-centered viewpoint: Words and situations are considered in terms of human experiences and values.

No brakes
No pedaling
Slowing down

Is there a force on the bike?

Student's view: "No, not really, because he is not pedaling or anything." (9, 11, 13)

Physicist's view: Reaction

Friction   (decelerating)

Pull of gravity

(c) Nonobservables don't exist: A physical quantity is not present in a given situation unless the effects of that quantity, or the quantity itself, is observable.

Golf ball
falling freely

Is there a force on the golf ball?

Student's view: "No, it's dropping freely." (7, 9, 11, 13)

Physicist's view:

Air friction

(accelerating)

Pull of gravity

(d) Endowing objects with charateristics of humans or animals: Objects are endowed with feeling, will, or purpose. These statements are often not metaphorical

Box

The box is not moving.

Is there a force on the box?

Student's view: "I suppose there is a bit of force because it has got to force itself to stay up." (7, 9)

Physicist's view: Reaction

Friction   Zero (stationary)

Pull of gravity

(e) Endowing objects with an amount of a physical quantity: An object is endowed with a physical quantity which is given an unwarranted physical reality.

Golf ball

Is there a force on the golf ball?

Student's view: "Yes, because the man would be hitting the ball and there would be a force on the ball which would be getting less as it goes up." (9, 11, 13, 15, 19)

Air friction

Pull of gravity   (Accelerating)

Figure 13.3 Five types of understanding of words identified by IAI's.

compare the utility of the IAI technique with the Kelly Repertory Grid technique (Kelly 1955). Construct validity has been investigated for demonstration-type interviews (e.g. Archenhold 1980) with respect to Piagetian stages. However, the significance of these results remains unclear while the mechanisms for the development of both word understanding and Piagetian stage promotion are uncertain.

In the meantime, it is only possible to check, using survey techniques, to see if the patterns of student understanding identified by interview means are replicated over large sample sizes. Watts and Zylbersztajn (1981) used a combined multiple-choice/explanation format for such an investigation into forces associated with movement (see Figure 13.4). A summary of some results obtained using this

A person throws a tennis ball straight up into the air just a small way.

The questions are about the total force on the ball.

If the ball is on the way up, then the force on the ball is shown by which arrow?

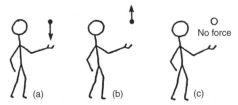

If the ball is just at the top of its flight, then the force on the ball is shown by which arrow?

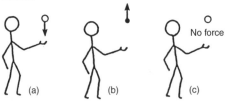

If the ball is on the way down, then the force on the ball is shown by which arrow?

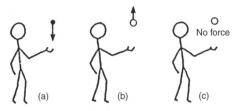

Figure 13.4 Multiple-choice questions used to check representativeness of interview data.

Table 13.1 Summary of some results obtained in the United Kingdom and New Zealand

Response pattern[a]	Response patterns for student age groups (%)				
	13–14 (N = 254)	14–15 (N = 195)	15–16 (N = 174)	16–17 (N = 147)	17–18 (N = 75)
bca	46	53	66	53	52
bba	4	1	1	—	—
baa	14	11	9	4	1
aca	9	5	5	14	21
aaa	5	11	6	2	19
acb	11	10	10	7	7
(other)	11	9	3	—	—

Note
a See Figure 13.4 for meanings of symbols.

approach in the United Kingdom and New Zealand is given in Table 13.1. In this case, the combined results confirmed the impressions gained from the analysis of transcript data.

The future use and development of the IAI approach

Some attempt is being made in present research to study children's egocentric views about the world through their own scientific conceptions, and in particular to investigate how their views change, and at what points in the child's development such changes occur. Additional research is under way which is aimed at examining

1   the retention of intuitive ideas in children despite formal teaching of those ideas;
2   the language used by students in discussing scientific concepts, and the way in which this language reflects their understanding;
3   the extent to which and the ways in which attitudes and orientation towards science can help in the development of scientific concepts;
4   the effects of different learning styles on the development of concepts;
5   the relationship between concepts, which reflects the alternative framework (Zylbersztajn 1983) developed by the students;
6   the possible existence and influence of sex differences in concept development.

However, we are most concerned that these results can be communicated to teachers. Thus we have developed three forms of a practical workshop (focusing respectively on force, particulars matter, and living) each of which runs for about half a day and which can accommodate primary, secondary, and tertiary teachers simultaneously (Gilbert and Osborne 1980a). We are also concerned with investigating the implications of these results for teaching (Gilbert *et al.* 1982). On the development side, we need to supplement the present use of the IAI technique as a research tool by exploring its diagnostic value in the hands of teachers, and its potential as a self-monitoring device for use by students of different ages and at different levels of conceptual development.

References

Archenhold, W.F. (1980). An empirical study of the understanding by 16–19 year old students of the concepts of work and potential in physics. In W.F. Archenhold, R.H. Driver, A. Orton, and C. Wood-Robinson (eds), *Cognitive Development Research in Science and Mathematics* (pp. 228–238). Leeds: Centre for Studies in Science Education.

Ausubel, D.P. (1968). *Educational Psychology: A Cognitive View*. New York: Holt Rinehart, Winston.

Driver, R. and Easely, J. (1978). Pupils and paradigms: a review of literature related to concept development in adolescent science students. *Studies in Science Education*, 5, 61–84.

Gilbert, J.K. and Osborne, R.J. (1980a). From children's science to scientist's science: a workshop (physics) (chemistry) (biology). Surrey: IET University of Surrey.

Gilbert, J.K. and Osborne, R.J. (1980b). "I understand, but I don't get it": some problems of learning science. *School Science Review*, 61, 664–673.

Gilbert, J.K., Osborne, R.J., and Fensham, P.J. (1982). Children's science and its consequences for teaching. *Science Education*, 66, 623–633.

Helm, H. (1980). Misconceptions in physics amongst South African students. *Physics Education*, 15, 92–105.

Kelly, G.A. (1955). *The Psychology of Personal Constructs*. New York: Norton.

Lachenmeyer, C. (1971). *The Language of Sociology*. New York: Columbia University Press.

Osborne, R.J. and Gilbert, J.K. (1980). A technique for exploring students' views of the world. *Physics Education*, 15, 376–379.

Pines, A.L. (1980). Protocols as indications of cognitive structure: a cautionary note. *Journal of Research in Science Teaching*, 17, 361–362.

Quine, W. (1960). *Word and Object*. Boston, MA: MIT Press.

Tasker, C.R. (1980). Some aspects of the student's view of doing science. *Research in Science Education*, 10, 19–23.

Tiberghien, A. (1980). Modes and conditions of learning. In Archenhold, W.F., Driver, R.H., Orton, A., Wood-Robinson, C. (eds), *Cognitive Development Research in Science and Mathematics*. Leeds: Centre for Studies in Science Education.

Viennot, L. (1979). Spontaneous reasoning in elementary dynamics. *European Journal of Science Education*, 1, 205–221.

Warren, J.W. (1979). *Understanding Force*. London: John Murray.

Watts, D.M. (1980a). A transcription format. Unpublished manuscript, Surrey: IET University of Surrey.

Watts, D.M. (1980b). Exploration of the concept of Force. Unpublished manuscript, Surrey: IET University of Surrey.

Watts, D.M. (1980c). Suggestions for generating and designing cards for the Interview-about-Instances approach. Unpublished manuscript, Surrey: IET University of Surrey.

Watts, D.M. and Zylbersztajn, A. (1981). A survey of some children's ideas about force. *Physics Education*, 16, 360–365.

Zylbersztajn, A. (1983). *A Conceptual Framework for Science Education: Investigating Curricular Materials and Classroom Interactions in Secondary School Physics*. Unpublished doctoral dissertation, Surrey: University of Surrey.

CHAPTER 14

CONCEPTS, MISCONCEPTIONS AND ALTERNATIVE CONCEPTIONS
Changing perspectives in science education

Gilbert, J.K. and Watts, D.M. 'Concepts, misconceptions and alternative conceptions: changing perspectives in science education'. *Studies in Science Education*, 1983, 10: 61–98

Introduction

Since the early 1970s there has emerged, on a world-wide basis, a strong interest in research into content- and context-dependent learning in science. Starting with individual papers e.g. Doran (1972), the pace has quickened through the production of theses e.g. Driver (1973), the organisation of conferences e.g. Archenhold *et al.* (1980) and Sutton and West (1982), to the appearance of books e.g. Driver (1983). However, the field shows signs of still being, using Kuhn's (1970) terminology, in a pre-paradigmatic phase. There is no general agreement on the aims of enquiry, the methods to be used, criteria for appraising data, the use to be made of the outcomes. For any coherence to appear in the field (Gilbert and Swift 1982) some semantic knots will have to be untied. One of these concerns the epistemological and ontological status of the descriptors used for the outcomes of such studies. They are severally referred to as 'misconceptions' e.g. Helm (1980), 'preconceptions' (e.g. Novak 1977), 'alternative conceptions' (Driver and Easley 1978) 'children's science' (Gilbert *et al.* 1982).

In this paper we attempt to show the relationship of these descriptors to particular meanings for the word 'concept'. These meanings are themselves portrayed against a background of the influences that have born, and still bear, on research in science education. A broad range of studies are then reviewed, where we try to identify the epistemological traditions within which they were conducted and to identify general patterns across studies based on particular words. Finally, we consider the implications of these studies for the modelling and conduct of conceptual development.

Some influences on research in science education

Science education, being a field which straddles both science and education, draws on the practices in both. As we shall discuss later, the research tradition in education makes extensive use of the prevailing tradition in science. Thus research in science education is heavily influenced, both directly and indirectly, by philosophy of science, as well as other notions drawn from psychology and sociology. These are orchestrated into research traditions whereby particular meanings for the word 'concept', a central notion for the formulation of enquiry, are adopted. As these issues are a necessary, but not sufficient, condition for the structure of this review, they will be presented compactly, making use of secondary sources.

The long and massive influence of Baconian empirical-inductivism on philosophy of science is well recorded e.g. Quinton (1980). Between the two Elizabeths this realist perspective dominated the representation of scientific activity. Within realism, the world exists in an absolute sense such that absolutely true discoveries could be made by suitably trained observers. Empirical-inductivism, according to Chalmers (1978), involved the production of singular statements ('a particular occurrence or state of affairs at a particular place at a particular time') by faithfully recording observations made with an unprejudiced mind and using unimpaired sense organs. 'Universal statements' (which 'refer to all events of a particular kind at all places and at all times'), or concepts, as we would now call them, were thought to be obtained by inductive generalisation from a suitably large number of diverse singular statements.

Despite its temporary strengthening through the emergence of logical positivism (see Papineau 1979), empirical-inductivism had, by the late 1940s, weakened its grip on philosophy of science. It was attacked on these grounds: the emerging evidence that sensory input involves selective attention, so that observation is theory-laden; that if observation is driven by theory, then the quality of the observation is governed by the quality of the pre-existent theory; that induction cannot logically take place (Chalmers 1978). It has been followed by a series of perspectives e.g. due to Popper (1972), Kuhn (1970), Lakatos (1970) and Feyerabend (1978) which can be, to a greater or lesser extent, cautiously described as relativist. Despite many points of disagreement, what these views share is: a belief in the provisional nature of knowledge; an acceptance of the theory-ladenness of observation; an evolutionary approach to theory building; a disagreement over the nature of, or even existence of, scientific method. As we shall see, this shift from realism to relativism is reflected in the social sciences.

In psychology, the decades following the 1920s were dominated by behaviourism. Passivity of mind was emphasised, with the environment providing an input whose information is directly transmitted to, and accumulated by, the organism. The resulting behaviour is the 'output'. This 'black box' approach to human functioning led to an experimental approach initially closely paralleling empirical-inductivism in science. Work was conducted on lower organisms in the belief that conclusions could be transferred to describe the behaviour of higher organisms, which was seen as differing only in complexity. The aim was to produce a blueprint of human behaviour.

Whilst behaviourism reached its peak of influence during the Second World War, it declined progressively thereafter. Factors included: its failure to address major social problems and the emergence of experimental data that was embarrassing to its underlying tenets e.g. the questioning of drive reduction models of motivation (Berlyne 1960).

During the 1970s, the influence of cognitive theorists e.g. Bruner, Piaget, Ausubel, increased. They argued that the process of development is neither direct biological motivation nor direct environmental pressure, but a reorganisation of psychological structures resulting from organism–environmental interactions. They advocated a psychology which monitored such interactions by establishing the personal meanings attached to experience. Phenomenological approaches, such as that due to George Kelly (1955), rejected the 'outsider' methods of experimental psychology and the focussing on intellect to the neglect of emotion. The main concern of the phenomenological movement is the monitoring and examination of what is seen through the eyes of the actor in the situation, so that understanding (*verstehen*) of a complex set of events can be developed rather than explanation using causal factors (*erklären*).

A parallel movement can be detected in sociology. Fay (1975) points to the historically important theme of trying to discover and implement a social structure which will serve a technological society i.e. deal with conflicts of interests, facilitate rapid social change. Science, within the empirical-inductivist and logical positivist traditions, has been extensively used in this endeavour. It claims to produce a 'real' knowledge of events of enquiry based on obtaining intersubjectively evident observations using reproducible experimental technique. The identification of causal laws is supported by the public verification of outcomes. However, the approach has been criticised on several grounds: the 'best end' in social matters involves the exercise of value judgements which make no use of conventional 'science'; the 'best means' to achieve any given end involves a notion of 'efficiency' which embodies value judgements; the separation between 'ends' and 'means' is itself problematic – as Fay (1975) puts it 'every means is an end relative to the means required to achieve it, so that any given course of action may be either a means or an end depending upon the point of view which one adopts'.

In these circumstances there has emerged an 'interpretative' tradition of social enquiry, based on the use of 'action concepts' i.e. terms used to describe doings as opposed to happenings. The purpose of an enquiry is to describe actions so as to make clear the intentions of the actor. These intentional explanations fit actions into 'a purposeful pattern which reveals how the act was warranted, given the actor, his social and physical situation, his warrants and beliefs'.

Any action concept is only usable within the context of a set of social rules ('social practice', Fay 1975) which provide criteria for the performance of an action (Shotter 1982). The purposes of enquiry within this tradition are: to make explicit the social rules underlying a class of actions; to discover the constitutive meanings (shared assumptions, definitions, conceptions) which make possible the existence of a given social practice; to identify a worldview underlying a set of constitutive meanings. This approach is seen by its advocates to: see the world through the actor's eyes; show the actors what they are doing when acting; provide enlightenment for an individual, for a partial action will be seen in the light of a whole; to improve communication between individuals; to provide an awareness of alternative lifestyles.

Drawing together strands from science, psychology and sociology, two underlying traditions emerge: the *erklären* tradition, in which explanation is the goal; the *verstehen* tradition, in which understanding is the goal. The *erklären* tradition is realist in outlook, showing allegiance to an empirical-inductivist view of knowledge, with a firm belief in the value of a reductionist approach to phenomena, the use of replicable experimental methods in the search for causal mechanisms. The *verstehen* tradition is relativist in outlook, showing the influence of post-inductivist views of knowledge, with a belief in the value of a holistic approach to phenomena, seeking to perceive understanding as shown by the individual actors in any human situation without the overt pursuit of generalisations.

It is not surprising that these two traditions have become reflected in two Paradigms for research in education, and hence in science education. These we call paradigm 1 and 2. Paradigm 1, following the *erklären* approach is: 'traditional', in that it is of longer standing; 'scientific', in that it relates to the empirical-inductivist view of science; 'experimental', in that the notion of controlled situations is employed; 'reductionist', in that phenomena are subdivided and the divisions selectively paid attention to; 'prescriptive', in that the outcomes of enquiry are intended to determine future actions; 'quantitative', in that suitable sections of a general population are enquired into; 'nomothetic', in that general laws are sought.

Paradigm 2, on the other hand, following the *verstehen* approach is: 'non-traditional', in that it is of fairly recent standing; 'artistic', in that its view of science is closer to the relativist schools; 'naturalistic', in that the notion of natural occurring situations is employed; 'holistic', in that phenomena are studied in their entirety; 'descriptive', in that there is no overt intention of determining future actions; 'qualitative', in that it seeks to enquire into phenomena without undue regard to their typicality; 'idiographic', in that it relates to the study of individuals. These two broad traditions of enquiry, and their associated operational paradigms, carry with them expectations as to the nature and meaning of the concepts which are the focus of their application.

Concepts of concept

If the transformation of meaning of the term 'concept', between the *verstehen* and *erklären* traditions, is to be represented, a central ambiguity must be recognised. It stems from the fact that the term is used with (at least) two separate and distinct referents. It is applied with equal facility to both an individual's psychological, personal, knowledge structure and to the organisation of public knowledge systems. Harré (1982) argues persuasively for a dimension that runs between private and public 'displays' of meaning. Here we take that spectrum as running from the soliloquy of personal thought, to the precise, formal and socially agreed concepts of public (Ziman 1978) science. We see the central forms of research in the field of research under review as being towards the mid-point of the spectrum studying the personal meanings of students as they confront the concepts of science. Harré calls this the 'toehold principle': the exploration of idiographic domains of personal meanings made intelligible by (with an interpretive effort) the members of a society, in our case the society of science educators.

To accede to such a dimension of conceptual display presents problems. However, the problems are more tractable than those posed by ignoring it and, as in traditional studies, treating psychological conceptions as if they were formal, strictly logical, scientific concepts e.g. Bruner *et al.* (1956). Such traditional studies (or abstractionalist studies as Bolton (1977) calls them) are not to be taken as a unified school. In 1965, before the emergence of the field under review, Wallace rued the absence of a consensus definition of concept or conceptual activity that would allow him to review the many studies he confronted. Claxton (1980) has made a similar point by listing the multitude of 'isolated' schools of thought and study concerning concepts. Freyberg (1980), too, has pointed up the many applications of the term as used by educationalists.

The traditional schools, nevertheless, do have an outlook in common which Smith and Medin (1981) characterise as the 'classical view'. To this we would add two others: the 'actional view' and the 'relational view'. These are now reviewed in turn.

The 'classical' view of concept

This outlook supports the view that all instances of a concept share common properties and that these properties are necessary and sufficient to define the concept. Although it is a view that dates back to Aristotle, it has many drawbacks. Primarily, it is a gross oversimplification (as Johnson-Laird and Wason (1977) point out).

An all-or-nothing, syllogistic concept bears little resemblance to the rather messy actuality of conceptual activity in either science subject-orientated terms, or in everyday life. As Pines and Leith (1981) suggest, it is inappropriate to see concept development in terms of single, bound entities acquired in 'all-or-none' integral steps. It is, as Markova (1982) describes, a Cartesian perspective. She characterises it as conceiving of concepts as 'logical atoms', a hierarchical subdivision of knowledge which Kelly (1963) has called an 'accumulative fragmentalist' view. An underlying assumption is that knowledge is stored in the mind in hierarchical layers which can be decomposed into smaller parts and studied independently.

Transferred to the area of educational instruction these assumptions imply that the process of acquisition of knowledge is decomposable into elementary steps that can be represented by hierarchical decision trees, so that progress in knowledge is dependent on whether the previous step is mastered in its entirety: whether it is right or wrong. Such a model is that used by Ausubel (1968), Gagné (1970) and Shayer and Adey (1981) for example. Public and private knowledge are seen as isomorphic and by implication are part of a static, logical and organised system: one which Piatelli-Palmarini (1980) characterises as a 'crystal'. White (1979a), for instance, considers concepts to be 'units of cognition' in a cognitive system that is 'more static than fluid'.

Such implications give rise to the notion of misconceptions: a flaw in the system. For example, Markle and Tieman (1970) argue that it is possible to clearly delineate such concepts as 'insect' or 'element' in terms of 'critical' attributes that define them. A non-example is then anything that excludes one or more of these attributes: to treat a non-example as an example is to display a misconception. A similar argument is developed by Herron *et al.* (1977) albeit whilst developing a taxonomy of difficulty in perceiving attributes, and by Stead (1980a) and Billeh and Khalili (1982) in terms of 'concrete' and 'formal' concepts.

Within science education research, the programme of detecting and illuminating misconceptions can be seen as an initial step in an effort to isolate, eliminate or repair and restore the failures, or 'bugs' (Brown and Vanlehn 1980) in the analytical framework in order to continue accretion on 'solid foundations'. In the light of what has been said earlier, such studies will be undertaken within Paradigm 1 and the *erklären* outlook.

The 'actional' view of concept

The group of studies which most strongly contrasts with the 'classical' view is the 'actional' view of concept. For Neisser (1976) as an example, conceptualising is 'a kind of doing'. It is a view of concepts as active, constructive and intentional. Freyberg and Osborne (1981) present concepts as 'ways of organising our experiences' so that new experiences do not leave concepts intact but that 'all cognitive learning involves some degree of re-conceptualising our existing knowledge'.

Others (like Schon 1963, 1979) contend that concepts are reconstructed with *every* change of detail if only in some minimal way. Thus, as Kelly (1963) would argue: conceptual development can be seen as a continuous, active, creative process of differentiation and integration of local conceptual domains. Nothing remains static and unchanged. In opposition to the crystal model, Piattelli-Palmarini (1980) proposes a 'flame'; often stable yet continuously dynamic.

This perspective leads to a research programme which is sharply different to that associated with the classical view. Rather than a search for invariants (or 'universals') in either conceptual content or development, it becomes one of mapping the 'topography' of local domains of understanding, or mini-theories (Claxton 1982), and of charting changes in frames of reference so that the durability, stability, coherence and consistency of conceptual constructions become the point of departure. This is done within Paradigm 2 of educational research, and in the *verstehen* tradition.

One crucial aspect of the 'actual' view of concept is the high epistemological status accorded to an individual's personal conceptions. Thus the phrase 'children's science' (Gilbert *et al.* 1982) accords to children's views the status that society frequently gives to science, and recognises that such views are developed using some notion of 'scientific method'. This view contrasts strongly with some e.g. McClelland (1982) who feel that young people's explanatory disquisitions are somehow unworthy of the terms 'concept' or 'theory'.

Other phrases have been used which are supportive of the notion of valuing, or at least respecting, personal meanings. Thus Rowell and Dawson (1980) have developed the notion of 'theories-in-action' (Karmiloff-Smith and Inhelder 1977) in terms of the commitment of young students to their own conceptions, as do Strike and Posner (1982), Posner and Hoagland (1981) and Hewson (1980) for university students. Driver and Easley's (1978) phrase 'alternative frameworks' falls within this category, and we shall use it as the basis for the two sections following the next section of this review. 'Alternative conception' (Driver and Easley 1978) also does so.

This view of concept has one immediate consequence. Students' 'errors' are then recognised as being natural developmental phenomena – personally viable constructive alternatives – rather than the result of some cognitive deficiency, inadequate learning, 'carelessness' or poor teaching. This outlook is supported by Kelly's (1955) theoretical position (especially his use of the metaphor man-the-scientist) in that it is an essential, unavoidable and *desirable* feature of personal experience that individuals generate their own varied conceptions for phenomena.

The 'relational' view of concept

Intermediate to the 'classical' and 'actional' views of concept lies the 'relational' view. Smith and Medin (1981) argue for a relational view of concepts in two parts. It is a composite view of concepts as containing both probabilistic and exemplar components. That is to say, that instances can be judged in terms of their degree of membership to a particular concept (their 'probability' of membership), and the concept judged in terms of its relationship to other concepts. It is the kind of work associated with the 'semantic networkers' Collins and Quillan (1969), Rips *et al.* (1973) and Norman and Rumelhart (1975). Maichle (1981), e.g. uses this approach to consider students' concepts in electricity; Champagne *et al.* (1981) in dynamics; Kempa (1982) describes similar studies with chemistry concepts; Schaeffer (1979) with biological ones.

In this tradition, concepts lose their bound syllogistic nature and admit to borderline cases, or degree of membership. It is limited to current work on 'fuzzy sets', (Zadeh 1965 discussed in Pope and Keen 1981). It is an improvement on the classical view in that it allows for a mixture of forms for concepts and moves towards a more naturalistic consideration of concept development. Enquiry here would seem to be based on an uncomfortable mixture of the Paradigm 1 and Paradigm 2 traditions.

Unlike the classical view it stresses the importance of the relational organisation of a concept within a network as well as the possession of a class of defined or distinctive features. It may well be that the phrase 'pre-conception' fits into the 'relations' view for these reasons. Driver and Easley (1978) point out that to have a pre-concept is to have something naïve, immature and more under-developed than a concept. Yet both 'pre-concept' and 'concept' can be seen to fit into the same network, separated only by the quality of features possessed.

Some difficulties with the relational view have been made by Claxton (1980), e.g. as being difficult to test empirically. Gilbert (1982) has argued against such approaches because of their reliance upon word-association tests, which permit little data concerning the *reasons* for associations to be collected. Markova (1982) criticises the philosophical assumptions of a relational view in that it emphasises a passive learner: it is static and atomistic. As a person's experiences accumulate, a concept changes in the number of exemplars and their degree of membership but remains essentially the same concept in terms of its internal features and its external links. For example, Clark (1973) claims that a child gradually adds new semantic features to those already present so that any (youthful) overextension of meaning is gradually reduced in time to adult meaning. For some (Nelson 1974, 1977; some of Piaget's writings as discussed in Piattelli-Palmarini 1980; and Neisser 1976, e.g.) it is not an adequate description of individual conceptual development simply to describe the process by which the person adds new attributes of a concept, or new semantic members to these that he or she has already acquired. Changing some elements whilst others remain untouched does not account for often radical or inconsistent changes in knowledge structure, of transfer or non-transfer between concepts, in either the personal or the public areas of knowledge.

On comparing the three views of concept presented earlier, it seems unlikely that a tight division can be maintained between them. The classical view shows much sympathy with a positivistic philosophy of science, which conceives of theories as organised by the logic of mathematics and taxonomy, as ideal logical structures (Harré 1972). This also suggests a fixed set of immutable concepts utilised by all (Newton-Smith 1981). Alongside the continued philosophical debates on science has come a reconsideration of scientific concepts and their role in scientific theorising. Feyerabend (1979) for instance declines the possibility of 'abstracting' concepts from a succession of observation statements, so that they are unsuitably theory-laden and mutable. Lakatos (1970) rejects a 'passivist' view of knowledge (that we 'live and die in the prison of our conceptual framework') and argues instead for a revolutionary activist view: 'it is *we* who create our "prisons" and we can also, critically, demolish them'. The point to their objections (Feyerabend's in particular) is that there are no (and can be no) straightforward rules to conceptualisation that work for all situations.

Conceptions, categories and frameworks

This use of the phrase 'alternative frameworks' as a descriptor for the outcome of Paradigm 2 enquiry is limited because there is little consensus in current literature to help explicate a common usage. Engel and Driver's (1982) suggestion is that 'frameworks' can be regarded as trans-contextual in small local domains, i.e. are demonstrated in the context provided by more than one interview question. They argue, too, that a framework ought to be a description of a perspective from which a prediction of events can be made. The act of describing a framework (by the researcher), they suggest, is influenced by the reporting of similar outlooks in other research studies, i.e. from outside even of the immediate research context. Contextual boundaries and prediction, then, are two useful markers to help distinguish terms.

Our proposal here is that 'conception' be used to focus on the personalised theorising and hypothesising of individuals. Our contention is that each person's knowledge is unique (though not infinitely diverse), which thus limits the generalisability of the single case study. In this sense Hewson (1980) talks about a conception as being a reflection, for an individual person, of how the world really is. Conceptions are accessed by the actions (linguistic and non-linguistic, verbal and non-verbal) of the person, often in response to particular questions.

To generalise beyond the individual is to construct groupings of responses which are construed as having similar intended meanings. This is to construct a category of response commonly in the context of single, or a specific set of, questions. Osborne (1980) e.g. makes categories of one-line statements from interviews which he calls 'viewpoints'. Engel (1982) limits categories to one response per question for each student participant. She notes, too, the commonplace difficulties of containing responses to one category so that mixed protracted responses are difficult to handle – a point echoed by Gilbert and Pope (1982) when considering students' responses about energy. Similarly Zylbersztajn and Watts' (1982) classification of students' responses about colour filters are broad classifications (in this case) of propositional data under convenient headings. Thus, categories are not individualised and represent an interpretation of statements at a more general, but functional, level.

The case to be made here is that alternative frameworks can profitably be seen as generalised non-individual descriptions. That is, their relation to the data base is one level further removed than that of a category of response. They can be seen, then, as short summary descriptions that attempt to capture both the explicit responses made and the construed intentions behind them. They are thematic interpretations of data, stylised, mild caricatures of the responses made by students. The distinction between the terms is illustrated in Figure 14.1.

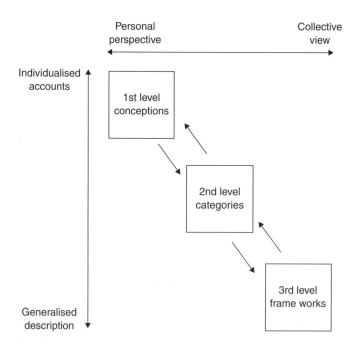

Figure 14.1 Representation of the levels of interpretation of data.

These represent levels of interpretation of the data and there is no suggestion that any one is preferable over the other. Each has its own focus and limits. Conceptions focus on the personalised accounts of individuals whereas categories interpret multiple data. They represent functional subdivisions of bulk data according to some features ascribed by the researcher, within a local context. Frame works focus upon a characterisation of responses and bridge small local changes in context.

Few of the studies in the next section are concerned either with misconceptions in the classical view or strictly relational concepts as described earlier. They will mostly be of the 'alternative frameworks movement' in that they, in Erickson's (1979) words, concern the 'substance of the actual beliefs and concepts' held by students. It needs to be said that the movement towards studying students' alternative frameworks has not developed in a neat and orderly way or followed a single straight line. Nor is it always possible to glean a clear view of the metaphysical or epistemological assumptions that lie behind the various research reports. On the whole they are what Marton (1981) and Säljö (1982) call second-order perspectives in that they are statements about peoples' ideas about the world, rather than normative performance-oriented (first-order) perspectives.

A review of the outcomes of some recent studies

It has been suggested (Champagne *et al.* 1981) that interaction between 'prior-knowledge' and formal instruction is more pronounced in the study of mechanics than for other science subjects. The argument is that all students begin the formal study of mechanics with an experientially verified set of principles, that allow them to predict the motion of objects, because such practical principles are necessary for coping with moving objects in everyday life. This proposal can be questioned on two counts. On a pragmatic basis it is to be expected that students also have experientially verified principles concerning: sound and auditory phenomena; light, vision and colour; temperature change and hot bodies; magnetic and electrical phenomena and so on, and that these are encountered in everyday life. From the theoretical position presented for the 'actional' view of concepts, it is to be expected that people develop individually constructed conceptions of experience, no matter what it is. There seems no obvious reason why moving objects should be different from other aspects of the dynamic stream of experience in a normal day. It is certainly the case that the major bulk of recent research in the body of work reviewed has been in mechanics, although other areas in science are now being considered as well.

Here we concentrate on the outcomes of studies in a small number of concept areas traditionally part of physics, i.e. force, gravity, energy, heat, electricity and light. In chemistry the concepts addressed are the mole, equilibrium and particulate matter; in biology, living. The clustering of studies into the traditional divisions of science probably reflects the backgrounds of research workers and the availability of students within the common divisions of curriculum time in schools.

It will be clear that many papers will not fit singularly into our various divisions and sometimes lie between and within separate categories. It is to be expected of such pre-paradigmatic work that underlying assumptions and modes of research conduct are not always specified explicitly or completely. Table 14.1 is an attempt to place where, in our opinion, certain exemplar studies are to be found both in terms of their discussion of concepts and their level of relation to the data. These, and other studies are discussed in further detail under the separate headings.

Table 14.1 Location of exemplar studies

| Topic area | Paradigm 1 | Paradigm 2 | |
	Classical	Relational	Actional
Force	Helm (1978) Za' Rour (1975)	Champagne *et al.* (1981)	Clement (1978) C Sjoberg and Lie (1981) CR Minstrell (1982) AF
Gravity		Jira and McCloskey (1982)	Moorfoot (1983) C Sneider and Pulos (1981) CR Watts and Gilbert (1983a) AF
Energy	Warren (1982)	Duit (1981)	Kreyenberg (1974) C Stead and Osborne (1980) CR Watts (1983b) AF
Electricity		Maichle (1981)	Shipstone (1982) CR Osborne (1981a) AF
Heat	Warren (1972)	Albert (1978)	Crookes (1982) C Andersson (1979) CR Erickson (1980) AF
Light			Stead and Osborne (1981) CR Guesne (1981) C Jung (1981) AF
Particles	Doran (1972)	Kircher (1981)	Selley (1981) AF
Mole	Duncan and Johnstone (1973)	Lazonby *et al.* (1982)	
Living			Stead (1980a) Brumby (1981) CR

Note
C: conception; CR: category of response; AF: alternative frameworks.

Force

The outcomes of many of the studies of force have concerned the strong association of force with motion; its interchangeability with the word energy; and its separation from the concept of weight. Some of the earliest work on 'misconceptions' was concerned with force. For example Helm's (1978, 1980) work, which attracted popular attention (Maddox 1978), shows that not only do students 'fail' conceptual mechanics questions but so do their teachers. That the latter are seen to be as 'bad' or 'worse' than their students prompted Maddox's title of 'the blind leading the blind up a blind alley'. The sense is unequivocal: misconceptions are 'dead ends' and a student would have to retrace his/her steps (or unlearn a 'misconception') in order to find the correct route (learn the 'correct concept'). In this same vein Aiello-Nicosia and Sperandeo-Mineo (1980) say of Italian high-school students that their misconceptions can have a 'negative' effect: the students draw 'misleading information' from common experience and then generalise it using 'wrong connections'. Similarly, Za' Rour (1975) and Ferbar (1980) equate mistakes and misconceptions with 'erroneous beliefs' about force.

In terms of alternative frameworks, force–motion frameworks are widely reported. Some of these can be summarised as follows:

(i) If a body is moving there is a force acting on it in the direction of movement.
(ii) Constant motion requires a constant force.
(iii) If a body is not moving there is no force acting on it.

These are not the only frameworks, simply the most commonplace. For example, there are some frameworks that would include forces as being involved with non-movement in contrast to (iii) above. Erickson and Hobbs (1978) report students who make a distinction between 'pulling forces' which seem to connote movement and 'holding forces' which do not. Watts (1983a) reports a similar distinction between 'Motive' forces and 'Configurative' ones. The association of force and motion, however, has had a wide press. Viennot (1979) instigated an important study with physics undergraduates using paper-and-pencil questions about oscillating masses; Watts and Zylbersztajn (1981) used a similar approach over a variety of situations with 13-year-old students. Both studies stress the importance of the direction of the force, which is commonly seen as being in the direction of motion. McCloskey *et al.* (1980) note that the direction need not necessarily be in a straight line. Thus, an object set in motion in a curved path (inside a hollow tube) will be predicted as continuing to travel in a curved path afterwards (when it emerges from the tube). Similar directional points have been made by Leboutet-Barrell (1976), Osborne (1980), Sjoberg and Lie (1981), Narode (1981), Clement (1982) and in developments of the Watts and Zylbersztajn work by Thomaz (1983) and Wright (1982).

The requirement for constant motion to have a constant force is widespread amongst students. Langford and Zollman (1982), for instance, comment on the strength and persistence of the notion that a force must continue to act on an object if it is to continue in motion even under (simulated) friction-free conditions. Where an object does not receive a constant force, then the force that causes its motion is 'used up' during movement until other forces (like gravity) take over. These kinds of frameworks are reported by Jira and McCloskey (1980), Lawson *et al.* (1980, 1981), Champagne *et al.* (1981), Sjoberg and Lie (1981), Watts and Zylbersztajn (1981) and Watts (1983a).

The third aspect of the force–motion framework concerns objects in their 'at rest' condition (Minstrell 1982). Here the absence of motion precludes the possibility of force, since (as students say) 'nothing is moving'. As Helm (1981) points out, since in physics a force is deemed to 'act' or 'be exerted' it seems only reasonable to look for the source of such action or exertion. The corollary to this is that the search is pointless where there *is* no action. Osborne (1980), Watts and Zylbersztajn (1981) and Watts (1983a) again raise similar issues.

These studies all span a variety of ages and aptitudes and yet it is difficult to glean clear developmental age-related changes. Erickson and Hobbs (1978), for instance, find few age-related differences, whilst Langford and Zollman (1982) comment on the similarities across aptitude levels. Sjoberg and Lie (1981) suggest there *are* differences but that 'The only "development" which seems to occur is that some of the mistakes are more clearly expressed by the more mature students. One may argue that they then have the "tools" to express views already developed earlier...'

Gravity

Some of the central features of studies in gravity to emerge are that

(i) gravity is concerned with and extends only as far as the earth's atmosphere;
(ii) gravity is differential: i.e. it relates to specifically different objects and varies according to circumstances;
(iii) gravity is associated with downward falling objects.

An early study by Kreyenberg (1974) used six case studies with 13-year-old students to explore theories on the cause and effect of weightlessness in space. A number of theories are posited: that it is *air* that has the force of attraction, so that astronauts are weightless on the moon because there is no atmosphere. Satellites, too, are weightless, being outside of the 'earth's force of gravity'. Later studies have shown similar results. Stead and Osborne (1980a), Watts and Zylbersztajn (1981), Watts (1982), Watts and Gilbert (1983a) and Moorfoot (1983) have indicated this close connection between gravity and the atmosphere so that gravity is not present in space, on the moon or, for some, underwater.

Stead and Osborne (1980a) point out, too, that gravity is seen to operate differently in different circumstances – for instance it is often seen to increase with height (Driver 1983). It operates to bring 'heavier' objects to the surface of the earth more rapidly (Gunstone and White 1980; Watts 1982) and is selective. In this sense gravity is seen to act differently on different objects like, e.g. an astronaut's 'heavy' boots (Stead and Osborne 1980a; Watts 1982) and not at all on hydrogen balloons (Watts and Gilbert 1983a).

The other main aspect of gravity reported is that it is commonly seen as being solely related to falling objects. Watts and Zylbersztajn (1981), e.g. note that for some students, gravity only begins to operate at the maximum height of a projectile's flight, where upon it acts to bring it down to the surface. Jira and McCloskey (1980) also suggest that some students saw a ball projected horizontally off a cliff as traversing a path resembling an inverted 'L'. That is, the ball would continue to travel horizontally until its force 'ran out', when gravity would 'take over' and the ball would fall vertically downward.

Others have noted how this local, vertical gravity bears upon students' picture of a spherical earth in space. Sneider and Pulos (1981), e.g. maintain that youngsters are presented with intuitive difficulties when required to recognise the part-to-whole relationship between visible ground and the whole earth. They present four models (or 'comographies') which show a progression towards an acceptance of the apparently absurd idea of the existence of people who live under our feet on the other side of the world. It is based upon an earlier work by Nussbaum and Novak (1976) who in turn note that though children would say the earth is 'round', they gave varying reasons for believing this. Some say that this referred to curved roads or mountains, some pictured the earth to be a circular island around which people could sail or fly, whilst yet others believed it to be a planet in the sky where only astronauts go. A fourth idea was that the earth is shaped like a ball but people live on a 'flat part in the middle'. Nor are all of these responses peculiar to young children as Vincentini-Missioni (1981) (working with non-science-specialist adults) has shown.

In only some studies do students respond to relate gravity with weight. Fleshner (1970) reports that the two terms are separate and that weight is attributed to the act of weighing, or the 'numerical result of the action of weighing, whilst

Moorfoot (1983) finds no evidence for this. For his students, no-gravity implies weightlessness.

Energy

The general framework that can be derived from a review of energy studies might be summarised as

(i) energy is to do with living and moving things
(ii) energy makes things work
(iii) energy changes from one form into another.

The first of these represents responses concerning energy that are described as anthropocentric and vitalistic. Stead (1980), e.g. discusses the responses of 8–13-year-old New Zealand students in terms of their strong tendency towards anthropomorphism. She suggests that the youngsters relate the word energy to living things by way of being energetic. Solomon (1982, 1983) makes similar points, noting too that students discuss human energy as being 'rechargeable' through food or by resting. Watts (1983b) calls this a 'human-centred' energy framework, which Gilbert and Pope (1982) also noted during the video-taping of discussions with younger (12-year-old) science students. All of these studies also comment on the emphasis given to energy and motion. It seems difficult to imagine any inanimate object as having a lot of energy whilst at the same time being stationary.

The exceptions to this last idea are often couched in terms of 'energy stores' that are concerned with making things work. Batteries, power stations, oil and coal are all seen as stores of energy, commonly intended for the benefit of humanity through technological appliances (Watts 1983b). As Duit (1981) comments, students would appear to suggest that 'for a life without technical aids no energy would be needed'.

There has been some recent controversy over the 'transformation' framework of energy since it is one that is sometimes explicitly taught. Warren (1982) denigrates this view of energy as emphasising the 'metamorphosis' of energy in a material sense. He suggests that it is appealing to young people who have become conditioned by stories of superman and 'fairy tales in which toads change into princes'. It is a view that sees energy as travelling through machines and wires and changing appearance at different points. In some cases it is thought to be stored in fuel or food. Whatever the merits of such a view (Richmond 1982; Schmid 1982) it is a common interpretation of students' responses when discussing energy (Duit 1981; Gilbert and Pope 1982; Watts 1983b).

In addition to these general research outcomes, one other feature is worth noting, Energy is frequently commented upon as being used by students interchangeably with other terms. Clement (1978), for instance, comments on the lack of differentiation between energy and other terms in college physics students in America. Viennot (1979) makes a similar point: that the concept of energy is 'inextricably mixed with the concept of force in a single undifferentiated explanatory complex'. Duit (1981), using word association tests, also notes the strong association between force and energy, an overlap of meaning that Watts and Gilbert (1983b) discuss in some detail. For Rhoneck (1981) the term energy is closely related to electric current; force being related to 'voltage' in an electric circuit.

Electricity

Much of the work on electricity has focussed upon simple battery-and-bulb circuits except for three more diverse studies by Preece (1976), Johnstone and Mughol (1978) and Rhoneck (1981). Preece's work consists of a 'relational' study of concepts concerned with electromagnetism and uses a word-association test to explore the semantic proximities of concepts for students who were graduate intending physics teachers. He suggests that the concept of 'electric current' is a pivotal one, in both the structure of the subject matter and within the structure implied from the students' responses. This suggestion is supported in some of the studies to be considered later. Johnstone and Mughol (1978) concentrate upon electrical resistance in symbolic form. They note that students (at all secondary school-age levels) are familiar with basic symbols of electrical circuits but that there is also considerable confusion (again at all ages) for the meaning of common electrical terms like voltage and power. Rhoneck's (1981) study considers such terms in more detail. Using a sequence of interviews, questionnaires and further interviews he explores secondary school students' meanings for electrical voltage and current. Responses indicate a close association between current and energy, voltage and force, and that, whereas voltage and current are related, they are generally not seen to be the same. He presents this relation by saying that a circuit is 'described by the current', and the voltage is 'closely associated' with the current. This kind of framework is supported by Maichle (1981) and is summarised here as:

(i) when an electric current 'flows' then 'voltage' should be present; when current is interrupted the voltage disappears.

With the exception of Riley *et al.* (1981) few of the other studies focus on the current–voltage aspects of a circuit, but concentrate on current itself and the components (commonly batteries and bulbs) in the circuit. Osborne and Gilbert (1979) piloted the Interview-about-Instances technique in exploring meanings for electric current. Like Johnstone and Mughol (1978) they comment on the similarities between a group of 7–14-year olds and a group of 16–18-year olds. They note that, for *some* sixth form physics students, intuitive ideas about electric current (and what happens to it) have not changed since childhood. Such studies on electric current give rise to the second general framework, which can be described as:

(ii) electric current 'flows' around a circuit in one direction and some of it is used up by each component it meets in turn.

Shipstone (1982) calls this a 'sequence model' and shows that (in his sample) it is very prevalent in the middle years of secondary schooling and persists amongst able students (who had been studying electricity for some four years) at A-level. Riley, Bee and Mokwa (1981) describe a similar model developed by students as they worked through sections of a USA Navy electricity and electronics course. These authors refer to it as a 'time-dependent' model of current flow. The same model also features as one of four proposed by Osborne (1981a, 1983). All of Osborne's models concern simple circuits (one battery and one bulb) and are, in turn, a 'unipolar' view that negates the need for a 'return' lead to the battery from the bulb; a bi-directional 'clashing currents' model; and an 'equal current', scientifically acceptable, model. Osborne has explored a wide age range of students and again finds that, despite teaching, many of them still use typical 'children's' models well into and, in some cases, at the end of the period of formal instruction. Zee *et al.* (1982) reinforce this point. They conclude, after interviewing nine physics undergraduate who had all completed a standard 'college electricity course', that all seemed to harbour the idea that current is 'used up' when it 'flows' through a light bulb: they do not intuitively 'conserve' current within circuits. One consequence of this, they suggest, is to invalidate the common assumptions that reside in the 'water flow' analogy, of water flowing through pipes and over waterfalls, etcetera. As an interesting aside they note that a comparable group who underwent an 'inquiry-orientated' course were better able to discuss such circuits and concepts 'without the dichotomy of conflicting interpretations'.

Some elements of Osborne's 'clashing' model are to be found in an extract provided by Driver (1983). In this model current 'flows' from both poles of a battery and meets inside the bulb in the circuit, producing light. In Driver's extract there are two bulbs in the circuit and the students (in this case 11-year olds) argue that the current from the battery will 'divide' (come out at each end) to produce light in each bulb. The interviewer attempts to probe this view by introducing a third bulb. Since there is not a third 'end' to the battery to service the third bulb, the students have to modify their responses to allow for something to go *through* the bulbs. Fredette and Lochhead (1980) probe similar issues. They suggest that students consider circuits such that some components (like batteries) must be in*cluded* within the circuit (have two leads) whilst others may be present as 'sinks' (have only one lead). It is not clear how widespread such responses are, but in many senses it may be a framework that complements the 'sequence' model. That is, rather than merely consuming *some* of the current 'flow' in a continuous circuit, some (specific) components may accept (and consume?) all in-flowing current.

Härtel (1982) argues that, because of such outcomes from recent research into electrical terms, it is advisable to talk, from the very outset of a teaching sequence, about single terms like current and resistance and at *the same time* about the complete system. In this way both the individual aspects and the circuit as a system can be broadened and specialised. There is a recurring problem, too, in that both students (and authors) discuss current 'flow' (an ambiguity in physical terms) rather than a flow of charge.

Heat

Studies in heat have been less numerous than those of the subject areas considered so far, making a collation of outcomes less easy. Early work, like Warren's (1972), has been concerned with a normative evaluation of university entrant students' definitions of heat. Here, for instance, Warren remarks that not one student gave a meaningful definition of heat, nor did any of the 150 respondents provide a statement of the first law of thermodynamics. He goes on to note that the most popular response was that 'heat is a form of energy', along with a collection of 'diverse' and 'vague' ideas.

In contrast, Tiberghien's (1979) work is more idiographic in that her stated aim is a particular interest in 'some of the representations or types of interpretations which children give concerning heat'. She notes a number of frameworks for 12–13-year olds, that heat is a fluid which moves over the surface of objects; that the 'transfer of heat' to the interior of solid objects is sometimes described by postulating holes for it to travel through; and that 'hot things heat; cold things cool', using heat as a verb in a framework that sees heat as the property of the material of an object.

Albert's (1978) study is with younger students over a range of 4–9-year olds. Some forty interviews are reported, conducted to a Piagetian model. In all she argues for six categories which she sees as being arranged developmentally according to chronological age. The last category, e.g. is a conceptualisation of heat as being produced by mechanical energy. This conceptualising is expressed at 8 or 9 years old, but not before. The six categories follow in sequence from entity, sometimes as a single dimension (8 years) and then energy as a source of heat, with some conceptualisation of temperature at 9 years old.

Erickson's work (1979, 1980) takes two separate looks at students' conceptions of heat. The initial study is interview based, first with 6–13-year olds and then a phase of ten in-depth, videotaped interviews with 12-year olds. Erickson presents his outcomes in terms of conceptual inventories, and one particular conception is prominent: the existence of cold as an opposite to heat. Like heat, cold is endowed with a material property as it is transferred from object to object. In addition, temperature is sometimes described as the measure of the mixture of hot and cold inside an object, a mixture that all objects have. Other examples describe heat as a substance rather like air or steam. Erickson's second study (1980) reports a subsequent survey of 276 students in three age groups: 9, 11 and 14-year olds. The results suggest that a 'caloric viewpoint' is well subscribed and remains constant over the three age bands. Changes occur in two other viewpoints (a 'children's' and a 'kinetic' viewpoint), which Erickson construes as a shift from a more perceptual bound, common sense explanation of heat phenomena to a more abstract perspective. However, he comments that there is no overall clear developmental pattern shown.

Both Stavy and Berkovitz (1980) and Andersson (1979) consider students' responses concerning temperature, in the first case about temperature mixing and in the second about boiling point. Crookes (1982) uses a case study approach to explore the conceptions of a single student as they develop in the face of discussion and discomfiting experience. For example, the student finds it difficult to come to terms with the realisation that common materials expand with a rise in temperature so that his previously 'solid' world alters all the time. Crookes paraphrases this worry by saying: 'If everything does expand and contract how can you survive in this suddenly flexible world'.

Engel's (1982) work is an extensive range of interviews over two age bands which are, in many cases, the same students interviewed two years later. Questions about heat cross a number of situations from boiling kettles, bath water, boiling potatoes, ice, to breathing 'clouds' on a cold day. Engel's analysis is in terms of frameworks and she notes that there are no clear general age trends and, in terms of heat frameworks, no age differences.

Light

There have been so few studies on light that it is difficult to discuss general, widespread findings. However, two particular frameworks are worth noting:

(i) A reflection is an image on a mirror or surface (and not a process by which light changes direction).
(ii) Light illuminates objects so that they are 'lit up' and can then be seen. The light is local to the scene and need not travel as far as the viewer.

One of the first studies, by Guesne (1981), asks students to respond to some optical experiments in what she calls 'directed' interviews. These are complemented by 'non-directed', or open ended questions in a separate set of interviews. She suggests that the students (13–14-year olds) talk in terms of reflection for a mirror but not for an illuminated sheet of white paper. Lenses are seen as making light bigger and that, in line with the second framework above, candies visible at a distance are not considered to be sending light as *far* as the observer, but are only capable of 'being seen'.

Stead and Osborne (1980) use a multiple-choice-test to consider the relative distances travelled by light at day and at night. For many interviewees (ranging between 9 and 16-year olds) the distance travelled depended first on the size of the source. The corollary to this is that a small source (like a candle) could not travel very far. Most 9-year olds did not consider light to travel more than a metre from a source during the day, and only slightly farther at night. Jung's (1981) paper comments on a sample of 12 and 15-year olds who suggest that their image is *upon* a mirror and not 'behind' it. A distinctive question asks about a completely black room with a mirror on one wall. A beam of light is directed at an angle onto the mirror whilst the interviewee is asked to imagine themselves at right angles to it. They are then asked if they would know if there was a mirror there or not. A typical response is that a small bright spot would be seen on the mirror, which Jung interprets as a 'de-coupling' between seeing the mirror (or light lying *on* the mirror) and receiving light *from* the mirror into the eye.

In a short study Zylbersztajn and Watts (1982) look at students' responses to the action of coloured filters placed into a beam of light from a torch. Many of these issues are taken up also by Andersson and Kärrqvist (1982) in a comprehensive study of students' interpretations of optical phenomena.

Particles

Some of the early work on the particulate nature of matter (Pella and Carey 1967, for instance) indicate age-dependence concerning students' conceptions. Later studies have been more equivocal so that, e.g. Novick and Nussbaum (1978) claim only that a student's change in conception from a primitive continuous to a particulate model is a major transition in outlook on the physical world. They find that many 13–14-year olds, after specific instruction do not conceive of empty space in ordinary matter, including gases. When questioned the students use a continuous matter interpretation or 'fill' space with more particles, dust, air and the suchlike. Moreover, these students find it difficult to interpret the constant motion of particles as intrinsic and are led to the view that there must be an agent of movement, commonly citing the air as the 'mover of particles'. In Doran's (1972) study, school students between ages 8 and 14 years old were questioned using 'alternate response questions', a choice between the accepted answer and a 'misconception' distractor. Doran notes that a continuous matter view is common amongst the students tested. Although he shows that a dynamic model is more widely opted for than a static model he does not directly question the source of the dynamism. Driver (1983) describes an extract from the work of 13-year olds, a piece of homework concerning the use of kinetic theory to explain the expansion of mercury in a thermometer during a temperature rise. She comments in particular about the notions of particles being embedded in a substance ('like raisins in a teacake') and of the particles themselves expanding. Selley (1981) remarks about the ingenuity and originality of youngsters in devising explanations for why iodine crystals dissolve and slowly diffuse upward in a potassium iodide solution, explanations they held to tenaciously in preference to the teacher's suggestions of a particulate model. During classroom discussion a group of 12-year olds suggest density differences, convection currents, heating effects, transportation by air bubbles and so on. As Selley comments, the particulate model is far from self-evident and cannot be inferred from phenomena by any logical route.

Kircher's (1981) extensive study is also equivocal about age-dependence and, as with Selley, notes the unconvincing nature of demonstrations like evaporation, dissolving crystals or crystallisation. This issue, which is essentially one of a mismatch between teacher and student of confirming classroom 'evidence', is also taken up by Tasker (1981).

The mole

Whilst there has been much journal correspondence concerning reasons *for* teaching the mole concept, and advice on *how* to teach it (e.g. Hudson 1976; Gower *et al.* 1977; McGlashan 1977; Whelan 1977), there has been less written by way of exploring students' understanding for the term. Three exceptions are Duncan and Johnstone (1973), Rowell and Dawson (1980) and Lazonby, Morris and Waddington (1982). Duncan and Johnstone's early study questioned students over an age range of 14–15 years old, from O-level classes. The authors summarise the responses by pointing out three 'difficulties' in teaching the mole, which are:

(a) the misapprehension that 1 mole of a compound will always react with 1 mole of another regardless off the stoichiometry of the reaction;
(b) balancing equations;
(c) manipulation of molarity solutions.

Rowell and Dawson (1980) describe an intervention study to teach apparently 'mismatched' students to understand the mole concept. They argue that spontaneous (Piagetian) formal thinkers are indistinguishable in test achievements from students who follow designed teaching sequence.

Lazonby *et al.* (1982) consider the individual aspects of the concept in terms of their chemical and mathematical logic and conclude that students (at O-level) do not find as much difficulty with most individual calculative items as they do with sequencing the parts together.

Living

This short section looks at just three papers on 'living'. All three are similar on a number of counts (beyond the topic of interest) and are by Tamir *et al.* (1981), Brumby (1981) and Stead (nee Bell) (1980a). In each case the authors address the study in terms of criterial (or critical) attributes in the manner of classical concept analysis and then produce categories of student response in terms of their inclusion (or exclusion) of such attributes. The authors point generally to the over-extension of a small number of criteria (usually movement and growth) to include, as living, such things as cars, fire and vacuum cleaners. These same criteria exclude seeds and embryos from 'living'. Tamir *et al.* (1981) suggest that these two indicators are used less (with age) as students progress towards more uniquely biological attributes.

Brumby (1981) does make some analysis of students' own explanations in terms of the frequency of categorised responses, and notes particularly the absence of any inclination towards experimentation. She comments, too, that explanations in terms of spontaneous generation, respiration and anthropomorphism were evident in a group of students even after six years of studying science.

As the alternative frameworks movement acquires research workers, it is likely that relatively neglected topics will be more widely addressed. Overviewing the work so far reported, there does seem to be a trend towards the use of an actional view of concept. It is of interest that new arrivals in the field (whose unpublished manuscripts are not extensively reviewed here) start work from the classic view and move progressively towards the actional view.

Conceptual development

The kinds of enquiry with which this review is concerned have proved of increasing interest to research workers and science teachers. Some of the reasons may have to do with the fact that, once basic research skills have been acquired (not a task to be undertaken lightly!), the opportunities for enquiry are immense, diverse and readily organised. However, the emergence of the field does seem also to have the potential for realising some long-felt societal needs regarding science education.

First, it offers the possibility of improving the quality of understanding achieved by would-be scientists, engineers and technologists. The elastic standards of the norm-based examination can then be replaced by the acrid estimations of the criterion-based examination. Second, the field offers hope of implementing a 'science for all' policy in schools. The Association for Science Education (1981), having pointed out the place of science in education as an intellectual discipline, through its applications, and as a cultural activity, states that

> The Association is convinced that the school curriculum should reflect each of the above contexts if it is to make an effective contribution to general education, prepare young people for their adult life in the community at large, and form a satisfactory base for post-school education in science, engineering and technology.

This view has been echoed by the DES (1982), albeit without mention, so far, of the resources involved. What: the 'alternative frameworks movement' may be able to do is to provide a starting place from which a science education for all can be constructed. Perhaps the greatest challenge, the third and the least addressed so far, is to provide science education within a multicultural society. Although general reviews on the education of minority groups have appeared e.g. Stone (1981), and the problems of teaching and learning science through a foreign language been addressed e.g. Strevens (1976), few articles have appeared which deal with the classroom organisation to meet these demands e.g. Jones and George (1981). Again the field does offer starting points for new departures.

The responses made by the personal meanings research and development field will be governed by the epistemological status accorded to those meanings. Work conducted with the classical view of concept will regard misconceptions as simply 'wrong'. An extreme response might be to merely ignore any evidence of misunderstanding' on the part of students, and to 'start from scratch' following a tabula rasa view of teaching (see Gilbert *et al.* 1982). There might be some temptation to merely repeat earlier teaching sequences, perhaps under the illusion that simple repetition or precisely formed phrases suffice to engender formal understanding. A more considered approach would be based on a notion of concept *formation* using the idea of 'attributes'. Sequences by which this may be done are suggested by Tennyson and Park (1980). An approach which uses the idea of 'attributes' to build on existing misconceptions (*sic*) is proposed by Stones (1979). As such it veers towards a relational view of concept. An approach to instructional design which appears to fit centrally into a relational view of concept is proposed by White (1979b).

It is our view that the more closely any study approaches an actional view of concept the more likely it is to contribute to a notion of conceptual *development*. By this we mean proposals that accept the existence, and value to their users, of alternative frameworks and seek to educate by the expansion of applicability of those frameworks or seek to modify them towards the consensus view of formal science. This will be no easy task, for the existence of alternative meanings for particular words (the central concern; of this review) will be compounded by the existence of alternative meanings for all facets of scientific enquiry, particularly those concerned with laboratory work and theory-building. The Learning in Science Project at the University of Waikato, New Zealand has shown the magnitude of the mismatch in interpretation of teachers' and students' expectations of 'practical work' (Tasker and Lambert 1981). Indeed, the evidence of 'alternative interpretations' has become so diverse as to lead Fensham (1980) to conclude that there is a need for some new objectives for science education. In view of the relevance of these objectives to this notion of conceptual development, a summary of them is included in Table 14.2.

Focusing initially on the facilitation of conceptual change by the teacher, West (1982) has identified two trends of development response to alternative frameworks: the 'revolutionary' and the 'evolutionary'. Revolutionary change involves arrangements for the deep restructuring of knowledge by the learner, what Strike and Posner (1982) call large-scale' change of 'accommodation', and is reflected in the work of, e.g. Erickson (1979) and Nussbaum and Novick (1981). Evolutionary change involves the facilitation of extensions in richness and

Table 14.2 Some new objectives for science education (after Fensham 1980)

Short description	Verbal description of objective
Definition	To introduce students to examples of how scientists have defined concepts in ways useful to them, but which conflict with common sense experience and usage
Accommodation	To make explicit the world views of natural phenomena that students hold and to relate these to world views held now and in the past by scientists
Over-simplification	To enable students to recognise that scientists *invent* general concepts which idealise and oversimplify real substances and phenomena (things)
Exemplification	To help students to relate a few scientific concepts to a wide variety of examples
Scientism	To help students take a natural phenomenon or piece of technology and identify several features with their corresponding scientific concepts
Representation	To help students to recognise and to use varieties of representation chemists and physicists use to describe substances and physical conditions

precision of meaning for students' frameworks, what Strike and Posner (1982) call 'small-scale' change or 'assimilation', and is reflected in the work of, e.g. Sutton (1982). Zylbersztajn (1983) has pointed out the parallels between revolutionary change and Kuhn's (1970) notion of revolutionary periods in science, and between evolutionary change and Kuhns (1970) notion of normal periods in science.

For teachers wishing to promote 'evolutionary' change, existing alternative frameworks may constitute a 'building block' (West 1982). The teachers' task is thus to form linkages between old and new knowledge, to help form cognitive bridges. Barnes' (1976) notion of the 'interpretative teacher' is valuable here, for such an individual will not see language only as an instrument by way of which meaning is received from others, but also as a tool to think with, by which meaning is constructed and interpreted by the knower, and knowledge reshaped. For the teacher wishing to promote 'revolutionary' change, existing alternative frameworks may constitute a potential 'stumbling block' (West 1982): stumbling block in that adherence to an earlier view may inhibit transition to a new view, and potential in that this may happen unless the limitations of the earlier view are directly addressed by the student. In this situation the teacher will wish to engender 'cognitive conflict' within a student with respect to the relative explanatory utilities of the old and new views.

Nussbaum and Novick (1981) have presented a design for learning activities which embodies a cognitive conflict strategy: students are expected to restructure their frameworks in order to accommodate results that present discrepancies when compared to predictions and explanations derived from their own ideas. This sequence is

1   The teacher creates a situation which requires students to invoke their frameworks in order to interpret it.
2   The teacher encourages the students to verbally and pictorially describe their ideas.
3   The teacher assists them, non-evaluatively, to state their ideas clearly and concisely.
4   Students debate the pros and cons of the different explanations that have been put forward. This will create cognitive conflict within many of those participating.
5   The teacher supports the search for the most highly generalisable solution and encourages signs of forthcoming accommodation in students.

Rowell and Dawson (1979) have presented three forms of resolution that may occur between pairwise conflicts: first, the logical force of one of the alternatives may be obvious to an individual student; second, the student may consider an analogous situation to the one under consideration and see one alternative as more effective there; third, neither of the alternatives seems entirely appropriate, so a unifying idea is constructed. There is no direct evidence that conflicts take place on a pairwise basis, but some success with conflict strategies has been reported, e.g. Stavy and Berkovitz (1980) and Gunstone *et al.* (1981), despite some evidence to the contrary, e.g. Bryant (1982).

Whether a teacher is following a revolutionary or evolutionary path to conceptual development, there are a number of general classroom strategies that may be valuable. If the proportion of 'student talk' in the classroom is increased (Barnes 1976) there is additional scope for more exploratory use of language, moving away from the teacher-dominated verbal-game tradition (Edwards and Furlong 1978) of teacher-opening: student-response: teacher follow-up. Lesson planning which is less tightly structured towards the teacher goals may be effective in this way (Sutton 1981), particularly where coupled to group work, although this latter does need to be brought to an orchestrated finale (Hornsey and Horsfield 1983; Sands 1982). Although practical work offers an obvious vehicle for the exploration of personal meanings, and a great deal does seem to be done (Beatty and Woolnough 1982), closer attention to the negotiation of purpose seems called for (Tasker and Lambert 1981). Metaphor and analogy has a major role to play in forming explanatory bridges between old and new knowledge (Pope and Gilbert 1983). Textbooks, because of their apparently central role in the organisation of teaching, have a potentially valuable part to play in promoting the exploration of frameworks, although this is rarely realised: the dominant paradigm presents a transition view of knowledge, an empiricist view of science and a truncated view of the history of science (Zylbersztajn 1983). Instructional systems which reflect these linguistic requirements have to be further developed (Osborne 1981b).

Looking at conceptual change from the students' point of view has been a contribution to the field by the Cornell–Witwatersrand group (Hewson 1981; Posner *et al.* 1982; Strike and Posner 1982). They see the process of, in West's (1982) terms, evolutionary change (conceptual change) and revolutionary change (conceptual exchange) as being rational decisions by an individual involving four considerations: *dissatisfaction* with an existing framework; the relative *intelligibility, plausibility* and *fruitfulness* of the competing frameworks. West and Pines (1983) doubt that any notion of rationality will be a dominant force in such change: non-rational forces like power, simplicity, aesthetic harmony and personal integrity will be controlling considerations.

Challenges to the alternative frameworks movement

It is our hope that this review will have pointed to many areas of this comparatively new field that require further research and scholarship. However, we do have our own foci of concern, which we would like to address here.

First, it is far from clear how representative of an individual's thinking a particular framework is. Indeed, given the manner in which such frameworks are obtained from interview (and other) data, it may well be that an individual's *conceptions* make use of several *frameworks*. In our own research (Watts and Gilbert 1983b) we have found that a one-to-one interview with a student can produce a fairly close adherence to one framework by a student, yet when students are left alone to discuss ideas amongst themselves (Gilbert and Pope 1982) an individual makes use of several frameworks. Moreover, given the educative nature of Paradigm 2 enquiry, it may well be that an elicited framework thereby becomes part of the intellectual history of the individual concerned: by confronting a personal understanding, or inventing one on the spur of the moment (as McClelland 1983 proposes), an individual may well modify it substantially. This would require a careful interpretation of successive re-enquiries into the frameworks of a particular word used by an individual over an extended period of time, e.g. several years.

Second, it is not yet agreed whether the focus of enquiry should be the individual as an isolate or the individual within a social group. In this review we have been mainly concerned with the study of individuals, believing that conceptions of ideas are a manifestation of personal culture. We see value in a phenomenological approach to psychology, particularly the Personal Constructivist ideas of

George Kelly (1955). These allow us to interpret social processes in cognition as a mutual construing of ideas, followed by a personal reappraisal. A complementary perspective is obtained from sociology, particularly as manifest in the word of Schutz and Luckman (1973). Their view point is expressed by Solomon (1982):

> This is a phenomenological approach to a taken-for-granted world of happening in which our experiences are assumed to be inter-subjective and built into a stock of knowledge which is shared. This sharing, by means of social exchanges, continually reinforces the meanings embedded in our thought and language.

The relative virtues of these two approaches as an interpretative frame has yet to be fully explored within the field under review.

Third, there is the question of the relation between the outcomes of research conducted with the assumption of context-independent learning (e.g. within the Piagetian tradition) and that conducted with the assumption of context-dependent learning (as is the theme of this review). Starting from a context-independent assumption, e.g. Shayer and Adey (1981), it seems possible to show an age-related graduation in the quality of understanding of specific content, i.e. in context-dependent learning. However, when the opposite assumption is made, i.e. that learning is context-dependent, as is the case in several of the studies reviewed here, no evidence of age-relatedness can be obtained. Is this a case of incommensurable paradigms, in the Kuhnian sense, or is there some underlying relationship? Driver (1983) suggests that

> Many substantive concepts in the sciences take their meanings not simply through the network of other substantive concepts to which they relate but through the nature or structure of the relationships between them. Content and structure should be complementary considerations in curriculum design.
>
> (p. 58)

This 'content plus process gives understanding' view may be a useful way of treating concepts in the classical view. However, it may be thought that, within the actional view of concept, content and process have become fused into one.

Lastly, there is the question of a model, or perhaps models, of conceptual development. We take Doran's (1978) point that appropriate models can form a focus and inspiration for future research. Three contenders are presented here:

The stepped-change model

This model (Figure 14.2) is only compatible with a classical view of concept, so that instantaneous progression between Euclidean point 'misconceptions' can take place. Second, it assumes that different 'misconceptions' can be hierarchically, and universally, arranged towards the 'true concept'. It also seems to assume that 'lower order' 'misconceptions' are lost on progression to 'higher order' 'misconceptions'.

The smooth-change model

This model (Figure 14.3) fits more comfortably within an actional view of concept, but the ontological status of frameworks is challenged, and a universally sequencing of progression between frameworks is maintained.

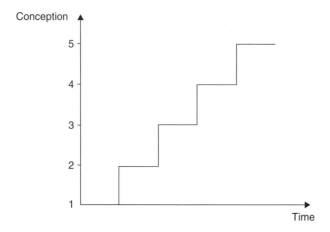

Figure 14.2 The stepped-change model.

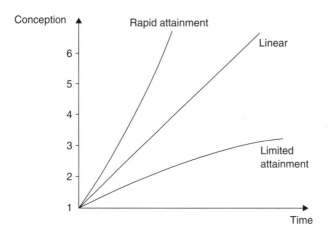

Figure 14.3 The smooth-change model.

A qualitative interpretation of this model can be attempted by extending Toulmin's (1972) analogy between Darwinian evolution and the Conceptual Evolution within a discipline. According to Darwin, species have specific characteristics which change gradually over time under the impact of selective pressures from the environment. The mechanism for this change involves the automatic random production of mutations, some of which provide an enhanced adaptation to a changing environment and which therefore convey greater breeding success. According to Toulmin, a discipline is composed of concepts which change gradually over time under the impact of selective pressures from the research environment. The mechanism for this change involves the continuous production of new conceptions, some of which show greater problem-solving power and therefore are retained through teaching. The extension of the analogy has an individual student constantly producing revised conceptions in an attempt to solve even more demanding problems: those conceptions that show the greatest potential are retained by frequent use.

One source of weakness in this chain of argument lies in doubts surrounding its Darwinian base. Gould has provided a modification to the theory (see Kemp 1982) called the punctuated equilibrium model: this suggests that sudden extreme changes in the DNA of specimens within an isolated breeding group can, particularly when associated with environmental changes, lead to rapid changes in the overall species. If the punctuated equilibrium model is used as the basis of an analogy of understanding of the conceptual changes that occur in an individual student, this suggests that change might be slow over extended periods of time but then accelerate rapidly before settling down again.

A catastrophe theory model

In its simplest form (which may not turn out to be the best for our purposes) a behaviour (in this case the conception displayed in conjunction with a given word) is governed by two Control Factors (which we suggest might be 'The Cost of Conceptual Change' and 'The Benefit of Conceptual Change').

Starting from point A, (see Figure 14.4) an individual's conception would gradually change under the impact of the Control Factors until point B is reached. Here, a point of Divergence, a catastrophic change to that conception represented by point C model occurs. Dependent on the operation of the Control Factors, the new conception could gradually change (towards, say, point F) or move towards

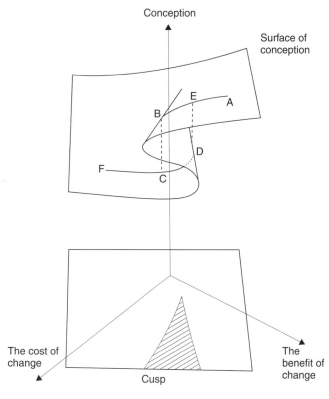

Figure 14.4 A catastrophe theory model.

point D, at which position the reverse catastrophe (to point E) would occur. Points E and B, representing initially the same conception, are separated by a cost/benefit tension called the Delay. The hatched area within the fold of the behaviour surface represents a conceptional space which is not stable. The hatched area on the floor of the model is the cusp; any point within it represents two possible conceptions for one pair of values of the Control Factors.

The Cost Factor may be thought of as having face interlocked components. These represent the effort required to: construct a new conception; present it to others; defend it against criticism; evaluate it against other possibilities. The benefit derived from a new conception would include: effecting a linkage between past and present experiences; the economy afforded by grouping experience (or instances); the anticipation of predictive control over future events; the parsimony of explanatory forms that need to be retained and displaced.

Of the three Models of Conceptual change that have been presented, that based on catastrophe theory seems to have the greatest potential. It is based on a continuous, or constructivist, notion of conception (see Thom 1975) yet allows for that rapid change of conception which is called the 'Ah, ha' experience. It does allow for the slow, or rapid, repression of conception under the impact of a constant range and number of instances. Most of all, it does offer the possibility of manipulating psychological factors, which have real meaning to the individual student, as away of promoting conceptual change.

In the classical (*sic*) fashion, we conclude by remarking that only continued research within clearly defined philosophical frameworks will identify the most profitable notion of concept and, hence of 'conceptual change'.

References

Aiello-Nicosia, M.L. and Sperandeo-Mineo, R.M. (1980). Formal thinking, physics concepts and misconcepts: an experimental study. Paper presented at the International Conference on Education for Physics Teaching, Trieste, September.

Albert, E. (1978). Development of the concept of heat in children. *Science Education* 62(3), 389–399.

Andersson, B. (1979). Some aspects of children's understanding of boiling point. Paper presented at the Cognitive Development Research in Science and Mathematics Conference, University of Leeds.

Andersson, B. and Kärrqvist, C. (1982). Light and its properties. EKNA Report No. 8, EKNA Project, Department of Educational Research, University of Gothenborg, Sweden.

Archenhold, W.F., Driver, R.H., Orton, A. and Wood-Robertson, C. (eds) (1980). *Cognitive Development Research in Science and Mathematics*. Leeds: Centre for Studies in Science Education, University of Leeds.

Association for Science Education (1981). Education through science: a policy statement of the Association for Science Education. *School Science Review* 63, 5–52.

Ausubel, D.P. (1968). *Educational Psychology: A Cognitive View*. New York: Holt, Rinehart and Winston.

Barnes, D. (1976). *From Communication to Curriculum*. London: Penguin.

Beatty, J. and Woolnough, B. (1982). Practical work in 11–13 science: the context, type and aims of current practice. *British Educational Research Journal* 8(1), 23–30.

Berlyne, D.E. (1960). *Conflict, Arousal and Curiosity*. New York: McGraw Hill.

Billeh, V.Y. and Khalili, K. (1982). Cognitive development and comprehension of physics concepts. *European Journal of Science Education* 4(1), 95–104.

Bolton. N. (1977). *Concept Formation*. London: Pergamon.

Brown, J.S. and Vahlehw, K. (1980). Repair theory: a generative theory of bugs in procedural skills. *Cognitive Science* 4(4), 14–26.

Brumby, M. (1981). Learning, understanding and 'thinking about' the concept of life. *Australian Science Teachers Journal* 27(3), 21–25.

Bruner, J.S., Goodnow, J.L. and Austen, G.A. (1956). *A Study of Thinking.* New York: John Wiley & Sons.

Bryant, P. (1982). The role of conflict and of agreement between intellectual strategies in children's ideas about measurement. *British Journal of Psychology* 73, 243–251.

Chalmers, A.F. (1978). *What is this Thing Called Science?* Milton Keynes: Open University.

Champagne, A., Klopfer, E.L. and Gunstone, R.F. (1981). Cognitive research and the design of science instruction. Paper presented at the International Workshop on problems concerning students' representation of physics and chemistry knowledge, Pädagogische Hochschule Ludwigsburg, September.

Clark, E.V. (1973). What's in a word? On a child's acquisition of semantics in his first language. In: T.E. Moore (ed.), *Cognitive Development and the Acquisition of Language.* New York: Academic Press.

Claxton, G. (1980). *Cognitive Psychology: New Directions.* London: Routledge and Kegan Paul.

Claxton, G. (1982). School science: falling on stony ground or choked by thorns? (mimeograph). University of London, Chelsea College, Centre for Science Education.

Clement, J.J. (1978). Mapping a student's causal conceptions from a problem solving protocol. Amherst, MA: University of Massachusetts, Department of Physics and Astronomy.

Clement, J. (1982). Students' preconceptions in introductory mechanics. *American Journal of Physics* 50(1), 66–71.

Collins, A.M. and Quillan, M.R. (1969). Retrieval time from semantic memory. *Journal of Verbal Learning and Verbal Behaviour* 8, 240–247.

Crookes, J. (1982). The nature of personal commitment in changes in explanations (mimeograph). University of Leicester, School of Education.

Department of Education and Science (DES) (1982). *Science Education in Schools.* London: Her Majesty's Stationery Office.

Doran, R.L. (1972). Misconceptions of selected science concepts held by elementary school students. *Journal of Research in Science Teaching* 9(2), 127–137.

Doran, R.L. (1978). Models for science education research. *Journal of Research in Science Teaching* 15(5), 423–431.

Driver, R. (1973). The representation of conceptual framework in young adolescent science students. Unpublished PhD Thesis. Illinois: University of Illinois.

Driver, R. (1983). *The Pupil as Scientist?* Milton Keynes: Open University.

Driver, R. and Easley, J. (1978). Pupils and paradigms: a review of literature related to concept development in adolescent science students. *Studies in Science Education* 5, 61–84.

Duit, R. (1981). Students' notions about the energy concept – before and after physics instruction. Paper presented at the International Workshop on problems concerning students' representation of physics and chemistry knowledge, Pädagogische Hochschule Ludwigsburg.

Duncan, I.M. and Johnstone, A.H. (1973). The mole concept. *Education in Chemistry* 10, 213–214.

Edwards, A.D. and Furlong, V.J. (1978). *The Language of Teaching.* London: Heinemann.

Engel, E. (1982). An exploration of pupils' understanding of the concepts heat, pressure and evolution. Unpublished PhD Thesis. University of Leeds, School of Education.

Engel, E. and Driver, R. (1982). Children's interpretations of scientific phenomena – analysis of descriptive data. Paper presented at British Educational Research Association Annual Conference, University of St. Andrews, September.

Erickson, G.L. (1979). Children's conceptions of heat and temperature. *Science Education* 63(2), 221–230.

Erickson, G.L. (1980). Children's viewpoints of heat: a second look. *Science Education* 64, 323–336.

Erickson, G. and Hobbs, E. (1978). A developmental study of student beliefs about force concepts. A paper presented at the 1978 Annual Convention of the Canadian Society for the Study of Education, Ontario, June.

Fay, B. (1975). *Social Theory and Political Practice.* London: George Allen & Unwin.

Fensham, P.J. (1980). A research base for new objectives of science teaching. *Research in Science Education* 10, 23–33.

Ferbar, J. (1980). Some misconceptions of force and energy. Paper presented at the International Conference on Education for Physics Teaching, Trieste, September.

Feyerabend, P. (1978). *Science in a Free Society*. London: New Left Books.

Feyerabend, P. (1979). *Against Method* (2nd edn). London: New Left Books.

Fleshner, E.A. (1970). The mastery by children of some concepts in physics. In: E. Stone (ed.), *Readings in Educational Psychology*. London: Methuen.

Fredette, N. and Lochhead, J. (1980). Student conceptions of simple circuits. *Physics Teacher*, 18(3), 194–198.

Freyberg, P.S. (1980). When is a concept not a concept? Paper presented to New Zealand Association for Research in Education, November.

Freyberg, P. and Osborne, R.J. (1981). Who structures the curriculum: teacher or learner? SET Research Information for Teachers 2, item 6.

Gagne, R.M. (1970). *The Conditions of Learning*. New York: Holt, Rinehart and Winston.

Gilbert, J.K. (1982). A constructivist approach to chemical education. Paper presented at Annual Congress of Royal Society of Chemistry, University of Aston, March.

Gilbert, J.K. and Pope, M.L. (1982). Schoolchildren discussing energy (mimeograph). University of Surrey, Institute of Educational Development.

Gilbert, J.K. and Swift, D.J. (1982). Towards a Lakatosian analysis of the Piagetian and alternative conceptions research programme. (Submitted to Science Education.)

Gilbert, J.K., Osborne, R.J. and Fensham, P.J. (1982). Children's science and its consequences for teaching. *Science Education* 66(4), 623–633.

Gower, D.M., Daniels, D.J. and Lloyd, G. (1977). The mole concept. *School Science Review* 58(205), 658–676.

Guesne, E. (1981). Children's conceptions of light. Workshop presented at the International Conference of Physics Teaching, Trieste, September.

Gunstone, R.F. and White, R.T. (1980). A matter of gravity. Paper presented to Australian. Science Education Research Association Conference, Melbourne, May.

Gunstone, R.F., Champagne, A.B. and Klopper, L. (1981). Instruction for understanding: a case study. *Australian Science Teachers Journal* 27(3), 27–32.

Harré, R. (1972). *The Philosophies of Science: An Introductory Survey*. Oxford: Oxford University Press.

Harré, R. (1982). Personal meanings: semantic relations of the fourth kind. In: E. Shepherd and J.P. Watson (eds), *Personal Meanings*. The First Guy's Hospital Symposium on the Individual Frame of Reference. Chichester: John Wiley & Sons.

Härtel, H. (1982). The electric circuit as a system: a new approach. *European Journal of Science Education* 4(1), 45–55.

Helm, H. (1978). Misconceptions about physical concepts among South African pupils studying physical science. *South African Journal of Science* 74, 285–290.

Helm, H. (1980). Misconceptions in physics amongst South African students. *Physics Education* 15, 92–105.

Helm, H. (1981). Conceptual misunderstandings in physics. In: *Perspective 3*, University of Exeter, School of Education.

Herron, J.D., Cantu, U., Ward, R. and Srinivason, V. (1977). Problems associated with concept analysis. *Science Education* 61(2), 185–199.

Hewson, P.W. (1980). A case study of the effect of metaphysical commitments on the learning of a complex scientific theory. Paper presented at Australian Education Research Association, April.

Hewson, P.W. (1981). A conceptual change approach to learning science. *European Journal of Science Education* 3(4), 383–396.

Hornsey, M. and Horsfield, J. (1983). Pupils' discussion in science: a stratagem to enhance quantity and quality. *School Science Review* 63, 763–767.

Hudson, M.J. (1976): Introducing the mole. *Education in Chemistry* 13, 110 and 114.

Jira, D.K. and McCloskey, M. (1980). Students' misconceptions about physical motion (mimeograph). Baltimore, MD: Johns Hopkins University.

Johnson, A.H. and Mughol, A.R. (1978). The concept of electrical resistance. *Physics Education* 13, 46–49.

Johnson-Laird, P.N. and Wason, P.C. (1977). *Thinking: Readings in Cognitive Science*. Cambridge: Cambridge University Press.

Jones, A. and George, N. (1981). The subject teacher in a multiracial school. In: C. Sutton (ed.), *Communicating in the Classroom*. London: Hodder and Stoughton.

Jung, W. (1981). Conceptual frameworks in elementary optics. Paper presented at the International Workshop on the problems concerning students' representation of physics and chemistry knowledge, Pädagogisch Hochschule, Ludwigsburg, September.

Karmiloff-Smith, S. and Inhelder, B. (1977). If you want to get ahead, get a theory. In: P.N. Johnson-Laird and P.C. Wason (eds), *Thinking: Readings in Cognitive Science*. Cambridge: Cambridge University Press.

Kelly, G.A. (1955). *The Psychology of Personal Constructs*. Vols 1, 2. New York: W.W. Norton.

Kelly, G.A. (1963). *A Theory of Personality: The Psychology of Personal Construct*. New York: W.W. Norton.

Kemp, T. (1982). The reptiles that became mammals. *New Scientist* 92, 583.

Kempa, R.F. (1982). Learning theories and chemical education. Paper presented at the Annual Congress of the Royal Society of Chemistry, University of Aston, April.

Kircher, E. (1981). Research in the classroom about the particle nature of matter (grades 4–6). In: Proceedings of the International Workshop on problems concerning students' representation of physics and chemistry knowledge, Pädagogische Hochschule Ludwigsburg.

Kreyenberg, L. (1974). Bedingungsanalyse zum Themenbereich Raumfahrt in 8. Schuljahr der Hauptschule unter besonderer Berücksichtigung der damit verbundenen methodologischen und methodischen Probleme als Voraussetzun für eine adäquate Planung ven Unterricht. (Unpublished dissertation for Hochschulinternes Fernsehen der Universität Osnabrück.)

Kuhn, T. (1970). *The Structure of Scientific Revolution*. Chicago, IL: University of Chicago Press.

Lakatos, I. (1970). Falsification and the methodology of scientific research programmes, In: I. Lakatos and A. Musgrave (eds), *Criticism and the Growth of Knowledge*. Cambridge: Cambridge University Press, 91–195.

Langford, J.M. and Zollman, D. (1982). Conceptions of dynamics held by elementary and high school students. Paper presented at the annual meeting of the American Association of Physics Teachers, San Fransisco, CA, January.

Lawson, R.A., Schuster, D.G. and McDermott, L.C. (1981). Students' conceptions of force (mimeograph). University of Washington, Department of Physics.

Lawson, R.A., Trowbridge, D.E. and McDermott, L.C. (1980). Students' conceptions of dynamics (mimeograph). University of Washington, Department of Physics.

Lazonby, J.N., Morris, J.E. and Waddington, D.J. (1982). The muddlesome mole. *Education in Chemistry*, 109–111.

Leboutet-Barrell, L. (1976). Concepts of mechanics among young people. *Physics Education* 11(7), 462–466.

McClelland, G. (1982). Alternative frameworks in science revisited (mimeograph). University of Sheffield.

McCloskey, M., Carmazza, A. and Green, B. (1980). Curvilinear motion in the absence of external forces: naive beliefs about motion of objects. *Science* 210, 1139–1141.

McGlashan, M.L. (1977). Amount of substance and the mole. *Physics Education*, 12, 276–278.

Maddox, J. (1978). Physics: how blind lead blind up a blind alley. *Times Educational Supplement*, December 1, 11.

Maichle, U. (1981). Representation of knowledge in basic electricity and its use for problem solving. Paper presented at the International Workshop of problems concerning students' representation of physics and chemistry knowledge, Pädagogische Hochschule Ludwigsburg, September.

Markle, S.M. and Tiemann, P.W. (1970). 'Behavioral' analysis of 'cognitive' content. *Educational Technology* 10, 41–45.

Markova, I. (1982). *Paradigms, Thought and Language*. Chichester: John Wiley & Sons.

Marton, F. (1981). Phenomenography – describing conceptions of the world around us. *Instructional Science* 10, 177–200.

Minstrell, J. (1982). Explaining the 'at rest' condition of an object. *Physics Teacher*, 20, 10–40.

Moorfoot, J.J. (1983). An alternative method of investigating pupils' understanding of physics concepts. *School Science Review* 64, 561–566.

Narode, R. (1981). Catalogue of misconceptions from introductory mechanics: a personal account (mimeograph). Amherst, MA: University of Massachusetts, Department of Physics and Astronomy.

Neisser, U. (1976). *Cognition and Reality.* San Fransisco, CA: Freeman.

Nelson, K. (1974). Concept, word and sentence: interrelations in acquisition and development. *Psychological Review* 81, 267–285.

Nelson, K. (1977). Some evidence for the cognitive primacy of categorisation and its functional bases. In: P.N. Johnson-Laud and P.C. Wason (eds), *Thinking: Readings in Cognitive Science.* Cambridge: Cambridge University Press.

Newton-Smith, W.H. (1981). *The Rationality of Science.* London: Routledge and Kegan Paul.

Norman, D.A., Rumelhart, D.E. and the LNR Research Group (1975). *Explorations in Cognition.* San Francisco, CA: Freeman.

Novak, J. (1977). *A Theory of Education.* Ithaca, NY: Cornell University Press.

Novick, S. and Nussbaum, J. (1978). Junior high school pupils' understanding of the particulate nature of matter: an interview study. *Science Education* 62(3), 273–281.

Nussbaum, J. and Novak, J. (1976). An assessment of children's concepts of the earth utilising structured interviews. *Science Education* 60(4), 535–550.

Nussbaum, J. and Novick, S. (1981). Brainstorming in the classroom to invent a model: a case study. *School Science Review* 62, 771–778.

Osborne, R.J. (1980). Force. Learning in Science Project. Working Paper no.16. Hamilton, New Zealand: University of Waikato.

Osborne, R.J. (1981a). Children's ideas about electric current. *New Zealand Science Teacher* 29, 12–19.

Osborne, R.J. (ed.) (1981b). The framework: towards action research. Learning in Science Project, Working Paper no. 28. Hamilton, New Zealand: University of Waikato.

Osborne, R.J. (1983). Towards modifying children's ideas about electric current. *Research in Science and Technological Education* 4, 11–21.

Osborne, R.J. and Gilbert, J.K. (1979). *An Approach to Student Understanding of Basic Concepts in Science.* Guildford: University of Surrey, Institute for Educational Technology.

Papineau, D. (1979). *Theory and Meaning.* Oxford: Clarendon Press.

Pella, M.O. and Carey, R.L. (1967). Levels of maturity and levels of understanding for selected concepts of the particle nature of matter. *Journal of Research in Science Teaching* 5, 202–215.

Piattelli-Palmarini, M. (1980). *Language and Learning: The Debate between Jean Piaget and Noam Chomsky.* London: Routledge and Kegan Paul.

Pines, A.L. and Leith, S. (1981). What is concept learning in science? Theory, recent research and some teaching suggestions. *Australian Science Teachers Journal* 27(3), 4–7.

Pope, M.L. and Gilbert, J.K. (1983). The role of metaphor in explanation: some empirical questions. *European Journal of Science Education* 5(3), 249–261.

Pope, M.L. and Keen, T. (1981). *Personal Construct Psychology and Education.* London: Academic Press.

Popper, K. (1972). *Conjectures and refutations: the growth of scientific knowledge* (4th edn). London: Routledge and Kegan Paul.

Posner, G.J. and Hoagland, G. (1981). Development of an instrument for assessing students' beliefs about science. Paper presented at the Annual Convocation of the North Eastern Educational Research Association, Fallsview Inn, Ellenville, New York, October.

Posner, G.J., Strike, K.A., Hewson, P.W. and Gertzog, W.A. (1982). Accommodation of a scientific conception: towards a theory of conceptual change. *Science Education* 66(2), 211–227.

Preece, P.F.W. (1976). The concepts of electromagnetism: a study of the internal representation of external structures. *Journal of Research in Science Teaching* 13(6), 517–524.

Quinton, A. (1980). *Francis Bacon.* Oxford: Oxford University Press.

Rhoneck, C.V. (1981). Students' conceptions of the electric circuit before physics instruction. Paper presented at the International Workshop on students' representation of physics and chemistry knowledge, Pädagogische Hochschule Ludwigsburg, September.

Richmond, P.E. (1982). Teaching about energy. *Physics Education* 17(5).

Riley, M.S., Bee, N.V. and Mokwa, J.J. (1981). Representation in early learning: the acquisition of problem solving strategies in basic electricity/electronics. Paper presented at the

International workshop on problems concerning students' representation of physics and chemistry knowledge. Pädogogische Hochschule Ludwigsburg, September.

Rips, J.J. Shoben, E.J. and Smith, E.E. (1973). Semantic distance and the verification of semantic relations. *Journal of Verbal Learning and Verbal Behaviour* 12, 1–20.

Rowell, J.A. and Dawson, C.J. (1979). Cognitive conflict: its nature and use in the teaching of science. *Research in Science Education* 9, 169–175.

Rowell, J.A. and Dawson, C.J. (1980). Mountain or mole hill: can cognitive psychology reduce the dimensions of conceptual problems in classroom practice? *Science Education* 64(5), 693–708.

Säljo, R. (1982). Learning and understanding: a study of differences in constructing meaning from text. *Göteborg Studies in Educational Sciences* 41, University of Gothenburg.

Sands, M.K. (1982). Group work on science: myth and reality. *School Science Review* 62, 765–769.

Schaefer, G. (1979). Concept formation in biology: the concept 'growth'. *European Journal of Science Education* 1(1), 87–101.

Schmid, B.G. (1982). Energy and its carriers. *Physical Education* 17(5), 212–218.

Schon, D.A. (1963). *Displacement of Concepts*. London: Tavistock Publications.

Schon, D.A. (1979). Generative metaphor: a perspective on problem-setting in social policy. In: A. Ortony (ed.), *Metaphor and Thought*. Cambridge: Cambridge University Press.

Schutz, A. and Luckman, T. (1973). *Structure of the Life World*. London: Heinemann.

Selley, N.J. (1981). Children's understanding of atoms and molecules (mimeograph). Kingston: Kingston Polytechnic.

Shayer, M. and Adey, P. (1981). *Towards a Science of Science Teaching: Cognitive development and Curriculum Demand*. London: Heinemann.

Shipstone, D.M. (1982). A study of secondary school pupils' understanding of current voltage and resistance in simple D.C. circuits (mimeograph). University of Nottingham, School of Education.

Shotter, J. (1982). Growth to autonomy within a system of conventions. In: M. Braham (ed.), *Aspects of Education*. Chichester: John Wiley & Sons.

Sjoberg, S. and Lie, S. (1981). Ideas about force and movement among Norwegian pupils and students (mimeograph). Oslo: University of Oslo, Centre for School Science.

Smith, E.E. and Medin, D.L. (1981). *Categories and Concepts*. Cambridge, MA: Harvard University Press.

Sneider, C. and Pulos, S. (1981). *Children's Cosmographics: Understanding the Earth's Shape and Gravity*. Berkeley, CA: University of California.

Solomon, J. (1982). How children learn about energy, or does the first law come first? *School Science Review* 63, 415–422.

Solomon, J. (1983). Socially acquired knowledge: an enquiry into British children's notions about energy prior to teaching (mimeograph). London: University of London, Chelsea College, Centre for Science Education.

Stavy, R. and Berkovitz, B. (1980). Cognitive conflict as a basis for teaching quantitative aspects of the concept of temperature. *Science Education* 64(5), 679–692.

Stead, B.F. (1980). Energy, learning in Science Project, Working Paper no. 17. Hamilton, New Zealand: University of Waikato.

Stead, B.F. (1980a). Living. Learning in Science Project, Working Paper no. 15. Hamilton, New Zealand: University of Waikato.

Stead, B.F. and Osborne, R.J. (1980). Exploring students' concepts of light. *Australian Science Teachers Journal* 26(3), 84–90.

Stead, K.E. and Osborne, R.J. (1980a). Gravity. Learning in Science Project, Working Paper no. 20. Hamilton, New Zealand: University of Waikato.

Stone, M. (1981). *The Education of the Black Child in Britain*. London: Fontana.

Stones, E. (1979). *Psychopedagogy*. London: Methuen.

Strevens, P. (1976). Problems of learning and teaching science through a foreign language. *Studies in Science Education* 3, 55–68.

Strike, K.A. and Posner, G.J. (1982). Conceptual change and science teaching. *European Journal of Science Education* 4(37), 231–240.

Sutton, C.R. (ed.) (1981). *Communicating in the classroom*. London: Hodder and Stoughton.

Sutton, C.R. (1982). The origins of pupils' ideas. In: C. Sutton and L. West (eds), *Investigating Children's Existing Ideas about Science*. Leicester: University of Leicester, School of Education.

Sutton, C.R. and West, L. (eds) (1982). *Investigating Children's Existing Ideas about Science*. Leicester: University of Leicester, School of Education.

Tamir, P., Galchoppin, R. and Nussinonitz, R. (1981). How do intermediate and junior high school students conceptualise living and non-living. *Journal of Research in Science Teaching* 18(3), 241–248.

Tasker, R. (1981). Children's views and classroom experiences. *Australian Science Teachers Journal* 27(3), 33–37.

Tasker, R. and Lambert, J. (eds) (1981). Science activities: the problem. Learning in Science Project, Working Paper no. 47. Hamilton, New Zealand: University of Waikato.

Tennyson, R. and Park, O. (1980). The teaching of concepts: a review of instructional design research literature. *Review of Educational Research* 50(1), 55–70.

Thom, R. (1975). *Structural Stability and Morphogenesis: An Outline of General Theory of Models*. Reading, MA: Benjamin.

Thomaz, M. (1983). An analysis of students' understanding about the concept of force (mimeograph). Aveiro, Portugal: University of Aveiro, Department of Physics.

Tiberghien, A. (1979). Modes and conditions of learning – an example: 'some aspects of the concept heat'. Paper presented at the Cognitive Development Research in Science and Mathematics Conference, University of Leeds.

Toulmin, S. (1972). *Human Understanding*. Vol. 1. Oxford: Clarendon Press.

Vicentini-Missioni, M. (1981). Earth and gravity: comparison between adults' and children's knowledge. Paper presented at the International Workshop of problems concerning students' representation of physics and chemistry knowledge. Pädagogische Hochschule Ludwigsburg, September.

Viennot, L. (1979). Spontaneous learning in elementary dynamics. *European Journal of Science Education* 1(2), 205–221.

Wallace, J.G. (1965). *Concept Growth and the Education of the Child*. Slough: National Foundation for Educational Research.

Warren, J.W. (1972). The teaching of the concept of heat. *Physics Education* 7, 41–44.

Warren, J.W. (1982). The nature of energy. *European Journal of Science Education* 4(3), 295–297.

Watts, D.M. (1982). Gravity – don't take it for granted. *Physics Education* 17(5), 116–121.

Watts, D.M. (1983a). A study of schoolchildren's alternative frameworks of the concept of force. *European Journal of Science Education* 5(3), 381–402.

Watts, D.M. (1983b). Some alternative views of energy. *Physics Education* 27(3), 14–23.

Watts, D.M. and Gilbert, J.K. (1983a). Appraising the understanding of physics concepts: gravity (mimeograph). British Petroleum/Guildford: University of Surrey, Institute of Educational Development.

Watts, D.M. and Gilbert, J.K. (1983b). Enigmas in school science: students' conceptions for scientifically associated words. *Research in Science and Technological Education* 4, 36–52.

Watts, D.M. and Zylbersztajn, A. (1981). A survey of some children's ideas about force. *Physics Education* 15, 360–365.

West, L.H.T. (1982). The researchers and their work. In: C. Sutton and L. West (eds), *Investigating Children's Existing Ideas about Science*. Leicester: University of Leicester, School of Education.

West, L.H.T. and Pines, A.L. (1983). How 'rational' is rationality? (mimeograph). Melbourne: Monash University.

Whelan, P.M. (1977). Introduction of the mole in the teaching of ideal and real gases. *Physics Education* 12, 279–284.

White, R.T. (1979a). Achievement, mastery, proficiency, competence. *Studies in Science Education* 6, 1–22.

White, R.T. (1979b). Describing cognitive structure. Paper presented at Australian Association for Educational Research, Melbourne, November.

Wright, R.W. (1982). Students' misconceptions of some principles in physics (mimeograph). Kenya: Kenyatta University College, Department of Physics.

Zadeh, L. (1965). Fuzzy sets. *Information and Control* 8, 338–353.

Za' Rour, G.I. (1975). Science misconceptions among certain groups of students in Lebanon. *Journal of Research in Science Teaching* 12(4), 385–391.

Zee, E.H. van, Evans, J., Greenberg, D.W. and McDermott, L.C. (1982). Student conceptual difficulty with current electricity. Paper presented at the National Meeting of the American Association of Physics Teachers, San Francisco, CA, January.

Ziman, J. (1978). *Reliable Knowledge*. Cambridge: Cambridge University Press.

Zylbersztajn, A. (1983). A conceptual framework for science education: investigating curricular materials and classroom interactions in secondary school physics. Unpublished PhD Thesis. University of Surrey.

Zylbersztajn, A. and Watts, D.M. (1982). Throwing some light on colour (mimeograph). Guildford: University of Surrey, Institute of Educational Development, December.

CHAPTER 15

CHILDREN'S SCIENCE AND ITS CONSEQUENCES FOR TEACHING

Gilbert, J.K., Osborne, R.J. and Fensham, P.J. 'Children's science and its consequences for teaching'. *Science Education*, 1982, 66(4): 623–633

Many research studies in recent years have shown that children have beliefs about how things happen and expectations which enable them to predict future events (Driver and Easley 1978). Evidence is accumulating from a wide variety of sources (Leboutet-Barrell 1976; Nussbaum and Novak 1976; Clement 1977; Stead and Osborne 1980) to show that children, on the basis of their everyday experiences of the world, hold these beliefs and expectations very strongly. Moreover, children have clear meanings for words which are used both in everyday language and also in formal science (Gilbert and Osborne 1980; Osborne and Gilbert 1980a). Such views of the world, and meanings for words, held by children are not simply isolated ideas (Champagne *et al.* 1979) but rather they are part of conceptual structures which provide a sensible and coherent understanding of the world from the child's point of view. These structures may be termed children's science.

In the development of science curricula the existence of children's science has usually either been ignored or inadequately considered (Fensham 1980). The two different assumptions on which science teaching has been based, and one on which it could be based, can be readily identified.

The "blank-minded" or "tabula rasa" assumption

This approach, which by implication underlies many modern curricula (Fensham 1980), assumes that the learner has no knowledge of a topic before being formally taught it. The assumption is that the learner's "blank mind" can be "filled" with teacher's science (S_T). This is diagrammatically shown in Figure 15.1.

The "teacher dominance" assumption

The assumption here is that, although learners may have some conceptual view of a new science topic before being taught it, this understanding has little significance

Figure 15.1 Science teaching in which it is assumed that the learners have no theoretical views of the topic or phenomena under study.

for learning and can be directly and easily replaced. Thus, even if children's science views (S_{Ch}) exist, they are not strongly held in the face of science teaching. This is diagrammatically shown in Figure 15.2.

The "student dominance" assumption

This assumption recognizes children's science views as sufficiently strong that they will persist and interact with science teaching. The interaction is diagrammatically shown in Figure 15.3.

There is growing evidence that the learned amalgam $\{S_{Ch} \setminus S_T\}$ of children's science and teachers' science can coexist in varying proportions. "Successful" learners use teachers' science when required in tests and examinations, but still retain children's science in dealing with many every day situations.

If science curricula and teaching are to be based on the third assumption, rather than on either of the first two, it will be necessary for us to learn much more about children's science: to know how to explore it, to know about its nature, and to consider the various ways in which children's science may, or may not, be modified by learning experiences.

The exploration of children's science

A variety of methods have recently been developed for use in investigating children's science. White (1979) has analyzed the similarities and differences of some of these methods. Most involve in-depth interviews with children (see, e.g. Pines *et al.* 1978; Brumby 1979; Tiberghien 1980). This study used two such methods which we have called the Interview-about-Instances approach and the Interview-about-Events approach. The Interview-about-Instances approach (Osborne and Gilbert 1980b; Gilbert *et al.* 1981) explores children's meanings for words by means of taped individual interviews. For a particular word, e.g., work, force, living, up to 20 familiar situations, depicted by line drawings on cards, are presented to the child. Some of the situations present an instance of the scientific concept embodied in the word and some do not. Children are asked, for each situation in turn, whether they consider it an instance or not. The children's reason for the choice is then elicited. The interview situation allows children to ask questions, to clarify perceived or actual ambiguities before answering, and also gives flexibility in discussing reasons or lack of reasons, for a particular answer. The method

Figure 15.2 Science teaching in which it is assumed that learners may have theoretical views but that these are easily displaced by the views presented by teachers.

Figure 15.3 Science teaching which recognizes that learners often do hold strongly entrenched theoretical views that persist in the face of teaching.

has been used to explore children's meanings for many words: e.g. "work" (Osborne and Gilbert 1979), "electric current" (Osborne and Gilbert 1979; Osborne 1981), "force" (Osborne and Gilbert 1980a; Watts 1980), "light" (Stead and Osborne 1980), "living" (Stead 1980), "friction" (Stead and Osborne 1981a), "gravity" (Stead and Osborne 1981b), and "animal"(Bell 1981).

The Interview-about-Events approach (Osborne 1980) places more emphasis on eliciting children's views of the world within the overall framework of children's science. It involves an individual discussion with an interviewee about an articulated series of demonstrations. This discussion is tape recorded, transcribed, and subsequently analyzed. The interview is built around a scientific concept, e.g., "physical change." The events are practical demonstrations of situations to which the concept may be applied. The demonstrations, performed by the interviewee with minimum assistance, are articulated to produce a smoothly linked conversation. The method has been used to explore children's views on "physical change" (Cosgrove and Osborne 1980), "chemical change" (Schollum 1981), and the "particle nature of matter" (Happs 1981). In the Appendix a sequence of steps used to investigate children's views on physical change is provided.

Patterns in children's science

On the basis of the findings from research which has been carried out using the two investigatory techniques, referenced above, at least five different patterns of children's science can be described (Osborne and Gilbert 1980a). These patterns will be illustrated from the sequence of discussions on physical change (Appendix). These illustrations arise from interviews with 43 New Zealand school children spread evenly over the 10–17-year age range. (The 10–15-year olds were studying general science, the 16–17-year olds were studying physical science.) The pupils were selected by their teachers as being of average attainment in science (Cosgrove and Osborne 1981). Each quote given will be followed by the step in the discussion sequence to which it applied, and the age of the interviewee.

Everyday language

Many words in science are used in an alternative way in everyday language. Often a student can listen to, or read a statement in science and *make sense* of it by using the everyday interpretation of the word. The interpretation is not the one intended by the teacher or textbook writer. For example:

> The air is made up of small particles (is anything else made up of small particles?) glass…they are made out of small particles of sand which have been turned hot…turned clear and then sort of take them out…and put them between two pieces of metal when they have been hardened and when they take it off they find that they have a clear surface called glass.
>
> (Step 7; age 11)

The word "particle" is commonly used in science classes to mean atom, molecule, or ion. In everyday use it refers to a small, but visible, piece of solid substance. The everyday meaning has been applied to air. The interviewee has apparently presumed that the "particle" size in sand is retained in glass. A parallel has been drawn between glass and air based on appearance.

Self-centered and human-centered viewpoints

Many very young children have very egocentric views of the world. By age 9 or 10 most children no longer adopt this strictly egocentric view but they still interpret and consider things in terms of human experiences and commonly held values. For example:

> Ice is just frozen water (what's the difference between frozen water and ordinary water?) You can't drink it very good.
>
> (Step 7; age 10)

Properties as a drink govern the evaluation made by this child of ice and water. A second example is:

> I think I said it was oxygen in the bubbles... but if you put your face over (the steam) and breathe in... it doesn't seem you can breathe too well... so I don't think there is much oxygen... it be more hydrogen.
>
> (Step 2; age 17)

Steam has been evaluated here by its capacity to support breathing, oxygen being known to be effective. In both cases, simple human concerns have governed the interpretation made of phenomena.

This different focus on how and why things behave as they do can result in children viewing situations in quite a different way to the more analytical, and impersonal, view of science. Answers given by children in science classrooms are sometimes apparently "off the track" hoped for by the teacher because of this difference in perspective of science teacher and student. The anthropocentric view often takes the form of some widely held beliefs – heavier objects do fall faster, things do get lighter when they are burnt, animals are things you take to the vet – and these human-centered views are reinforced by everyday language to some extent.

Nonobservables do not exist

To a number of children, and some learners despite formal teaching, a physical quantity is not present in a given situation unless the effects of that quantity or the quantity itself is observable. Some examples are: "If you cannot feel an electric current it is not present" (Osborne and Gilbert 1979); "if the effects of the presence of light, for example, flickering on a wall, are not observable the light is not present" (Stead and Osborne 1980).

> Oh, it has evaporated. (What does that mean?). Well it has not gone into the steam form because it doesn't look as if it has gone up in the water state... it must have split up because you couldn't sort of see steam or anything rising. (What do you mean split up?) The hydrogen and the oxygen molecules.
>
> (Step 5; age 16)

The student has presumed that, on all occasions, the visibility of water is maintained on the transition from the liquid state to the vapor state. When this visibility is not maintained, an explanation is presented in terms of elements known to be invisible and constituents of water, i.e., hydrogen and oxygen, commonly encountered in their gaseous form.

Endowing objects with the characteristics of humans and animals

Children often endow objects with a feeling, a will, or a purpose. This is partly related to children's view of living things being much broader than the biologists' viewpoint (Stead 1980), but it is also reinforced by the use of metaphor in both common language and even in the teaching of science. Teachers make statements like "the electric current chooses the path of least resistance," "the positive ion looks out for a negative ion." However, it would appear that, not surprisingly, children do not always consider such statements to be metaphoric. For example:

> It's cold in there and the chill's coming to the outside...the coldness just...um...oh, it's cold in there and it's just trying to get out...and it's somehow got out.
>
> (Step 6; age 13)

"Cold" is thought to move towards the outside of the jar under the effect of an implied will.

Endowing objects with a certain amount of a physical quantity

It is not uncommon for children to endow an object with a certain amount of a physical quantity and for this quantity (e.g. force, momentum, energy) to be given an unwarranted physical reality. For some physical quantities (e.g. force, coldness, etc.) this tendency of children leads to considerable difficulties in learning, particularly in appreciating the abstract nature of these quantities and their relationship to other quantities. For example:

> The heat makes the air bubble come out of the element.
>
> (Step 2; age 12)

The implication here is that heat is a physical entity. It is thought to physically force the air bubble to come out of the heating element in the kettle. Both the nature of heat and the source of air bubbles have been unconventionally understood. A second example is:

> The coldness of the ice could have brought the water... but that's a bit funny.
>
> (Step 7; age 12)

Here "coldness" is thought to have a physical identity.

Teachers' views of science

Just as by *children's science* we mean those views of the natural world and the meanings for scientific words held by children before formal science teaching, so *scientists' science* (S_S) means the consensual scientific view of the world and meaning for words. Ideally the view of science presented to children by teachers, or directly through curriculum material, will closely relate to scientist's science. However, this may not always be so. Teachers undoubtedly have a wide variety of viewpoints, (S_T), ranging from almost children's science to scientists' science, but often different from both these in distinguishing less clearly between the objects of science and the concepts that relate to them (Fensham 1979). This teacher's view of science interacts with the science curriculum and its materials as he/she prepares for teaching. This may or may not modify this view in the direction of scientists' science as shown in Figure 15.4. The resultant is the *viewpoint presented* by the teacher to the pupils. It is the interaction of children's science and their teacher's science that will have profound implications for the outcomes of teaching.

The consequences of children's science for teaching

A further consideration of the data collected using the Interview-about-Instances and Interview-about-Events techniques suggests that for children who have been taught science there are at least five patterns of outcomes from these interactions. The five outcome patterns will again be illustrated from protocols using the same Interview-about-Events sequence in Appendix (Cosgrove and Osborne 1981).

The undisturbed children's science outcome

Some children have an undisturbed viewpoint despite formal teaching. Reasonably common among this pattern of learners are those who now incorporate some language of science to describe the viewpoint, but whose viewpoint is essentially unaltered. The following is an example of undisturbed children's science despite teaching:

> (Where have you used the word particle?) In the science lab. (Are there particles in the jar of ice water?) Yes, I suppose so. (Which are the particles to you?) The ice blocks. (Has the water got anything to do with particles?) Oh, they melted into the water.
>
> (Step 7; age 13)

The children's science view was that a visible piece of ice is a particle. The language of science, using "particle" to mean molecule of water, has had little impact on this view. This type of interaction is presented in Figure 15.5. Similarly:

> The water has melted it...it has become part of the water...but there are parts of it left that you can't see...the taste of sugar.
>
> (Step 4; age 11)

Figure 15.4 Strongly held teachers' views of science may persist or interact with the views in science curricula.

Figure 15.5 A prelearning or children's view of science can persist unchanged by science teaching.

The children's science view, that taste is separate from material substance, has not been modified by contact with the phenomenon of dissolving.

The two perspectives outcome

It is possible for the student to basically reject the teacher's science as something that can be accepted in terms of how to view the world, but to consider it as something that must be learned, e.g., for examination purposes. The student, therefore, has two views, but the learned science viewpoint is not one that has been adopted for use outside the formal learning situation. For example:

> It is dry...the water has evaporated...the water has gone (where to?) well...the teachers tell me that it has gone you know...that it makes up the clouds, you know in the sky and that sort of thing. (I see, it has gone up to the sky?) it is meant to have (where do you think the water that was on the saucer has gone?) I don't know...I don't think about it (it is not still on the plate dried up is it?) no, I don't think so...(how does it get from here to the clouds?) I don't know (magic?) no...it's sort of a gas there...not magic (where did you learn about clouds and evaporation?) in about fourth grade (9–10 years)...around there somewhere (oh, well they wouldn't have talked about it in much detail at that sort of level would they?) no (and all that you can sort of remember is that when water evaporates it goes into the clouds?) yet (but you don't have a picture of how that goes on?) no, except for little arrows that point up (I see, what were those arrows do you think?) can't remember (so you have got this sort of picture of water, arrows and clouds?) yes, and it sort of comes down as rain.
>
> (Step 3; age 14)

This student has the view that water disappears from a place into the air. However, the standard explanation, concerning evaporation and using diagrams, has proved less than believable to the student. Nevertheless, it has been learned but is not used willingly to explain phenomena. This type of interaction is presented in Figure 15.6.

The reinforced outcome

The dominance of the students' prior understandings and meanings for words can, as suggested earlier, often lead to quite unintended uses of what is being taught. One

Figure 15.6 Science teaching can result in a second view being acquired for use in school but the original children's view persists elsewhere.

Figure 15.7 The original children's view is strengthened by science teaching which now is misapplied to support it.

common example of the outcomes of this is the confusion between physical quantities. Quantities defined in science in a particular way can be misinterpreted to mean something quite different. In Figure 15.7, the children's science viewpoint is being maintained following teaching but now scientific concepts are put forward to explain or underpin a particular viewpoint. For example, the statement by a younger student:

> It would come through glass.
>
> (Step 6; age 10)

becomes, for an older student

> Through the glass...like diffusion through air and that...well it hasn't got there any other way (a lot of people I have talked to have been worried about this water...it troubles them) yes, because they haven't studied the things like we have studied (what have you studied which helps?) things that pass through air, and concentrations, and how things diffuse.
>
> (Step 6; age 15)

The notion of diffusion, learnt in connection with movement through air and water, has been applied to explain movement through glass. The children's science idea of "movement through air" has been transformed into "diffusion through glass."

The mixed outcome

In many cases, scientific ideas are learned, understood, and appreciated by learners. However, the interrelationships of these ideas are manifold and at any one time only a limited amount can be learned. Often this results in students holding ideas that are not integrated and may be self-contradictory. In this outcome the learners' views are a mixture of amalgam of children's science views and teachers' views, Figure 15.8. For example:

> I think it is the same atoms in the ice before and now they are unfrozen in the water (what else is in there besides the atoms? the stuff that freezes?) no...I don't know...yes...no...it's all atoms but the atoms are just frozen.
>
> (Step 7; age 14)

The idea of the conservation of matter between physical phases has been learned. However, the microscopic change in structure is being interpreted as a general change in the properties of microscopic components, i.e., atoms (*sic*).

The unified scientific outcome

The aim of all science education is that a learner should obtain a coherent scientific perspective (S_S) which he understands, appreciates, and can relate to the environment

Figure 15.8 Science teaching resulting in a mixed outcome where children's science and teachers' science now coexist together.

Figure 15.9 Science teaching which extends children's science and teachers' science to a more unified science view.

in which he lives and works. Students can be found who have this view in relation to specific words and viewpoints that we have investigated. In some of these cases, the learned viewpoint is in fact more closely aligned to scientists' science than to the teacher's views of the science. This outcome is represented in Figure 15.9. A typical example of the coherent scientific perspective:

> It is wet on the outside… 'cos the jar's cold… 'cos the ice is inside it and therefore the water molecules that are in the air moving around…although we can't see them…when they hit the cold jar…that makes them cold…and therefore they group together again in their groups of molecules and then they become water again because they've been cooled down.
>
> (Step 6; age 15)

It is the outcome that all teachers would wish to arise from their interaction with students.

Conclusion

This paper suggests, by argument and example, that the view which children bring with them to science lessons are, to them, logical and coherent and that these views have a considerable influence on how and what children learn from their classroom experiences. Our conclusions from a variety of studies support the view of Wittrock (1977) that people tend to generate perceptions and meanings that are consistent with prior learning. Learning can be anticipated and understood in terms of what the learners bring to the learning situation, how they relate the stimuli to their memories, and what they generate from their previous experiences.

We have also attempted to suggest, by argument and example, that the aim of science teaching and learning can be viewed as the development of children's science. Traditionally, the goal of the development is scientists' science. This has proved to be an immense task that is often very incomplete even among so-called successful learners. As happens in many present science classes, we may have to be satisfied with largely undisturbed children's science as our outcome. A more modest and manageable goal in these cases would be to make these learners aware that there *is* another viewpoint, the scientists' viewpoint, which is useful to scientists and may have more general use also. Only by adapting our teaching to make these two views explicit is this new goal likely to be achieved. This approach may also facilitate the development process on its way. Such a development will only occur in a genuine and nonsuperficial way if the scientific perspective appears to students to be at least as logical, coherent, useful, and versatile way of viewing the world than their present viewpoint.

Whatever the goal, it would seem that teachers need to be aware of children's science and to encourage students to express their views. We all need, as teachers, to listen to, be interested in, understand and value the views that children bring with them to science lessons. It is only against that background of sensitivity and perception that we can decide what to do, and how to do it. This is a major challenge for science teaching.

Appendix

Interview-about-Events outline schedule for physical change

Step 1 The interviewee is presented with a screw-top jar containing ice and is invited to dry the jar thoroughly. The jar is then set aside.

Step 2 The interviewee is invited to observe the water coming up to, and boiling, in an electric kettle. Preliminary questions are "What is happening?" and "What are the bubbles made of?"

Step 3 The interviewee holds a saucer in the steam and is invited to comment on what is observed and why it has happened. After these questions, it is put, inverted, to one side.

Step 4 Some of the hot water (from Step 2) is put in a cup. The interviewee puts some sugar in it and stirs the mixture. The preliminary question is again "What is happening?"

Step 5 The inverted saucer (see Step 3) is now reconsidered. The dryness is discussed through "What has happened to it?" and "Why is that?"

Step 6 The jar (see Step 1) is now reconsidered. It now has water on the outside. The interviewee is asked "Is that different to when you had it before?" and "Can you tell me about that?"

Step 7 The lid of the jar (see Step 6) is removed, and some water and ice extracted on a spoon. The questions begin with "What is happening here?"

References

Bell, B.F. (1981). When is an animal not an animal? *Journal of Biological Education*, 15, 3.

Brumby, M. (1979). Students' perceptions and learning styles associated with the concept of evolution by natural selection. Unpublished doctoral dissertation, University of Surrey, UK.

Champagne, A., Klopfer. L., and Anderson J. (1979). Factors influencing learning of classical mechanics. Paper presented at AERA Meeting, San Francisco, CA, April.

Clement, J. (1977). Catalogue of students' conceptual models in physics. Working Paper, Department of Physics and Astronomy, University of Massachusetts.

Cosgrove, M. and Osborne, R.J. (1980). Physical Change, L.I.S.P. Working Paper No. 26. Hamilton, New Zealand: University of Waikato, Science Education Research Unit.

Driver, R. and Easley, J. (1978). Pupils and paradigms: a review of the literature related to concept development in adolescent science students. *Studies in Science Education*, 5, 61–84.

Fensham, P.J. (1979). "Conditions for co-operation and strategies for innovation." In *Co-operation between Science Teachers and Mathematics Teachers*. H.G. Steiner (ed.), Institute für Didaktik den Mathematik der Universität Bielfeld, pp. 553–580.

Fensham, P.J. (1980). A research base for new objectives of science teaching. *Research in Science Education*, 10, 23–33.

Gilbert, J.K. and Osborne, R.J. (1980). "I understand, but I don't get it": some problems of learning science. *School Science Review*, 61(218), 664–674.

Gilbert, J.K., Watts, D.M., and Osborne, R.J. (1981). Eliciting student views using an interview-about-instances technique. Symposium paper presented at the AERA Conference, Los Angeles, CA, April.

Happs, J.C. (1981). Particles, LISP. Working Paper No. 18. Hamilton, New Zealand: University of Waikato, Science Education Research Unit.

Leboutet-Barrell, E.M. (1976). Concepts of mechanics among young people. *Physics Education*, 11(7), 462–466.

Nussbaum, J. and Novak, J.D. (1976). An assessment of children's concepts of the earth using structural interviews. *Science Education*, 60, 535–550.

Osborne, R.J. (1980). Some aspects of students' views of the world. *Research in Science Education*, 10, 11–18.

Osborne, R.J. (1981). Children's views on electric current. *New Zealand Science Teacher*, 27, 12–19.

Osborne, R.J. and Gilbert, J.K. (1979). An approach to student understanding of basic concepts in science. Guildford, Surrey, UK: University of Surrey, Institute for Educational Technology.

Osborne, R.J. and Gilbert, J.K. (1980a). A technique for exploring students' views of the world. *Physics Education*, 15(6), 376–379.

Osborne, R.J. and Gilbert, J.K. (1980b). A method for the investigation of concept understanding in science. *European Journal of Science Education*, 2(3), 311–321.

Pines, A., Novak J., Posner, G., and Van Kirk, J. (1978). The clinical interview: a method for evaluating cognitive structure. Ithaca, NY: Department of Education, Cornell University.

Schollum, B. (1981). Chemical Change, LISP. Working Paper No. 27. Hamilton, New Zealand: University of Waikato, Science Education Research Unit.

Stead, B.E. (1980). The description and modification of some students biological concepts. Unpublished MEd thesis, University of Waikato, New Zealand.

Stead, B.E. and Osborne, R.J. (1980). Exploring science students' concepts of light. *Australian Science Teachers Journal*, 26(3), 84–90.

Stead, K.E. and Osborne, R.J. (1981a). What is friction? Some children's ideas. *Australian Science Teachers Journal*, 19, 41–52.

Stead, K.E. and Osborne, R.J. (1981b). What is gravity? Some children's ideas. *New Zealand Science Teacher*, 30, 5–12.

Tiberghien, A. (1980). Modes and conditions of learning. An example: the learning of some aspects of the concept of heat. In *Proceedings of Cognitive Development Research Seminar*, F. Archenhold *et al.* (eds), Leeds, UK: University of Leed Centre for Science Education, pp. 288–309.

Watts, D.M. (1980). An exploration of students' understanding of the concepts "force" and "energy." Paper presented at the Conference on Education for Physics Teaching, Trieste, September.

White, R. (1979). Describing cognitive structure. Paper presented at the Australian Association for Research in Education Conference, Melbourne, Australia, November.

Wittrock, M.C. (1977). Learning as a generative process. In *Learning and Instruction*, M.C. Wittrock (ed.), Berkeley, CA: McCutcheon, pp. 621–631.

CHAPTER 16

CONSTRUCTIVE SCIENCE EDUCATION

Pope, M.L. and Gilbert, J.K. 'Constructive science education'. In F. Epting, and A. Landsfield (eds), *Anticipating Personal Construct Psychology*. Lincoln, NE: University of Nebraska Press, 1985, pp. 111–127

Psychology bases for science education

Behavioristic psychology has traditionally had a large influence on the design and conduct of science education in many countries; however, things are changing. In 1979, the major association of science educators in Britain (Association for Science Education 1979) suggested that alternative models of psychology, such as that of Kelly (1955), should be considered for their implications with respect to science education. That consideration, which has historical parallels in psychology itself, is gaining momentum.

Davisson (1978) noted that Kelly's theory received little recognition when it was originally published in the United States. The US psychological community were, relative to their British counterparts, slow to accept and use the insights provided by personal construct theory. Among the reasons given by Davisson is that the theory was neither "scientific nor humanistic enough in the 1955 sense of the terms for Kelly's views to become a significant factor in American psychological thinking." Since the 1950s, there has been considerable debate as to just what justifies the label "scientific," and indeed this remains an open question. However, since psychology is no longer dominated by the scientistic conception of science underpinning behaviorism, personal construct psychology is now attracting a new interest within the United States and continues to develop in Britain. Although much of this development has been within the sphere of clinical psychology as an extension of Kelly's original work in this area, his views are gaining increasing recognition among educators.

Within science education this recognition is a recent phenomenon. Current pedagogical practice in science education seems to be dominated by a cultural transmission approach to teaching, and its allied view of knowledge neglects the role of personal experience in the construction of knowledge (Pope and Gilbert 1983). Piaget's views have had a considerable impact on science education, but science educators using Piagetian models have restricted their focus to notions of the fixity of stages of cognitive development at the expense of the constructivist thrust of Piaget's epistemology.

Since 1980 we have directed a group of researchers whose common commitment is the recognition of the role of personal construction in knowledge, hence the name Personal Construction of Knowledge Group (PCKG). The work of the group spans a range of disciplines. However, in this paper we are restricting ourselves to the area of science education that forms the predominant focus of the group. This group forms part of a growing invisible college of science educators who believe that the teaching of science should acknowledge current philosophies of science (i.e. the role of personal construction in the development of scientific knowledge). The psychological perspective of George Kelly provides a framework that is compatible with many of the metaphysical commitments of these educators. Within the PCKG we have found that the spirit of his work not only relates to our views on teaching but also has been a powerful influence in the conduct of our research (see Pope 1981).

An essential part of our research has been, within the Kellyan tradition and similar to that of the ethnomethodologists (e.g. Garfinkel 1967; Cicourel 1974), to find out how the people *within* a situation view it. In the case of describing behavior in a classroom, it has been vital to have the comments of the actors – the teachers and students – when describing classroom events. This constructivist emphasis can be seen as an alternative to the naive empiricist-associationistic theories which have dominated science education and which have led to a passivist approach to knowledge.

Passivist and activist theories of knowledge

Lakatos (1970) noted an important demarcation between passivist and activist theories of knowledge:

> "Passivists" hold that true knowledge is Nature's imprint on a perfectly inert mind: mental *activity* can only result in bias and distortion.... "Activists" hold that we cannot read the book of Nature without mental activity, without interpreting them in the light of our expectations or theories. Now *conservative "activists"* hold that we are born with our basic expectations; with them we turn the world into "our world" but must then live forever in the prison of our world.... But *revolutionary activists* believe that conceptual frameworks can be developed and also replaced by new, *better* ones; it is *we* who create our "prisons" and we can also, critically, demolish them.

The positivist, empiricist–inductivist conception of science is in sympathy with an absolutist view of truth and knowledge, and thus if teachers hold to that conception of science, then curriculum content and the manner in which students are taught will place little or no emphasis on the student's own conceptions and active participation. The model of the learner as "impotent reactor" might be appropriate here. The passivist theories of knowledge are rejected by personal construct psychologists. They also reject a conservative–activist position, such as that described by Lakatos, since it implies that the "limiting cages" of personal constructions are permanent limitations. Piagetians who stress the limiting aspects of stages of development are adopting a conservative–activist view, and indeed there is much within Piaget's genetic epistemology which is similar to Lakatos's description of a conservative–activist position. A Kellyan philosophy is akin to Lakatos's description of the revolutionary activist's views. Rather than see constructions of reality as "prisons," Kelly would see them as dynamic frames which can limit conceptual development *if* an individual chooses not to exercise responsibility for the creation of such frames. Individuals are not inevitably limited by their world views.

For Kelly, the construction of reality is a subjective, personal, active, creative, rational, and emotional affair. If we are to believe modern philosophers of science, then similar adjectives can be applied to scientific theorizing and methodology. However, students' experiences of science in schools and colleges do not appear to be developing this viewpoint. The ideas of "children's science" and "students' science" have proved useful in interpreting the understandings that young people demonstrate (Gilbert *et al.* 1982b), yet their interaction with "teachers' science," although capable of recognition (Gilbert *et al.* 1982a), has been largely ignored in science classrooms (Zylbersztajn 1983). It is worthwhile to speculate on why this is so.

A challenge to traditional views of science education

Within the Kellyan tradition, person-the-scientist and scientist-the-person are both engaged in a process of observation, interpretation, prediction, and control. In our work we have been interested in the personal theorizing of the young person-scientist and have advocated that science educators should pay attention to the personal meanings of their students (Gilbert and Pope 1982; Watts *et al.* 1982; Pope and Gilbert 1983). This advocacy presents a challenge to traditional views of science education. The key seems to lie in the notion of science which underlies science education.

In putting forward his philosophy of constructive alternativism, Kelly was sensitive to the possible problems entailed by its relativist nature. At the time he was writing, philosophers of science such as Popper, Kuhn, and Lakatos had not had the impact on thinking in science that these theorists have today. Relativity of knowledge received scant attention. As Kelly pointed out, people, like the scientists of his day, may find it hard to accept that their personal models are not the world as it is but are constructed realities and that they are not soundly based in absolute truths. When faced with the challenge of constructive alternativism, people may be unwilling to accept the responsibility that goes along with the acknowledgment that is *they* who construct their own world views. For many, it is more accept-able to believe that their views are imposed on them by "the way things really are." A Baconian view of science still informs much of science education (Cawthron and Rowell 1978; Swift *et al.* 1983).

Kelly (1970a) recognized the challenge of constructive alternativism as follows:

> A person who spends a great deal of his time hoarding facts is not likely to be happy at the prospect of seeing them converted into rubbish. He is more likely to want them bound and preserved, a memorial to his personal achieve-ment. A scientist, for example, who thinks this way, and especially a psychol-ogist who does so, depends on his facts to furnish the ultimate proof of his propositions. *With these shining nuggets of truth in his grasp it seems unneces-sary for him to take responsibility for the conclusions he claims they thrust upon him.*
>
> To suggest to him at this point that further human reconstruction can com-pletely alter the appearance of the precious fragments he has accumulated, as well as direction of their arguments, is to threaten his scientific conclusions, his philosophical position, and even his moral security. No wonder, then, that, in the eyes of such a conservatively minded person, our assumption that all facts are subject – are wholly subject – to alternative constructions looms up as culpably subjective and dangerously subversive to the scientific establishment (our italics).

The challenge of relativist views of science – for example those due to Kuhn, Lakatos, and Feyerabend – when coupled to a constructive alternative view of science education has only partially been accepted by science educators.

In the early 1970s, a number of books emphasized the role of people's active construction of experience in the creation of knowledge and the necessity of revolution for its development. By revolution we are referring to the day-to-day questioning of their views as people develop their concepts about any topic in addition to the well-documented revolutions that have occurred in, for example, the area of science when the dominant paradigm of the day was challenged by those who stepped outside the limits of present theory and engaged in what Kuhn called "extraordinary science" (Kuhn 1970). Despite the considerable emphasis on constructivism, however, practice in science classroom does not adequately reflect this concern.

Part of the reason for the lack of implementation may be ignorance. However, it is just possible that philosophies of constructive alternativism such as those put forward by Kuhn, Lakatos, and Feyerabend pose a threat to a core construct held by many people – that is, "the world as they see it is the world as it is." Any other conception would, for such people, be met with Kellyan hostility. No doubt that hostility exists within the ranks of science educators, but the climate of the times is such that the epistemology inherent in constructive alternativism is more readily accepted. Given such acceptance, we have sought to encourage science educators to explore some of the major tenets of personal construct psychology and their implications for the practice of science education.

The role of the teacher in students' conceptual change

For Kelly, whether or not a person will change his or her constructs depends on the permeability of constructs and on the success or otherwise of predictions entailed by the constructs; and the extent of change will depend on the nature of the interrelationships between constructs with the person's repertoire. Kelly, like Piaget (1971) and Werner (1957), saw conceptual development as an evolutionary process which involved the progressive differentiation of conceptual structures (groups of constructs) into independently organized substructures and the hierarchic integration of these substructures at progressively higher levels of abstraction. The functional differentiation of structures enhances the range of convenience of an individual's construct system. However, hierarchic integration of these differing substructures is necessary for the integrity of a person's construct system. Having a metatheory which provides a linkage among a number of minitheories can allow the individual to make a wider range of cross-references than is possible within a very differentiated system. Bruner (1977) saw this as an essential aspect of the development of intelligence.

Although adequate cognitive functioning requires integration through the deployment of superordinate constructs, differentiation or fragmentation can have its merits. One implication of Kelly's fragmentation corollary is that the constructive alternativist can test new hypotheses without having to discard the old hypotheses or constructs. Because constructs are hypotheses, we can hold on to constructs that are incompatible. Kelly (1970b) saw this as a feature of human thought which was especially noted in children: "The nice thing about hypotheses is that you don't have to believe them. This, I think, is a key to the genius of scientific method. It permits you to be inconsistent with what you know long enough to see what will happen. Children do that. What is wonderful about the language of hypothesis is its refreshing ability to free the scientists from the entangling consistencies of adulthood. For a few precious moments he can think again like a child, and, like a child, learn from his experience."

Kelly clearly valued the children's, at times inconsistent, theorizing as scientific, while the adults' tightly ordered constructs could limit their theorizing and thus stunt the development of new knowledge. We also believe that children's alternative and often inconsistent frameworks should be recognized for their epistemological status.

Kelly's theory is in a sense a metatheory – a theory about theories. His individuality corollary may help to explain the differing philosophies of science held by the philosophers: philosophers' science. However, the premises embedded in the theory are relevant to the constructions of science held by children, students, and science educators. Children, even before they meet school science, are scientists – that is, they have their personal theories and indulge in experimentation. Through their direct experiences with the physical world and informal tuition, the child will have evolved a set of personal theories to explain events: children's science.

At this stage, it may be useful to take a look at an illustration of the personal theories of a young scientist by drawing on an example given by Claxton (1982) – one of the members of the invisible college who recognizes the role of personal construction in knowledge.

As Claxton points out, we do not have *a* personal theory we have *many* personal theories – "each one of these 'mini-theories' being defined by the domain of experience to which it applied" – in other words, each minitheory has what Kelly called a focus of convenience. These minitheories or constructs will be linked to others and will have varying ranges of convenience, positions within the system, degrees of permeability, and so forth. A minitheory will change through extension or restriction of its range and its position and entailment with other minitheories. Claxton suggests that a situation can arise when a number of originally separate minitheories are found to contain a common subtheory. He gives an example of a child who has developed separate minitheories to deal with elements such as drinks, bath time, boats, and the local pond. Eventually the child may "come to see that there is a reliable and predictable trans-situational component of all of them called *water*" and subsequently develop a trans-situational minitheory about water.

Claxton suggests that one way of looking at science teaching is that it is analogous to "trying to civilise a land that is already densely populated with people who have their own, perhaps 'primitive' but certainly workable culture." Teaching can be designed so that these so-called primitive beliefs can be understood and respected. These beliefs or constructs are an integral part of a person's world view, and since the person has found them workable for his or her purpose, they may be resistant to change. *Recognition should be given to the existing knowledge of the child because this is the basis upon which the child constructs experience of the formal science lesson.* As Claxton notes, if a child is confronted in a science lesson with the question, "What will happen if we drop this weight into this measuring cylinder of water," the child may not possess a trans-situational construct about water. However, the child may have minitheories about drinks, baths, boats, and beds as a repertoire of existing knowledge which could be used to generate a prediction because

> each of them has something to say about the situation of lowering a solid object onto or into a "squishy" medium. If he chooses *drinks*, he may say that the water level will remain unchanged – because when he puts a spoonful of sugar to a cup of tea, no change in the overall volume is noticed. If he extends

boats, he will get the same answer because neither the sea nor the local pond visibly rises or falls when a real or a toy boat is launched. If he uses the *beds* mini-theory, however, he may predict that the water level will go down – because his bed goes down when he lies on it. And, finally, if he chooses to base his prediction on *baths*, then he will say that the level will go up, because that is what happens when he gets into the bath.

In the example given, only the minitheory *baths* generates the correct prediction. In order to understand the alternative predictions or misconceptions of the child who extends a boats, drinks, or beds minitheory, the teacher could elicit an explanation of the child's prediction. Claxton gives the following examples of the types of explanations that may be given for the alternative predictions: "The *drinks* theory supports its prediction by generating 'The weight will just appear, like the sugar.' *Boats* predicts no change because 'The weight will just sit in the water like a ship.' *Beds* gives us 'The water level goes down because the weight squashes it.' "

In our own work we have sought to encourage science teachers to recognize the importance of seeking the pupil's explanation. The explanations here are each "a perfectly rational deduction from an unappropriately applied theory" (Claxton 1982). From a Kellyan point of view, the teacher needs to recognize that these minitheories have been viable in their contexts, that some may be very firmly embedded in a system of interrelationships with other minitheories, and that initially the child may not readily accept invalidation of an inappropriately applied theory. Indeed, in cases where a correct prediction is made, one should not assume that the explanation given will necessarily correspond to the symbolic science explanation.

The teacher will want to gain some understanding of the elaboration of the minitheory which results in the correct prediction. What is the pupil's reasoning as to why the level of the water goes up when a person gets into the bath? By further probing, the teacher will gain an insight into the interrelationship of ideas that the child can bring to bear on problems such as the one described here. Indeed, by talking to children about their minitheory or personal constructs, the teacher may find that a particular child has a construct which assumes that the level of water will go up when he or she gets into the bath but has not extended the range of convenience of this construct to apply to the event of dropping weights into measuring cylinders of water.

One task of a science teacher, as seen from a constructivist viewpoint, is to develop situations for learners whereby their personal constructs or minitheories can be articulated, extended, or challenged by the formal constructs of the currently accepted scientific view or to make a bridge between a construct from personal experience which can be extended to events in the domain of science.

Recent interest in learners' constructs is not confined to the teaching of school children. Posner (1981) stresses the need to consider college students' preconceptions, purposes, values, and conceptions of past experiences which they bring to particular curricular tasks. He points out that college students' preconceptions can be resistant to change because they have been acquired through interaction with the physical world without formal instruction and are "therefore very functional in and adaptable to most circumstances." Unless ways are found to develop or make bridges between these preconceptions and the formal concepts of science, Posner suggests that students may simply compartmentalize their knowledge "claiming that the problem is a physics problem and therefore does not have anything to do with the 'real world'." A constructivist curriculum must include methods which militate against excessive compartmentalization and fragmentation.

Adopting a Kellyan perspective would require the teacher to recognize pupils' or students' personal scientific constructs as having both important epistemological value and high educational status. This is not an assumption shared by those who uphold the traditions of science education who stress that children's beliefs must be directly and efficiently overcome by the strength and knowledge of "true" science. When noticed, children's constructs are not to be built and developed, but corrected. This view of education is increasingly under attack (see, e.g. Barnes 1974; Donaldson 1978). However, science teachers rarely adopt a phenomenological perspective and seek to elicit and utilize the pupils' views within their teaching (Zylbersztajn 1983). For Donaldson, *it is important that the teacher comes to an understanding of the premises used by the child.* Just telling the child the correct answer couched in the language of high science has been shown to be ineffectual (Driver 1973). The child may be able to use the language of the teachers' science but, in Kellyan terms, the teachers' science is only incorporated into the person's system of constructs in a fragmented fashion and is unconnected in any meaningful way with the person's more central or core constructs.

A Kellyan approach to the teaching of science would suggest that, as a matter of classroom policy, differences among the learner's personal meanings, those of the teacher, and the formal concepts of scientists' science should be dealt with in an open forum where the differences are valued for what they are (i.e. constructive alternative ways of seeing).

Although the focus of his theory was the individual, Kelly (1970a) did not totally ignore the social context, since he suggested that we may limit our potential for extending our view of the world for fear of the consequences within a social setting: "Novel ideas, when openly expressed can be disruptive to ourselves and disturbing to others. We therefore often avoid them, disguise them, keep them bottled up in our minds where *they cannot develop in the social context,* or disavow them in what we believe to be loyalty to the common interest. And often, against better judgment, we accept the dictates of authority instead, thinking thus to escape any personal responsibility for what happens" (our italics).

Kelly saw other people as important in the validation or invalidation of a person's constructs and recognized that a person may have a construct about the importance of authority or the importance of being loyal to the common interest and that these constructs may be paramount and will, if a person holds them, restrict certain experimentation. Thus if a child perceives the social context of school as one in which the teacher's knowledge is the only one to be valued, then there may be a reluctance to experiment with his or her personal views, and knowledge may be limited to "what the teacher tells one." Proctor and Parry (1978) reject the way many sociologists tend to view "the individual merely as a passive occupant of socially determined roles." They prefer Kelly's view that cultural control is something that is "within the client's own construct system." There is an essential dialectic between the constraints of our culture and personal freedom. As Proctor and Parry put it, "Ideologies determine us to the extent to which we internalise the values of the culture." Nevertheless, we are able to recognize this process and, in doing so, stand above it. *If we recognize our personal agency in the construction of knowledge, we can begin to identify where we have constrained ourselves by adopting, uncritically, the prevailing social ideology.*

It is possible to think of theories in science as being determined by physical reality. It is also possible to view the child's conception of science as being determined by his or her social context (i.e. what they are told by teachers, parents, textbooks, or other pupils). Children may choose to construe knowledge as being constrained in this way. However, we would argue that throughout their schooling, pupils should recognize that their constructions are important, that they are active, rational, and emotional beings who develop their own world views, and that the progress of science itself depends on such personal meanings.

Knowledge in science can be seen as progressing from the personal construction of individual scientists seeking to make sense of their experiences and anticipate events toward some consensus of construing a topic by a community of scientists. Kelly's commonality corollary allows for overlap between personal views and therefore a partial consensus. Along the way, the journey taken by scientists is fraught with conflicts, personal and interpersonal, and in a sense the journey is one that will not end, because they must be open to transformations of the consensus. The body of formal knowledge, which we call science, should be seen as constructed by and related to the personal commitments of those who form the scientific community. The ideas held by a consensus of the scientific community are neither inviolate nor incorrigible. As Watts *et al.* (1982) suggest, Kelly's epistemology and theory of personal constructs allow one to blur the distinction between personal meaning and formal knowledge of science: "Rather than treating all concepts as if they are distinct clear-cut entities of physics, we propose to view the process through which scientists structure their domains as being similar to the way that people deliberately construct their own world-views. Accepting Kelly's model of man-the-scientist emphasises that we see people as essentially inquisitive and constructively scientific, whilst at the same time casting science as yet another form of frail and fragile human activity."

This view of science is reflected in the writings of modern philosophers of science. Indeed, an activity that engendered a lot of discussion and eventually a monograph written by five members of the PCKG resulted from the conjecture, "Man-the-scientist but what kind of scientist?" We examined the implication of Kelly's original metaphor embracing respectively the philosophers Bacon, Popper, Kuhn, Lakatos, and Feyerabend (Swift *et al.* 1983).

Kelly was concerned with helping his clients recognize that it is the "spectacles" which they choose to see the world through that may be limiting their effective functioning. By recognizing one's own potential for changing the constructs which one holds, one is opening the door to creativity. As has been noted, Kelly does not deny the existence of reality but chooses to focus on the importance of coming to an understanding of a person's constructions of reality. It is possible for there to be an indefinite, even infinite, number of ways of construing some aspect of reality, and since none of us can directly view reality, there is no way that one construal of it can be seen as absolutely true while all others are absolutely wrong. However, since reality does exist, there will be some wrong construals in the sense that an individual's ideas do not map to reality as it exists. Over a period of time the scientific community will come up with a way of viewing reality which accounts for more phenomena than previous theories. This can be seen as a relatively better theory. Following a Kellyan viewpoint does not mean that children should be left to construct their world views without having such relatively better theories presented to them. However, what is essential is that such formal knowledge is presented as conjectural and that it should be open to the personal reconstruction and appraisal of the learner. It would be recognized that if the learner does not see the formal knowledge as being intelligible, plausible, fruitful, and in keeping with his or her fundamental metaphysical commitments, the learner has no reason for conceptual change (Hewson 1980).

By following Kelly's view on the nature of a personal construct system, the teacher would wish to find out about the learner's framework of ideas and recognize that these ideas will have an emotional significance for the individual and that some of the minitheories which are held are deeply entrenched (core constructs) and therefore, like the hard core of a Lakatosian research program, will be highly resistant to change. *If* a teacher wishes to encourage conceptual change and acceptance of the received view of a scientific concept, then experiences have to be arranged within a supportive relationship whereby the learner has his or her views challenged. Readers might recognize the implications of Kelly's fixed-role therapy for such a process.

It would be seen as a teaching technique that encourages pupils to challenge their own world views in order to promote conceptual change. The received view takes its place alongside the pupils' constructions, and the onus is on the young personal scientists to evaluate the alternative theories for themselves. This scientific activity should lead to a committed choice, albeit tentative, among the alternative theories and indeed may generate a new theory. The possibility will exist that one of the student's theories may go beyond the received view and provide

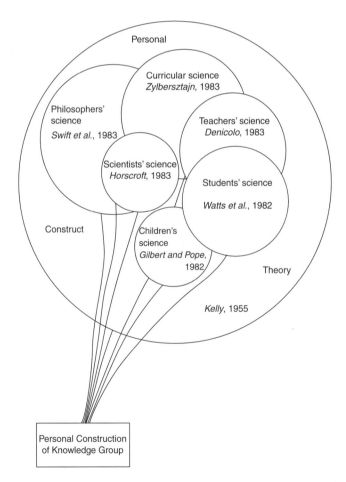

Figure 16.1 Some publications by the Personal Construction of Knowledge Group.

a more useful explanation of events. The progress of science requires such revelations. The recognition that our ways of categorizing and explaining the world can be an impediment is just as necessary for the creation in science as it is for the personal growth of a client in therapy. As Kelly (1969) put it, " 'Hardening of the categories', a common affliction amongst scientists, usually marks the end of the creative phase of a distinguished career."

Conclusion

Through our work we have sought to soften the categories of hostile science educators. As a research program we have examined philosophers' science, scientists' science, curricular science, teachers' science, students' science and children's science. Each of these can potentially be radically different from each other. We have demonstrated differences among these versions of scientific concepts and have also pointed to the range of alternatives that can be found within any one of them. Our research program has raised implications for the teaching of science, and personal construct psychology has been a useful meta-theory which anticipates the existence of the alternatives we have found. In Figure 16.1 we have diagramed examples of papers from the PCKG, each of which has as its focus an identified area. It should be noted, however, that a major strength of our research program, as we see it, is the interlinkages with the work of individuals within the group. Much of this cohesion is related to the use of the epistemological base and constructivist research methodologies implied by personal construct psychology. Readers may care to deepen their understanding of our work through one or more of the papers identified.

References

Association for Science Education. (1979). *Alternatives for Science Education: A Consultative Document*. Association for Science Education, Hatfield, Herts, England.

Barnes, D. (1974). *From Communication to Curriculum*. London: Penguin Books.

Bruner, J. (1977). *The Process of Education*. New York: Harvard University Press.

Cawthron, E.R. and Rowell, J.A. (1978). Epistemology and science education. *Studies in Science Education*, 5, 31–59.

Cicourel, A.V. (1974). *Language Use and School Performance*. New York: Academic Press.

Claxton, G. (1982, April). *School Science: Falling on Stony Ground or Choked by Thorns*. Paper presented at Symposium on Investigating Children's Existing Ideas about Science, University of Leicester.

Davisson, A. (1978). George Kelly and the American mind. In F. Fransella (ed.), *Personal Construct Psychology 1977*. London: Academic Press.

Denicolo, P. (1983, April). *Metaphor in the Teaching and Learning of Chemistry – an Empirical Study*. Paper presented at the Annual Conference for Postgraduate Psychology, St. Andrews.

Donaldson, M. (1978). *Children's Minds*. London: Fontana.

Driver, R. (1973). *The Representation of Conceptual Frameworks in Young Adolescent Science Students*. Unpublished doctoral dissertation, University of Illinois.

Garfinkel, H. (1967). *Studies in Ethnomethodology*. Englewood Cliffs, NJ: Prentice-Hall.

Gilbert, J.K. and Pope, M.L. (1982). *Schoolchildren Discussing Energy*. Unpublished manuscript, University of Surrey, Institute of Educational Development.

Gilbert, J.K., Osborne, R., and Fensham, P. (1982a). Children's science and its consequences for teaching. *Science Education*, 66, 623–633.

Gilbert, J.K., Watts, D.M., and Osborne, R. (1982b). Students' conceptions of ideas in mechanics. *Physical Education*, 17, 62–66.

Hewson, P.W. (1980, April). *A Case Study of the Effect of Metaphysical Commitments on the Learning of a Complex Scientific Theory*. Paper presented at AERA, Los Angeles, CA.

Horscroft, D. (1983). *The Notions of Reactions Rate and Equilibrium used by a Group of Practising Industrial Chemists*. Unpublished manuscript, University of Surrey, Department of Educational Development.

Kelly, G.A. (1955). *The Psychology of Personal Constructs*. New York: Norton.

Kelly, G.A. (1969). Ontological acceleration. In B. Mayer (ed.), *Clinical Psychology and Personality: The Selected Papers of George Kelly*. New York: Wiley.

Kelly, G.A. (1970a). A brief introduction to personal construct theory. In D. Bannister (ed.), *Perspectives in Personal Construct Theory*. London: Academic Press.

Kelly, G.A. (1970b). Behaviour as an experiment. In D. Bannister (ed.), *Perspectives in Personal Construct Theory*. London: Academic Press.

Kuhn, T.S. (1970). *The Structure of Scientific Revolutions* (2nd edn). Chicago, IL: University of Chicago Press.

Lakatos, I. (1970). Falsification and the methodology of scientific research programmes. In I. Lakatos and A. Musgrave (eds), *Criticism and the Growth of Knowledge*. New York: Cambridge University Press.

Piaget, J. (1971). *Genetic Epistemology*. New York: Norton.

Pope, M.L. (1981). Personal experience and construction of knowledge in science. In K. Abrahamsson (ed.), *Co-operative Education, Experiential Learning and Personal Knowledge*. Stockholm: National Board of University and Colleges.

Pope, M.L. and Gilbert, J.K. (1983). Personal experiences and the construction of knowledge in science. *Science Education*, 67, 193–203.

Posner, G. (1981, April). *Promising Directions in Curriculum Knowledge: A Cognitive Psychology Perspective*. Paper presented at AERA, Los Angeles, CA.

Proctor, H. and Parry, G. (1978). Constraint and freedom: the social origins of personal construct. In F. Fransella (ed.), *Personal Construct Psychology*. London: Academic Press.

Swift, D., Gilbert, J.K., Pope, M.L., Watts, D.M., and Zylbersztajn, A. (1983). *Philosophies of Science and Science Education*. Unpublished manuscript, University of Surrey, Institute of Educational Development.

Watts, D.M., Gilbert, J.K., and Pope, M.L. (1982, February). *Alternative Frameworks: Representing of School Children's Understanding of Science*. Paper presented to First International Symposium on Representing Understanding, Guy's Hospital, London.

Werner, H. (1957). *Comparative Psychology of Mental Development* (3rd edn). New York: International University Press.

Zylbersztajn, A. (1983). *A Conceptual Framework for Science Education: Investigating Curricular Materials and Classroom Interactions in Secondary School Physics*. Unpublished doctoral dissertation, University of Surrey.

INDEX